A MIND OVER MATTER

A MIND OVER MATTER

Philip Anderson and the Physics
of the Very Many

Andrew Zangwill

Georgia Institute of Technology

OXFORD
UNIVERSITY PRESS

OXFORD

UNIVERSITY PRESS

Great Clarendon Street, Oxford, OX2 6DP,
United Kingdom

Oxford University Press is a department of the University of Oxford.
It furthers the University's objective of excellence in research, scholarship,
and education by publishing worldwide. Oxford is a registered trade mark of
Oxford University Press in the UK and in certain other countries

First Edition published in 2021

Impression: 3

Published in the United States of America by Oxford University Press
198 Madison Avenue, New York, NY 10016, United States of America

British Library Cataloguing in Publication Data

Data available

Library of Congress Control Number: 2020944518

ISBN 978–0–19–886910–8

Printed and bound by
CPI Group (UK) Ltd, Croydon, CR0 4YY

Contents

Prologue

In 1990, the US House of Representatives authorized a total federal expenditure of $5 billion dollars to construct a giant proton accelerator called the Superconducting Super Collider (SSC). The purpose of this machine was to test a sophisticated theoretical description of subatomic particles and to announce to the world that the United States was not prepared to cede leadership in high-energy particle physics research to Europe.

Some scientists and science administrators not involved in particle physics feared that the construction and maintenance costs of the SSC would siphon off government funds from their own areas of research. As a result, the scientific community did not speak with one voice when the budget for the project came up for review every year by Congress. Two Nobel laureates emerged as the principal spokespersons for and against the SSC. The particle physicist Steven Weinberg supported the project; the condensed matter physicist Philip Anderson opposed it.

Weinberg was an expert in the physics of the very small—one of the creators of the theoretical "standard model" of subatomic particles that the SSC was designed to test. He believed that the most important problems in science aimed to discover the physical laws obeyed by the minutest particles in the cosmos. Knowing these microscopic laws, one could derive (in principle) the macroscopic laws obeyed by larger objects like nuclei, atoms, molecules, solids, plants, animals, people, planets, solar systems, galaxies, etc.

Anderson was an expert in the physics of the very many—one of the creators of condensed matter physics, the science of how vast numbers of atoms interact with each other to produce everything from liquid water to sparkling diamonds. He agreed that the standard model was interesting, but he denied the assertion that the laws of elementary particle physics had anything useful to say about famously difficult and unsolved problems like: Why is there such a thing as window glass? How does turbulence develop in fluids? How does the brain learn? He feared that building and maintaining the SSC would inevitably consume the majority of the funds the government allocates to support scientific research of all kinds.

The debate over the SSC was a unique forum where the differing scientific philosophies of Weinberg and Anderson intersected with a hard decision Congress had to make about spending many federal dollars on a single scientific project. For that reason, the two theorists provided testimony to Congress on several occasions. But only once, at a 1993 hearing, did they testify together in person.[1] Portions of their verbatim testimony follow.

DR. WEINBERG: I am grateful to the chairman to allow me to come here to talk about the Super Collider. In essence, the Super Collider is a machine for creating new kinds of matter, particles that have existed since the Universe was about a trillionth of a second old. To produce these particles requires an energy about twenty times higher than the energy of the largest accelerators that now exist, which is why the Super Collider is so big and therefore why it is so expensive.

This little statement that I have made really does not do justice, however, to what the Super Collider is about because particles in themselves are not really that interesting. . . . If you have seen one proton you have seen them all. We are not really after the particles, we are after the principles . . . that govern matter and energy and force, and everything in the Universe.

Culminating around the mid-1970s, we developed a theory called the standard model which encompasses all the forces we know about, all the different kinds of matter that we can observe

---

[1] *Superconducting Super Collider*, Joint Hearing 103–85 before the Committee on Energy and Natural Resources and the Subcommittee on Energy and Water Development of the Committee on Appropriations. United States Senate. August 4, 1993, pp. 48–60.

with existing laboratories. We know that [this theory] is not the last word [because] it leaves out things that are pretty important, like the force of gravity...In addition, the particles that we know, quarks, electrons and so on all have mass...But [the theory] does not know exactly what [these masses] are. This is the question that the Super Collider is specifically designed to answer.

But there is a sense, nevertheless, [that] this kind of elementary particle physics is at the most fundamental level of science. That is, you may ask any question, for example, how does a superconductor work, and you get an answer. You get an answer in terms of the properties of electrons and the electromagnetic field and other things. And then you ask, well, why are those things true? And you get an answer in terms of the standard model....And then you say, well, why is the standard model true? And you do not get an answer. We do not know. We are at the frontier. We have pushed the chain of why questions as far as we can, and as far as we can tell we cannot make any progress without the Super Collider. Thank you.

CHAIRMAN: Thank you very much Dr. Weinberg. Our next witness is Professor Philip Anderson from the Department of Physics, I think that is Applied Physics, at Princeton University.
DR. ANDERSON: Thank you. For the record, I am not an applied physicist. I like to call myself a fundamental physicist as well; I am just fundamental in a somewhat different way.
CHAIRMAN: Was that because of the W particle?
DR. ANDERSON: No, it was the Higgs boson that I helped invent.

Now, on several occasions over the years I have testified against the SSC and against other big science projects, and in favor of funding a wider variety of fundamental science on a peer-reviewed basis through institutions such as the National Science Foundation and the National Institutes of Health, which have good records of responsible distribution of funds. I will try to be as brief as possible and in any case I do not think I can be anywhere near as eloquent as my colleague here, Steve Weinberg.

DR. WEINBERG: You can try.

DR. ANDERSON: The point of my testimony is priorities. The physics being done by the SSC is in a very narrow specialized area of physics with a very narrow focus. It focuses on the very tiny and very energetic sub-sub-substructure of the world in which we live. Most of that substructure is well understood in a very definite sense. Nothing discovered by the SSC can, for the foreseeable future, change the way we work or think about the world and cannot change even nuclear physics.

Perhaps a couple of hundred theorists (too many for such a narrow subject in my opinion)...and a few thousand experimentalists work in this particular field of science. That is less than ten percent of the research physicists in the world....Yet the budget of [the SSC] dwarfs the budget for all the rest of physics. The fact is that particle physicists are funded, on average, ten times as liberally as other physicists...In this sense, the SSC is not a very efficient jobs program, at least for physicists.

At least two books and many articles have been published recently trying to justify the special status for this particular branch of physics as somehow more fundamental than all other science. That so many particle physicists have time to write such books and articles may tell you something about the real interest in the field; it has not made much progress lately, and so they do not have anything else to do.

There are many other really exciting fundamental questions which science can hope to answer and which people like myself are, on the whole, too busy to write books about. There are questions like: How did life begin? What is the origin of the human race? How does the brain work? What is the theory of the immune system? Is there a science of economics?

All these things have in common that they are manifestations not of the simplest things about matter—the elementary particles—but of the complexity of matter and energy as we ordinarily run into them. These manifestations of complexity do not...have any possibility of being affected by whatever the SSC may discover....On the other hand, the future seems to me to belong to these subjects, to these questions, rather than to the infinite regression of following the tiny substructure of matter. Perhaps you should think which fundamental questions are easier and less expensive to solve. Thank you.

Congress cancelled the SSC two months later. Many particle physicists blamed the demise of the SSC on the testimony and lobbying skill of the outspoken Anderson. He had broken ranks and given public voice to a dispute best handled quietly within the family of physicists. Who was this condensed matter physicist and how had he become so influential?

Acknowledgments

My former Georgia Tech colleague Paul Goldbart suggested I write this book. I had never met Phil Anderson, so Paul volunteered to pitch the idea to him through a mutual friend. The prospective subject was not enthusiastic. He thought his life was too dull for anyone to write about.

I introduced myself to Anderson electronically and explained why I thought a biography would be both timely and interesting. Eventually he agreed and I am grateful for his cooperation. He submitted to several interview sessions and answered the many questions I posed in subsequent emails. He also supplied documents, letters, and photographs without restriction. I regret only that I was not able to complete this book before his death.

Formal interviews of Anderson were indispensable to this book. The website of the American Institute of Physics maintains transcripts of oral history interviews of Anderson conducted by Lillian Hoddeson (1987–1988), Alexei Kojevnikov (1999–2000) and Premela Chandra, Piers Coleman, and Shivaji Sondhi (1999–2002). Author Laurence Gonzales also kindly provided me with a tape of an interview he conducted with Anderson more recently for a history of the Santa Fe Institute.

It is a pleasure to thank the more than one hundred people who spoke or wrote to me about their interactions with Anderson. This group included family (particularly his daughter Susan), friends, acquaintances, students, collaborators, and competitors. I also thank historian Michael Gordin, who was very encouraging at an early stage.

For their help with primary documents, I am grateful to the archivists at Harvard University, the Library of Congress, the Massachusetts Institute of Technology, Princeton University, Rutgers University, the University of California at San Diego, the University of Cambridge, the University of Illinois, the University of Oxford, Wabash College, and Yale University. I am also grateful for the help I received from the American Institute of Physics Center for History of Physics, the Aspen Center for Physics, the AT&T Archives and History Center, the Chemical Heritage Foundation, the Niels Bohr Archive, and the Santa Fe Institute.

Several individuals read draft chapters of this book and I thank Bert Halperin, Volker Heine, David Joffe, Tony Leggett, Joseph Martin, Marc Merlin, Núria Munoz Garganté, Rob Phillips, Wayne Saslow, and Richard Werthamer for their constructive criticism. I am grateful also to Joshua Weitz for suggesting the title. Special thanks go to two people. Glenn Smith was present when this project began and he patiently provided feedback, advice, and friendship throughout. Dan Goldman read the final manuscript and his comments led me to produce a much improved, truly final manuscript. Of course, all errors and obscurities are my responsibility alone.

Lastly, I am grateful to Sonke Adlung and the entire editorial and production staff of Oxford University Press for their excellent work in bringing this book to fruition.

1

Introduction

History will judge Philip W. Anderson (Figure 1.1) to have been one of the most accomplished and influential physicists of the second half of the twentieth century. His name is not widely known to non-scientists because his accomplishments do not involve the physics of the *very small* (quarks and string theory) or the physics of the *very distant* (supernovae and black holes).

Anderson's expertise was the physics of the *very many*, primarily very many atoms and/or very many electrons. How many? A typical question in his field might ask for the energy required to disassemble one grain of sand into its constituent atoms. The number of these atoms is about equal to the number of grains of sand in the Sahara desert.[1] Special methods and talents are needed to answers questions of this kind.

During his nearly sixty-year career at Bell Telephone Laboratories, the University of Cambridge, and Princeton University, Anderson played a dominant role in shaping the character and research agenda for *solid-state physics*. This is the subfield of physics that deals with ordinary matter like iron, wood, glass, and pencil lead. It also provides the basic understanding which supports the semi-conductor industry, computers, lasers, smart phones, fiber optics, magnetic resonance imaging, and most of the other drivers of our technological society.

Important as they are, these applications of solid-state physics did not direct Anderson's personal research. His preference to focus on basic principles led him to study phenomena with

---

[1] This number is about one hundred quintillion (10^{20}).

exotic-sounding names like superconductivity, antiferromagnetism, the Josephson effect, superfluidity, the Kondo effect, spin
glasses, Mott insulators, liquid crystals, heavy fermions, and resonating valence bonds. His share of the 1977 Nobel Prize for Physics
recognized his theoretical discovery of a phenomenon now called
Anderson localization, which describes how a propagating wave can
be stopped in its tracks by a disordered medium.

Over the years, Anderson earned a reputation for his ability to
identify and then tackle very difficult solid-state physics problems
and for the deep, seemingly magical, intuition he brought to bear
on them. The type of questions that engaged him were often easy
to state but very difficult to answer. Why are some solids rigid
while others are not? Why do electrons move easily through
some solids but not at all through others? What are the fundamental mechanisms responsible for magnetism and superconductivity?

Anderson's intuition often led him to reach conclusions
instinctively rather than by conscious deduction. Some part of
this ability comes from a breadth and depth of knowledge that
permitted him to weave multiple strands of information together
into a single coherent story. But at least some of his intuition—
which was so often correct—came from a place that remained a
mystery to even his closest friends and collaborators.

As a theoretical physicist, one of Anderson's greatest strengths
was his uncanny ability to strip away the details from a complicated problem and identify its key elements. He would then construct a mathematical model (description) which retained only
those elements. Invariably, the models he developed were simple
enough to analyze in detail, yet complex enough to exhibit the
physical behavior he hoped to understand.

Anderson's nearly 500 scientific papers provide a clear guide to
his research achievements. However, his story is compelling
beyond his individual accomplishments because, more than any
other twentieth-century physicist, he transformed the patchwork of ideas and techniques formerly called solid-state physics

into the deep, subtle, and intellectually coherent discipline known today as *condensed matter physics*.

This was not merely a cosmetic change of name. To a great extent, Anderson and a few other like-minded physicists abandoned the prevailing methodology of concentrating on the differences between solid substances and devoted themselves to discovering, exploiting, systematizing, and educating others about the *universal* properties of solids, i.e., those properties that always appear when 10^{23} particles interact strongly with one another. This so-called *many-body problem* fascinated Anderson endlessly.

In a solid, the relevant particles are electrons and an important part of solid-state physics resembles a chess game where every chess piece is an electron. We know the rules obeyed by the pieces, but their vast number generates a huge number of possible arrangements for them. At the highest level of achievement, which is where Anderson operated, the insight and skill of a grandmaster are required to gain an understanding of the true behavior of the electrons.

Anderson wrote a book where he identified a handful of fundamental organizing principles and showed that many seemingly disparate phenomena in condensed systems are actually different manifestations of these few principles. His book is not easy to read, but it had a profound effect on many of the leaders of the next generation of theoretical physicists. Important ideas spread quickly and a glance at the current textbook literature shows that Anderson's perspective now permeates the *gestalt* of the entire condensed matter community.

The concept of *broken symmetry* is one of the ideas that the physics community identifies most closely with Anderson's personal research. He discovered its importance at an early stage and applied it over and over with great success to a variety of problems. Symmetry breaking also describes a recurring feature of Anderson's life where he deliberately—and often contrarily—disassociated himself from the behaviors or beliefs of others. This pattern had many consequences, not least in producing the circumstances for a deep disappointment which settled on him during the final phase of his long career.

Anderson made it his business to influence the culture and politics of American science. He sought and found profound ideas in condensed matter physics as part of a deliberate effort to challenge a high-energy physics community that had spent decades claiming the intellectual high ground for its own activities. By arguing strenuously for the fundamental nature of his own field, Anderson hoped to blunt the influence particle physicists had long enjoyed with government officials and science journalists. The former kept the money flowing to build ever-larger particle accelerator machines. The latter breathlessly reported the cosmic significance of every newly discovered subatomic particle while noting that the latest research by solid-state physicists might produce a better toaster.

Anderson vented his frustrations in a 1972 article where he pointed out that symmetry breaking generated novel properties in large many-particle systems. Moreover, he insisted, these novel

Figure 1.1 Philip Warren Anderson in 1988 at age 65. Source: Donn Forbes.

properties are impossible to predict knowing only the properties of the individual constituent particles and their mutual interactions. He used this idea to attack the claims of particle physicists who asserted that the essential job of science was to discover the laws governing subatomic particles because all other "laws" of Nature were ultimately derivable from them.

To the contrary, Anderson argued, the hierarchical structure of science (e.g., from physics to chemistry to biology to psychology) is not merely a convenient way to divide research practice. Rather, it reflects the existence of fundamental laws at each level that do not depend in any significant way on the details of the laws at lower levels. The higher level laws must be *consistent* with the subatomic laws, but the likelihood that one can derive the former from the latter is essentially zero. Anderson inspired a small intellectual renaissance among philosophers because (unbeknownst to him) his ideas revived a concept called *emergence* which had been proposed a century earlier.

In the 1980s, Anderson helped found the Santa Fe Institute, a think tank devoted to developing strategies to study complex systems as dissimilar as turbulent fluids and the US economy. Anderson knew that some physical systems produced complex behavior starting from very simple rules of engagement. This led him to suggest that the mathematics used to analyze these systems might be useful to analyze complexity in other situations. With some success, he lobbied practitioners in fields as diverse as finance, neuroscience, economics, computer science, operations research, physiology, and evolutionary biology to adopt his approach.

Keeping up the pressure on particle physicists, Anderson was a lightning rod for controversy when, as described in the Prologue, he testified in Washington to oppose plans by the US government to build the Superconducting Super Collider. The project was eventually cancelled and some members of the physics community never forgave him for breaking ranks and publicly exposing disagreements inside the larger community of US physicists.

The arc of Anderson's career reveals a shy mid-western boy who learned the more sophisticated ways of the larger world as a college and graduate student at Harvard University. Eschewing an academic career, he went to work at an industrial laboratory, rose quickly to the top rank of theoretical physicists, and stayed there for thirty-five years. His name became synonymous with success in condensed matter physics and the breadth of his ideas and his skills as a polemicist gave him influence well outside the traditional community of physicists.

Anderson's novel theory of high-temperature superconductivity in the late 1980s should have been the crowning jewel of his career. However, it proved difficult to work out the predictions of the theory in sufficient detail to compare them with the results of experiments. As time went on and other physicists offered alternative theories, Anderson sometimes became dismissive and combative towards them. This behavior damaged his reputation and drove some young theorists away from the problem. In the end, twenty years of effort failed to convince the majority of his colleagues that his basic idea was correct. This experience left a bitter taste in his mouth.

My original conception of this project was to use Anderson's career as a vehicle to discuss the intellectual history of condensed matter physics. The impossibility of this task soon became apparent. Entire books could be written to trace the history of the community's efforts to understand magnetism, superconductivity, the Kondo effect, the Hubbard model, and dozens of other topics. For that reason, I was forced to adopt an extremely Anderson-centric perspective and leave unmentioned the contributions of a great many other excellent scientists unless they bore directly on his involvement.

I did not meet Anderson in person until I interviewed him for this biography. He was no *tabula rasa* because he had been diligent to curate his own history. Personal historical commentary appears in a volume of his collected essays, in the annotations to the papers included in two volumes of his selected scientific works, in the transcripts of three oral histories, and even in the text of many of his technical papers.

This is not a textbook, so I have aimed to make my discussions of Anderson's physics as descriptive as possible. I use virtually no equations and diagrams do most of the heavy lifting. The main requirement is that the reader be able to follow logical arguments of the sort used in college science and engineering courses. Technical terms are unavoidable, but all of them are defined, and when one re-appears later in the text, the reader will lose virtually nothing by skipping over it lightly as a non-musician might skip over a technical musical term when reading a biography of a great composer.

This leads to a broader discussion of the political and cultural aspects of Anderson's career. It was, in fact, an issue of science politics which put him on the path that led to his interests in emergence and complexity mentioned above. Important sources of information here are the many non-technical essays and book reviews he wrote over the years. These feature his opinions about religion, education, computers, journalism, statistics, the culture wars, and the history, practice, sociology, and philosophy of science.

The mathematician Mark Kac once contrasted the "ordinary genius," who was someone simply "many times better" than his colleagues, with the "magician," for whom "even after we understand what they have done, the process by which they have done it is completely dark."[2] Phil Anderson has always struck me as a magician in this sense. I cannot pretend to have completely discovered how he got that way, so I have attempted to understand how this characteristic influenced his scientific trajectory and the effect it had on his students, coworkers, his community, and on the enterprise of physics.

---

[2] Mark Kac, *Enigmas of Chance: An Autobiography* (Harper and Row, New York, 1985), p. xxv.

2

Son of the Heartland

Philip Warren Anderson was a winter baby, born December 13, 1923. He grew up in an academic family deeply rooted in the American Midwest. His father, Harry, was a professor of plant pathology at the University of Illinois at Urbana-Champaign. His mother, Elsie (née Osborne), was the daughter and sister of professors of mathematics and English, respectively, at Wabash College in Crawfordsville, Indiana.

Anderson's parents were natives of Crawfordsville and a very pregnant Elsie insisted on a just-in-time road trip to ensure that her son was born on Indiana soil. Later, holiday and long summer visits kept Phil connected to his extended Indiana family. Until the age of thirty, he spent almost every Christmas at his maternal grandfather's home in Crawfordsville. These visits exposed him to the traditional Hoosier values of pugnacity, skepticism, patriotism, and sensitivity.[1] There is a grain of truth in all regional stereotypes and the reader can judge the extent to which these traits appear in some of the behavior of the mature adult.

Phil and his sister Eleanor Grace (older by four years and always called Graccie by the family) engaged with science from an early age.[2] Their father encouraged them to collect insects and ask questions, just as he had done as a child. Harry's professional interest in horticulture led him to encourage his children to

---

[1] *Readings in Indiana History*, edited by Oscar H. Williams (Indiana University, Bloomington, IN, 1914), p. 259.
[2] Graccie is pronounced "Gracky."

learn about this subject and he largely succeeded. Both became passionate and knowledgeable gardeners as adults.

Harry set up a small chemistry laboratory in his home where his grade school-aged son managed to synthesize hydrogen. The boy failed to produce a working firework in this laboratory, but a child-like enthusiasm for skyrockets and Roman candles survived far into adulthood.[3] The Anderson kids also learned to love the outdoors. Graccie was a tomboy and she and young Phil spent many summers carousing with their cousins on the farms still owned by their parents' families.

Elsie was the guardian of academic standards and she was quite unhappy if her children ever brought home a grade less than an "A." The nurturing example of Harry, fueled by pressure from Elsie, made it almost inevitable that their children would dream about careers in science. Harry subscribed to the weekly magazine *Science* and the Anderson kids always made a stab at reading it as high school students. Phil was good at math and he thought he might become a mathematician. Graccie planned to become the Marie Curie of biochemistry.

The Saturday Hikers

Anderson learned about the world beyond Illinois and Indiana from his father's membership in an institution unique to the University of Illinois called the Saturday Hikers.[4] This group of 15–30 male faculty members drove out to a river or lake in the countryside outside Urbana every Saturday morning to hike, canoe, play softball, swim, battle chiggers, and enjoy a campfire cookout. Afterward, there would be singing, university gossip, and spirited political discussion.

---

[3] Interview of Claire and David Jacobus by the author, Princeton, NJ, May 5, 2016.

[4] P.W. Anderson, "Growing Up with the Illinois Faculty's Saturday Hikers" (unpublished, 2016).

The politics of the Saturday Hikers was mostly leftist at the national level and strongly interventionist when it concerned Europe and Asia. During the 1930s, they debated President Herbert Hoover's plan to relieve farmers facing mortgage foreclosures, the appropriateness of adopting the "Star Spangled Banner" as the national anthem, the hunger strike of Mahatma Gandhi, the rise of the Nazi party in Germany, the Spanish Civil War, and President Franklin Roosevelt's plan to add justices to the Supreme Court of the United States.

At the time, the left-wing orientation of the Hikers was not common in the Agriculture and Engineering Colleges of the University of Illinois. This meant that the members of the group tended to come from other parts of the campus. Among these people, Harry Anderson was particularly friendly with the chairman of the physics department, Wheeler Loomis, the psychologist Coleman Griffith, and the political scientist Clarence Berdahl.

The creator and leader of the Saturday Hikers for thirty-five years was William Abbott Oldfather, a distinguished professor of classics. Oldfather was a fearless scholar who expressed his socialist opinions "vigorously and often vituperatively."[5] The Andersons and Oldfathers were quite close. They vacationed together in the Teton Mountains and spent two weeks sharing an isolated cabin in Ontario, Canada. On that occasion, Oldfather read aloud from a copy of Thorstein Veblen's famous economic and sociological analysis of consumerism, *Theory of the Leisure Class* (1899).

No faculty children or spouses attended the Saturday Hikes. However, the Hikers often trooped out again on Sundays, this time with their families invited. The serious hiking and political debate simply picked up from where it had ended the previous day. Anderson remembered these occasions as his happiest hours as a child and adolescent.[6]

---

[5] Winton U. Solberg, "William Abbott Oldfather: Making the Classical Relevant to Modern Life" in *No Boundaries: University of Illinois Vignettes*, edited by Lillian Hoddeson (University of Illinois Press, Urbana, IL, 2004), pp. 69–87.

[6] Biography of Philip W. Anderson, Nobel Foundation, 1977.

With the encouragement of his parents, Phil honed his out-door skills and adopted the political views of the Hikers. Often, the Andersons hosted a rotating dinner/dance called the "Indoor Yacht Club" where progressive political talk was a major feature. Phil and Graccie were not invited, but they stayed out of sight and absorbed all that was said. Both remained committed liberal Democrats their entire lives.

Formal Education

Eleven-year-old Phil Anderson entered the Laboratory High School of the University of Illinois in the fall of 1935. The quality of the instruction at this small private high school (nicknamed "Uni") was very high. A unique feature—exploited by Anderson—was the ability to begin with a "sub-freshman" year which con-solidated the seventh and eighth grades. The annual tuition was $25 and the students were mostly children of University of Illinois faculty and wealthy businesspeople.[7]

Graccie was a senior at Uni when Phil entered as a sub-freshman. She excelled academically and socially, serving as both the vice-president of the junior class and president of the senior class. Later, she majored in chemistry at the University of Illinois, served as a meteorologist for the US Navy, and earned a PhD in biochemistry from the University of Wisconsin. The birth of her children interrupted her plans to work in that field but she later enjoyed a long and successful career as a scientific librarian, biog-rapher, and translator.[8]

Graccie and her brother rarely quarreled. However, as one might expect of two close and very smart siblings, they main-tained a healthy intellectual rivalry all their lives. When they both lived in New Jersey, they raced each other to be the first to complete a difficult word puzzle which appeared every week in the Sunday *New York Times*.

[7] Interview of Henry P. Noyes by the author, September 20, 2015.
[8] Interview of Andrew Maass by the author, June 18, 2017.

Figure 2.1 Early influences. Phil Anderson's mathematics teacher, Miles Hartley (left) and his physics teacher, Wilber Harnish (right) in 1939, when Anderson was a junior. Source: *U and I*, the yearbook of the University High School, University of Illinois.

Anderson's favorite teacher at Uni was Miles Hartley, a University of Illinois PhD who taught plane geometry and algebra to the juniors and solid geometry and advanced algebra to the seniors (Figure 2.1). He was a stickler, but he was also a pedagogical innovator who used plywood and dowels to construct models to illustrate theorems in solid geometry.[9] Inspired by Hartley, Anderson cemented his plan to major in mathematics in college.[10]

The physics teacher at University High School, Wilber Harnish (Figure 2.1), was the anti-Hartley. Harnish's background was in education, not physics, and he avoided quantitative deductions by focusing on student experiments. This meant that Phil and his classmates

[9] Miles C. Hartley, "Models of Solid Geometry," *The Mathematics Teacher* **35** (1) 5–7 (1942).

[10] Jeremy Bernstein, *Three Degrees Above Zero: Bell Laboratories in the Information Age* (Cambridge University Press, Cambridge, 1984), p. 121.

worked with vacuum pumps and electric motors, but they were not taught the laws of physics that made them work. Harnish discussed gravity in connection with pulleys and falling objects, but he neglected to point out that gravity also governs the motions of the planets. His teaching provided not a hint of the unity of the subject.[11]

Phil compensated by borrowing popular science books from the school library. His two favorites were *The Einstein Theory of Relativity* (1936) by Lillian and Hugh Lieber, and *Mr. Tompkins in Wonderland* (1940) by George Gamow. The Liebers were expert at using cartoons and geometrical diagrams, but they included serious algebraic manipulations also.

George Gamow was a world-renowned theoretical physicist who used fiction to introduce modern physics to a popular audience. His hero, Mr. Tompkins, was a bank clerk who attended physics lectures at a local university. Each night, he dreamed of a fantastical world where the usually unseen effects of special relativity, quantum mechanics, and the curvature of space due to gravity become apparent during the course of daily life. Clever cartoons and a clear presentation of the physics are notable features of Gamow's book also.

The Lieber and Gamow books leave the reader with a vivid impression of the interplay between theory and experiment in physics.[12] In light of Phil's later insistence on the importance of experiment to guide and inform theory, it is not surprising that these particular books never left his memory. On the other hand, much of the material he read from the Uni library differed so much from what he saw in Harnish's physics class that he was a college freshman before he realized they were all part of the same subject.[13] The search for connections would be a characteristic feature of his research for his entire career.

[11] Interview of Henry P. Noyes by the author, September 20, 2015.

[12] Both books are still in print: Lillian R. Lieber and Hugh Gray Lieber, *The Einstein Theory of Relativity: A Trip to the Fourth Dimension* (Paul Dry Books, Philadelphia, PA, 2008); George Gamow, *Mr. Tompkins in Paperback* (Cambridge University Press, Cambridge, 2012).

[13] Interview of PWA by Alexei Kojevnikov on March 30, 1999, Niels Bohr Library & Archives, American Institute of Physics, College Park, MD.

Anderson was a diligent student, and the editors of the school year-book used the title of an Oscar Wilde play, *The Importance of Being Earnest*, as the caption for his 1940 graduation photo. The accompanying thumbnail biography reveals that he acted in the school play every year, wrote and read the senior class history at commencement, and participated in the biology and chess clubs. At graduation, he ranked first in his class, tied with "three others, one a girl" as his transcript put it. He earned a grade of "A" in every course except typewriting and physical education. The latter probably reflects disinterest because he won the school tennis championship as a junior, competed in the state track meet as a miler, and was a talented speed skater.

The Krebiozen Affair

Elsie Anderson stressed education to her children, but she also put great emphasis on the importance of self-respect and respect for others.[14] There were many opportunities to communicate this message, particularly because the ravages of the Great Depression were grimly apparent on the streets of Champaign and Urbana during her children's school years. One day, an out-of-work man came to the back door of the Anderson residence looking for food. Elsie treated him with kindness and respect and it was made clear that no other behavior was acceptable.[15]

Anderson also looked to his father for advice and example. Harry's success as a researcher was a clear model for a life devoted to science.[16] Less apparent, but perhaps more important, Phil put great value on his father's personal integrity. A striking example

[14] Letter from PWA to Liberty Santos, November 17, 1986. *Anderson, Philip W.*; Faculty and Professional Staff files, Subgroup 13: P, AC107.13, Princeton University Archives, Department of Rare Books and Special Collections, Princeton University Library.
[15] Interview of Andrew Maass by the author, June 18, 2017.
[16] Harry Anderson's *Diseases of Fruit Crops* (McGraw-Hill, New York, 1956) continues to garner citations in the scientific literature, more than sixty years after its original publication. Google Search, September 26, 2018.

is the role Harry played in the notorious Krebiozen affair nearly a decade after his son left Urbana.[17]

In 1949, a man named Stevan Durovic came to Chicago from Argentina to meet Dr. Andrew Ivy, a physiologist then serving as Vice-President of the University of Illinois. The two men shared a belief that the human body could be stimulated to fight tumors. Durovic convinced his host that a substance he had synthesized from horse plasma called Krebiozen was the stimulant they sought. Ivy arranged for clinical tests and announced at a crowded press conference that Krebiozen was "an agent for the treatment of malignant tumors."

The American Medical Association (AMA) examined the clinical data and concluded otherwise. The Chicago press clamored for increased funding for Ivy and Durovic but the President of the University of Illinois cited the AMA statement and demurred. Under pressure from Chicago politicians, the Illinois State Legislature held hearings on Krebiozen throughout 1953.

A letter entered into evidence at the hearings by an Argentine physician stated that he and Prof. Harry Anderson (who was in Argentina to attend a scientific conference) had visited the facility where Durovic claimed to produce Krebiozen. They found an abandoned building with no laboratory facilities. Back in Chicago, a lawyer for Dr. Ivy attacked Anderson saying he had not submitted the letter under oath. Anderson offered to do so, but the hearing chair deemed it unnecessary.

Phil was livid when he heard that a lawyer had publicly impugned his father's testimony. To his son, Harry Anderson was incapable of lying because "he embodied integrity just by being."[18] Phil admired this trait in his father and sought to emulate it all his life. As we will see, he took a principled stand not to participate in military consulting work, he was embarrassed personally when

[17] George D. Stoddard, *Krebiozen: The Great Cancer Mystery* (Beacon Press, Boston, 1955); Patricia Spain Ward, "Who Will Bell the Cat? Andrew C. Ivy and Krebiozen," Bulletin of the History of Medicine **58**, 28–52 (1984).

[18] Interview of PWA by the author, October 7, 2015.

he failed to detect scientific misconduct at Bell Labs, and he immediately disowned an entire book he wrote when he realized that the theory at its core was incorrect.

The Illinois Legislature concluded its hearings by endorsing Krebiozen. Durovic and Ivy (who had by now left the University) began a ten-year campaign to build public enthusiasm and political support for the substance. Negative reports issued in 1963 by the Federal Drug Administration and the National Cancer Institute did little to dampen the hope of desperate patients. Krebiozen continued to be manufactured and sold in Illinois until 1973 when the state criminalized those activities.

Informal Education

Peers were important to Anderson and several played a continuing role in his life. One of these was (Henry) Pierre Noyes, a bright fellow who attended elementary school, high school, and college with Phil. Pierre became a theoretical physicist also. At some point before high school, Phil and Pierre began to question the logic and historicity of the stories presented in the Bible. It was not long before they rejected religion and embraced atheism, a decision abetted by both of their fathers. Anderson's father helped him resist his mother's entreaties to attend church services. Years earlier, Harry had abandoned organized religion in reaction to his own father's hellfire and brimstone form of faith.

Pierre's father (a chemistry professor at the University of Illinois) did his part by giving the boys a copy of *Heavenly Discourse* (1927) by C.E.S. Wood. Wood was a prominent attorney who defended anarchists like Emma Goldman and other political radicals. His satirical book records conversations between heavenly figures like God, Saint Peter, Jesus, and Satan, and historical figures like Voltaire, Joan of Arc, Thomas Jefferson, Charles Darwin, Theodore Roosevelt, and Mark Twain. *Heavenly Discourses* aimed its satire at exposing the sanctimonious nature of religious zealots and decrying the use of religion to justify war.[19] Phil and Pierre took these messages to heart.

Three other high school students, Henry Swain, Philip Thompson, and Warren Goodell, joined Phil and Pierre to form a close-knit group of five friends. Thompson, who went on to a distinguished career in mathematical meteorology, recalled that:

> We all had a very strong scientific bent, particularly in mathematics...I think we all learned a great deal from each other because we were constantly stimulating each other. Through our late high school days until even in our college days, we had a kind of mathematical competition in which we would pose problems to each other...Phil Anderson was particularly good at this. He had a flair...to use any method that was available to solve a problem.[20]

Thompson also recruited Anderson to play violin (which he took up at Uni) in a string quartet comprised of the two of them, Henry Swain, and Henry's sister Martha. Years later, Martha's best friend, Joyce Gothwaite married Phil Anderson.

In March of 1937, soon after his thirteenth birthday, Anderson accompanied his father, mother, and sister on a five-month excursion to Europe. The occasion was a sabbatical leave of absence for Harry to visit foreign botanical research facilities. Because of its timing, this trip had a significant impact on Phil's education and maturation. The Spanish Civil War was raging and, just a month earlier, Adolf Hitler had declared that "the noblest and most sacred [task] for mankind is that each racial species must preserve the purity of the blood which God has given it."[21]

[19] Robert Hamburger, *Two Rooms: The Life of Charles Erskine Scott Wood* (University of Nebraska Press, Lincoln, NE, 1998).

[20] Interview of Philip D. Thompson by Joseph Tribbia and Akira Kashara, December 15–16 1987, American Meteorological Society, Oral History Project, accessed June 17, 2017.

[21] "On National Socialism and World Relations," a speech delivered by Chancellor Adolf Hitler in the German Reichstag on January 30, 1937.

The Andersons crossed the Atlantic on a cruise ship with their family automobile stored below decks.[22] They spent their first ten weeks visiting horticultural centers and sightseeing in England and France. In London, a night spent sitting on a curb in Hyde Park ensured an unobstructed view of the Coronation Procession of King George VI. In Paris, the just-completed painting *Guernica* by Pablo Picasso graced the Spanish pavilion of the Exposition of 1937.

The family devoted the next ten weeks to an extended road trip to European centers of horticulture in Utrecht, Heidelberg, Munich, Vienna, Budapest, Belgrade, Sofia, Sarajevo, Dubrovnik, and Trieste. Phil and Graccie angered their parents when they crossed from the Netherlands into Germany and raised their hands to give a mock "Heil Hitler" salute to the border guards.[23] Nothing came of it, but the family soon perceived a change in atmosphere. Some people in Germany would not talk to them; others whispered to the visiting Americans that they hated the regime. Later, they witnessed huge pro-Nazi demonstrations in the streets of Vienna. The family had read about the plight of the Jews in Germany and they were sympathetic.

Following their return to the United States, the long drive from New York City to Urbana gave Anderson plenty of time to reflect. Demagoguery was not something a kid from the American Midwest was used to seeing. Fifteen years later, the grown man had a visceral negative reaction to the same behavior in Senator Joseph McCarthy when he pursued his campaign to root out the supposed Communist infiltration of American institutions.

Champaign-Urbana in the 1930s

Like most people, the time and place of Anderson's upbringing shaped his world-view. His family's home was in a cobblestoned

[22] Grace Anderson, Log of 1937 European Trip. Courtesy of Andrew Maass.
[23] Interview of Andrew Maass by the author, June 18, 2017.

and tree-lined Urbana neighborhood adjacent to the university where most of the residents were faculty members. He saw these comfortable people every day. On the other hand, bus rides to tennis matches with high schools in neighboring small towns exposed him to people whose survival depended on a good crop from the seemingly endless corn and soybeans fields of Central Illinois.[24]

The two largest employers in the adjacent towns of Urbana (pop. 12,000) and Champaign (pop. 20,000) were the University of Illinois and the mechanical shops of the Illinois Central and Big Four railroad companies.[25] The shops employed workers of all races and Anderson gained some awareness of the small (5 percent) African-American population of Champaign and Urbana when he and his friends rode their bikes to the shops and poked around until they were shooed away.

Harry and Elsie taught their children to respect people of all races although the family's direct personal experience with minority groups was quite limited.[26] Black citizens of Champaign-Urbana had to sit in designated sections of movie theaters and the public swimming pools were off-limits to them.[27] Most of the hotels, restaurants, and barbershops in both towns refused to serve African-Americans, despite an explicit law in Illinois forbidding discrimination.

Anderson knew that southern-bound trains passing through Champaign had segregated passenger cars, but he did not really understand why his parents consistently refused to take family

---

[24] David Foster Wallace, *A Supposedly Fun Thing I'll Never Do Again: Essays and Arguments* (Little, Brown, and Company, New York, 1997). Wallace writes about his upbringing in Philo, Illinois, a small town ten miles from Urbana.

[25] Roger L. Geiger, *To Advance Knowledge: The Growth of American Research Universities: 1900–1940* (Oxford University Press, New York, 1986), p. 273. The Big Four was the popular name of the Cleveland, Cincinnati, Chicago, and St. Louis Railroad.

[26] The Anderson family employed an African-American woman when Phil was an infant and later when they needed occasional help with parties.

[27] Janet Andrews Cromwell, *History and Organization of the Negro Community in Champaign-Urbana*, Illinois, MS Thesis, Sociology, University of Illinois, 1934.

road trips to southern states. His real education in racial matters came in high school from one of his Indiana cousins, who passed on the attitudes he learned from his progressive aunt, the dean of students at Bennington College in Vermont.

Champaign and Urbana were politically conservative places at this time. The majority of the population was Protestant and church events were important to the social fabric of both towns. Anti-Semitism and anti-Catholicism existed, but not in minds of Phil and Graccie.[28] Their parents' disapproval of those who made bogeymen out of religious minorities was another constant of their childhood.

The left-wing politics of the Anderson family was not common and neither was the atheism embraced by Phil and Pierre Noyes. This divide burst into the open in 1945 when a woman named Vashti McCollum sued the Champaign Board of Education to prevent them from holding voluntary religion classes at her son's public school. Three years later, a landmark decision of the United States Supreme Court held that these classes violated the "establishment of religion" clause of the First Amendment of the Constitution.[29]

Anderson's late childhood and adolescence coincided with the years of the Great Depression. In the winter of 1932, nearly one third of Americans were unemployed. In Champaign, a few dozen lucky men found employment when the federal Works Progress Administration built an administrative head-quarters for the town.[30] By contrast, a 1938 analysis of nearly

[28] See, e.g., Winton U. Solberg, "The Early Years of the Jewish Presence at the University of Illinois," *Religion and American Culture: A Journal of Interpretation* 2(2), 215–45 (1992).

[29] The 1948 US Supreme Court case is *McCollum v. Board of Education of Champaign County*. Vashti McCollum and her husband John were atheists. John happened to be a junior faculty colleague of Harry Anderson in the Horticulture Department of the University of Illinois. Leigh Eric Schmidt, *Village Atheists: How American Unbelievers Made Their Way in a Godly Nation* (Princeton University Press, Princeton, NJ, 2016), pp. 268–71.

[30] Raymond Bial, *Images of America: Champaign* (Arcadia Publishing, Charleston, SC, 2008).

200 colleges and universities found that very few professors lost their jobs. Instead, they experienced an average salary cut of about 15 percent.[31]

Harry Anderson experienced a salary cut of just this magnitude in 1933. However, his salary that year of $3500 was already well above the national mean of $1970 and his 1940 salary of $4200 placed him among the top 15 percent of all American wage earners.[32] This income (about $77,000 in 2020 dollars) permitted Elsie to hire a live-in college girl to help with housework and child care. Harry bought a new car every few years and the family enjoyed regular summer vacations. Unlike many high school boys around the country, Phil did not have to contribute to the family income by working after school or during the summer. By the standards of a nation coping with the Depression, he led a privileged life.

Money was not a major issue for the Andersons until the time came to consider college options for Phil. Graccie wanted to pursue a PhD in biochemistry at the University of Wisconsin and Harry's salary was not quite enough to pay simultaneously for two children pursuing college degrees.[33] One solution was for Phil to attend the University of Illinois as his sister had. That would cost nothing. Another possibility was to attend inexpensive Wabash College where his grandfather and uncle taught. A third option—the one taken thanks to a generous scholarship—was Harvard College.

[31] Walter M. Kotschnig, "Depression, Recovery and Higher Education: A Review and Preview," *Bulletin of the American Association of University Professors (1915–1955)* **24**, 19–28 (1938).

[32] Transactions of the Board of Trustees of the University of Illinois, 1928–1930, 1934–1936, 1938–1940; *Statistical Abstract of the United States, 1934–1935*, No. 177. Individual Income Tax Returns: By Income Classes; *Statistical Abstract of the United States, 1944–1945*, No. 285. Income Tax Returns, Individual, Estate, and Trust, by Net Income Classes: 1935–1941.

[33] Grace Anderson did earn a PhD in biochemistry, but the birth of twin boys prevented her from pursuing the subject professionally. Later, she forged a successful career as a university science librarian. Interview of Andrew Maass by the author, June 18, 2017.

A Snapshot

Recommendation letters in his Harvard College application file provide a glimpse into Anderson's personality and temperament as a graduating senior (Figure 2.2). The letters come from Charles Sanford (Principal, University High School), Prof. Wheeler Loomis (Chair, Physics Department, University of Illinois), and Prof. William Oldfather (Chair, Classics Department, University of Illinois). The recommenders were not free from bias—their purpose was to help Phil gain admission—but their comments provide some insight nonetheless.

Figure 2.2 Phil Anderson at age 16 in the photograph he submitted as part of his application to Harvard College in 1940. Source: Harvard University, Office of the Registrar.

Principal Sanford remarked that "as a student, [Phil] is honest and responsible. He is courteous and pleasant but somewhat reserved and self-conscious." Sanford listed the young Anderson's outstanding personality characteristics as "persistence, intelligence, wit, originality, modesty, and sincerity." He attested to Phil's emotional balance and noted that "he seems to limit his friends to a very close circle."

The physicist Wheeler Loomis had earned all his degrees from Harvard and, as a Saturday Hiker, he played a key role in convincing Anderson to apply for admission and seek a scholarship. He stated that "everything I know about Phil is favorable" and then focused on the Anderson family "traditions" which he characterized as "reliability, perseverance, a sense of humor, a force of character, and an unusual breadth of culture."

William Oldfather, the 60-year-old leader of the Saturday Hikers, emphasized his personal knowledge of Phil. He judged him "a normal wholesome boy in every respect" who "is more widely read than any boy I have ever known". He praised Anderson's "alert and inquiring intelligence" and then made a prediction: "I should rate his promise of becoming a conspicuous figure in society as above that of the young James Tobin, also from this community." Tobin had graduated from Uni and gone to Harvard on a scholarship four years earlier. He won a Nobel Prize for Economics in 1981.

The most remarkable document in Anderson's Harvard file is a letter written by his father. The Harvard Dean of Freshmen had written to the parents of all incoming freshmen requesting a profile of their child to help Harvard with their advising process. Harry Anderson's clear-eyed letter begins, "as a parent, I am naturally not unbiased . . . but I believe Philip has always been happy in his home life." He remarks that "Philip is not religiously minded . . . and I would judge that he could be classified a mild radical in his political and social thinking."

Harry predicts that the Dean will find "Philip a good natured, even-tempered boy, tolerant of others' opinions but likely to defend his own stubbornly." He then candidly states that:

> Philip's greatest weakness is his inability to make friends easily. He is not at ease with people who do not interest him. He has never cultivated the art of getting along with people and appearing to be interested in them. I think he realizes this weakness and there is some evidence he is trying to remedy it. He needs training socially.

It is a common stereotype that many physicists need training socially. Nevertheless, later events testify to the truth of much of what Harry Anderson wrote about his college-bound son.

3

Making Waves

In early September 1940, sixteen-year-old Phil Anderson and his high school friend Pierre Noyes said goodbye to their families at the railroad depot in Urbana, Illinois. Twenty-five hours, two train changes, and a subway ride later, they lugged their suitcases through one of the elegant gated entrances of Harvard University. Their freshman dormitory, Matthews Hall (Figure 3.1), was only a few steps away.

Figure 3.1 Matthews Hall. Anderson's freshman dormitory at Harvard. Source: Rickinmar.

Anderson studied at Harvard for six years, earning first a bachelor's degree and then a PhD. The United States entered World War II in his sophomore year, and he served his country for the two years between graduation and the end of the war. Two actions Anderson took during these eight years turned out to have important professional consequences. As an undergraduate, he prepared for war-related work by taking mostly electronics and radio engineering courses rather than pure physics classes. After the war, he returned to Harvard and completed a PhD thesis in chemical physics rather than work in the hot new field of nuclear physics. Together, these decisions led him to a career in theoretical physics at Bell Telephone Laboratories.

College Accelerated

Anderson attended Harvard because a National Scholarship paid most of the costs of his tuition, room, and board. These scholarships had been created five years earlier by Harvard's president, James Bryant Conant, to make it possible for "boys with superior intellectual endowment" and a "high development of character and personality" to attend Harvard.[1] The scholarship program provided a (nearly) full-ride because the cost of tuition, room, and board at Harvard in 1940 was $924, well above the cost of most other colleges and well above the means of most American families.[2]

Sixty percent of the seniors who graduated from Harvard the year Anderson arrived came from the wealthiest 2.5 percent of the United

---

[1] *The Harvard College National Scholarships: A Descriptive Report at the End of Six Years*, (Harvard University Press, Cambridge, MA, 1949), pp. 12–16.

[2] The corresponding costs at the University of Pennsylvania and the University of Michigan were $520 and $590, respectively. *New York Times*, November 14, 1982; http://www.nytimes.com/1982/11/14/education/paying-for-college-is-working-your-way-through-still-possible.html?pagewanted=all. Accessed July 9, 2017; University History, Tuition and Mandated Fees, University Archives and Records Center, University of Pennsylvania. http://www.archives.upenn.edu/histy/features/tuition/1940.html. Accessed July 9, 2017; Bulletin of General Information, University of Michigan, 1940–1941, p. 15.

States population.[3] A typical class was very homogeneous in class, race, religion, and ethnicity.[4] Most were graduates of the Eastern preparatory schools favored by the Protestant elite who still dominated the major political and cultural institutions of the country.[5]

Harvard's National Scholarships were open only to high school students from seventeen states in the Midwest, South, and Far West. This restriction and the emphasis on students with "high character" were not accidents. They gave Conant the flexibility to continue Harvard's fifteen-year-old policy of limiting the percentage of Jews who attended the College to around 10 percent. The original and continuing purpose of that policy was to appease the patrician families (who paid full tuition) so they did not begin to abandon Harvard and send their sons to college elsewhere.[6]

The semi-autobiographical novel, *The Last Convertible*, by Anton Myrer provides a glimpse of the culture Anderson encountered when he arrived on campus. Myrer's alter ego enters Harvard as a freshman in 1940 and remarks,

> The last thing you could call me was sophisticated. I was on full scholarship; I owned exactly one sports jacket and two suits; my spending money was what I could earn [from campus jobs]. I was all too conscious of the gulf that separated me from the other students in a thousand and one ways...every nuance of social distinction and the hierarchies of privilege.[7]

[3] Jerome Karabel, *The Chosen* (Houghton-Mifflin, Boston, 2005), p. 159. The future American President John Fitzgerald Kennedy graduated from Harvard College the year Phil arrived.

[4] The Harvard acceptance rate at this time (85%) reflects the special consideration given to boys from wealthy and socially prominent families and "legacy" boys whose fathers were Harvard graduates. Official Register of Harvard University **39** (5) 1942.

[5] James D. Davidson, "Religion among America's Elite: Persistence and Change in the Protestant Establishment," *Sociology of Religion* **55** (4), 419–40 (1994).

[6] Jerome Karabel, *The Chosen* (Houghton-Mifflin, Boston, 2005), p. 109 and Chapter 6; Paul F. Zweifel, Norman J. McCormick, and Laurie H. Case, "Kenneth Myron Case 1923–2006: A Biographical Memoir," National Academy of Sciences, 2013.

[7] Anton Myrer, *The Last Convertible* (GP Putnam's Sons, New York, 1978), p. 31, 34.

Anderson's experience was very similar. He felt awkward and socially naïve around the prep school kids, most of whom did little to hide their disdain for scholarship students.[8] Harvard did not assign his high school friend Pierre Noyes to be his roommate and he found it difficult to make new friends. Often, he went to the dining hall alone. His finances were tight and he saved money by mailing his laundry home to Urbana rather than use local laundromats.[9] Speed skating at a rink in downtown Boston provided a rare but welcome release for the 5' 8", 150 pound freshman.

Anderson enrolled in five courses in his first year: math, physics, English composition, French, and European history. The history course was memorable because it sparked a lifelong fascination with the early medieval period and its architecture. As a prospective major in mathematics, Phil took Analytic Geometry and Calculus. However, on the advice of the Saturday Hiker and University of Illinois Physics Department Chair, Wheeler Loomis, the physics class he chose was the first course recommended for physics majors—Mechanics, Heat, and Sound.

The physics course lecturer was Wendell Hinkle Furry, the instructor Anderson later judged to be the single best teacher he experienced at Harvard. It was a small-world encounter because Furry was an Indiana native who had gotten his PhD in theoretical molecular physics at the University of Illinois, just a few blocks from the home of the then nine-year-old Phil Anderson.

Furry was a strict and challenging professor. At first, Anderson was disconcerted because he had no intuition for some of the concepts Furry discussed.[10] This changed in a few weeks, but it

---

[8] Private communication with Katia Noyes, October 9, 2018.

[9] Harvard dormitories did not have coin-operated washing machines and dryers at that time. The average yearly cost to use a local (Cambridge, MA) commercial laundry service was about $30 per year. Deirdre Clemente, *Dress Casual: How College Students Redefined American Style* (University of North Carolina Press, Chapel Hill, NC, 2014), p. 83. The cost to mail a package home to Urbana was $0.36 per pound. History of United States Postal Rates, accessed October 7, 2018.

[10] Interview of PWA by Lillian Hoddeson, May 10, 1988, Niels Bohr Library & Archives, American Institute of Physics, College Park, MD.

struck him forcefully that intuition was essential to his overall understanding of any problem he faced. It became his secret weapon, a fact noticed by many as his career evolved.

Anderson's second year at Harvard differed completely from his first. He and Pierre Noyes now roomed together in Winthrop Hall. Directly above them lived Harvey Lincoff, the son of a Jewish jeweler from Pittsburgh, and he and Phil soon developed a strong friendship that lasted a lifetime.[11] Phil and Pierre also found a group of boys with whom they could drink beer and try to meet girls. Academically, Anderson was able to coast because the brightest kids in his math and physics classes banded together to form a study group.

The academic star of the study group was Thomas S. Kuhn, the self-confident scion of a wealthy Jewish family from the suburbs of New York City.[12] A physics major, Kuhn became editor of *The Harvard Crimson* (the student newspaper) as a sophomore and he won election as a junior to Phi Beta Kappa, the academic honor society. Anderson and Kuhn took quite a few classes in common and, according to Phil, he "saw to it that his grade was always a point or two higher" than Kuhn's in those classes.[13] This competition continued when both pursued PhDs in physics at Harvard. Later in life, Kuhn's groundbreaking book, *The Structure of Scientific Revolutions*, made him one of the most prominent historians of science of the second half of the twentieth century.[14]

Life changed for everyone on December 7, 1941 when the Imperial Japanese Naval Air Service attacked the US Pacific Fleet at its base at Pearl Harbor, Hawaii. Within days, the United States was at war with Japan, Germany, and Italy. Harvard responded by

[11] Author correspondence with Ingrid Kreissig, July 31 2019.

[12] N.M. Swerdlow, "Thomas S. Kuhn, 1922–1996," *Biographical Memoir of the National Academy of Sciences* (National Academy Press, Washington, DC, 2013).

[13] Interview of PWA by Alexei Kojevnikov, March 30, 1999, Niels Bohr Library & Archives, American Institute of Physics, College Park, MD.

[14] Karl Hufbauer, "From Student of Physics to Historian of Science: T.S. Kuhn's Education and Early Career," *Physics in Perspective* 14, 421–70 (2012).

organizing a Radio Research Laboratory to develop radar countermeasures and an Underwater Sound Laboratory to improve
sonar systems and design torpedoes.

When Anderson returned to campus after the Christmas
recess, he discovered a host of other changes: a compulsory exer-
cise program for all undergraduates, a 12-week 'third semester'
offered during the summer, and a new set of applied physics courses
focused on vacuum tubes and radio waves.[15] Like many others
eager to contribute to the war effort, he switched majors to a new
and accelerated program that would permit him to graduate in
eighteen months with a bachelor's degree in 'Electronic Physics'.

Anderson's accelerated degree program began in the spring of
1942 with a course on Alternating Currents taught by the theoretical physicist John Van Vleck. Five years later, Van Vleck would
agree to serve as Phil's PhD supervisor. The summer of 1942 and
the entire 1942–1943 academic year were devoted to the study of
electron tubes, amplifiers, electric oscillations, radio waves,
antennas, and high-frequency electronics.

Eventually, most of these electronics classes passed out of the
physics department and into the curriculum of Harvard's Division
of Engineering and Applied Science.[16] But at the time, two
Harvard physics professors carried almost all the Electronic
Physics teaching load: E. Leon Chaffee and Ronold W.P. King.
Anderson took three courses from Chaffee and liked them all. He
also took three courses from King and disliked them all.

These reactions are an early display of an attitude Anderson
exhibited his entire career: a disdain for mathematical formalism
and a celebration of simple intuitive reasoning. In Phil's opinion,
Chaffee was an intuitionist and King was a formalist. It is interesting that Anderson's academic competitor, Tom Kuhn, recalled

[15] Official Register of Harvard University **41**, No. 23, September 26, 1944. Issue
Containing the Report of the President of Harvard College and Reports of
Departments for 1941–1942, pp. 92–3.
[16] Morton Keller and Phyllis Keller, *Making Harvard Modern: The Rise of America's
University* (Oxford University Press, Oxford, 2001), p. 106.

these electronics classes somewhat differently. To him, "Chaffee was an incredibly bad teacher, King was a very good teacher."[17]

At one point, Anderson acted on a bulletin board invitation and joined the 'Harvard Guerrilla Unit', a volunteer group whose members trained to create havoc behind enemy lines.[18] They learned how a German tank works, how to make Molotov cocktails, and how to neutralize enemy sentries.[19] Anderson also applied to an ROTC program designed to train Air Force navigators. The recruiters rebuffed him because he wore glasses.

Anderson graduated in May 1943 with a BS (summa cum laude) in Electronic Physics. It was a degree rooted in pre-twentieth century classical physics because, as he put it much later, "I heard not a word about quantum mechanics during my entire undergraduate experience."[20] Those of his classmates who were Physics (rather than Electronic Physics) majors fared slightly better because elementary quantum mechanics was used in one of the courses recommended for them: Introduction to Atomic Physics.

The difference between Physics and Electronic Physics turned out to be consequential because the recently created national Office of Scientific Research and Development sent recruiters to Harvard and other universities seeking graduating seniors and graduate students to work in war-related research facilities.[21] The recruiters sent the Physics majors to Los Alamos to work on the Manhattan

---

[17] Aristides Baltas, Kostas Gavroglu, and Vassiliki Kindi, "A Discussion with Thomas A. Kuhn," in *The Road Since Structure: Philosophical Essays 1970–1993 with an Autobiographical Interview*, edited by James Conant and John Haugeland (University of Chicago Press, Chicago, 2000), p. 267.

[18] Primus V. "Guerrillas in the Yard," *Harvard Magazine*, July–August 2011; http://harvardmagazine.com/2011/07/guerrillas-in-the-yard; accessed July 13, 2017.

[19] "175 Turn Out For Guerrilla Unit," *Harvard Crimson* July 31, 1942.

[20] Interview of PWA by P. Chandra, P. Coleman, and S. Sondhi, October 15, October 29, and November 5, 1999, Niels Bohr Library & Archives, American Institute of Physics, College Park, MD.

[21] Irvin Stewart, *Organizing Scientific Research for War: The Administrative History of the Office of Scientific Research and Development* (Little, Brown and Company, Boston, 1948), Chapter 18.

Project.[22] They sent the Electronic Physics majors like Phil, Pierre Noyes, and Tom Kuhn to radar research facilities like the MIT Radiation Lab and Harvard's own Radio Research Laboratory.

Wartime Service

From June 1943 to September 1945, Phil Anderson worked nine hours a day, six days a week in an antenna group at the Naval Research Laboratory (NRL) in Washington, D.C. Anderson's group focused primarily on "identify friend or foe" (IFF) radar systems designed to positively identify Allied aircraft. They also contributed to the development of countermeasures to disrupt enemy radar signals. Military events dictated these particular choices of tasks.

The afternoon of the Pearl Harbor attack, the aircraft carrier *USS Enterprise* sent out a squadron of Wildcat fighter planes to search for the Japanese fleet. With night approaching and the aircraft short of fuel, the *Enterprise* instructed six of the fighters to land on Ford Island in Pearl Harbor. This led to tragedy when American anti-aircraft gunners shot down five of the six fighters. In the darkness, the gunners lacked a reliable means to distinguish the Wildcats from Japanese fighters.

By the time Anderson arrived at NRL, the main goal of the IFF effort was to develop a radio-radar system to distinguish friendly aircraft from hostile aircraft that was fast and used separate frequencies for the challenge and reply signals.[23] Phil's team acquired components and expertise from specialists at Westinghouse, RCA, Bell Laboratories, Western Electric, General Electric, and DeForest Radio.[24] The quality and reliability of the components and advice

---

[22] The recruiters obfuscated the work that the Physics majors were expected to do at Los Alamos because the true nature of the Manhattan Project was a highly guarded secret.

[23] Louis A. Gebhard, *Evolution of Naval Radio-Electronics and Contributions of the Naval Research Laboratory* (US Naval Research Laboratory, 1979), p. 256.

[24] A. Hoyt Taylor, *Radio Reminiscences: A Half-Century* (US Naval Research Laboratory, Washington, DC, 1948), p. 164.

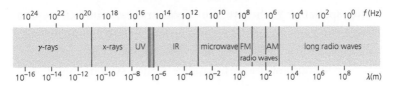

Figure 3.2 The spectrum (range) of electromagnetic waves that transport electric and magnetic fields through space at the speed of light, c. The different names refer to waves which differ only in their frequency f (measured here in Hz, i.e., cycles per second) or, equivalently, their wavelength, $\lambda = c/f$ (measured here in meters). The colored bars identify the visible portion of the spectrum. UV and IR stand for ultraviolet and infrared.

they received gave Anderson a lasting admiration for these technology companies.[25]

The NRL IFF system operated in the microwave portion of the electromagnetic frequency spectrum shown in Figure 3.2. Microwaves are like radio waves in every way except that their frequency is greater and their wavelength is shorter. This technology was absolutely essential to the Allied war effort because, compared to all previous radar systems, microwave radar sets had greater range, were easier to aim at a target, and were small enough to be mounted in aircraft and small naval vessels.[26]

The radar countermeasure work in the NRL antenna group sprang from another military incident. In August 1943, the German Luftwaffe began using radio-controlled glide bombs against Allied ships in the Mediterranean Sea. A mother plane released its bomb and then remained safely out of anti-aircraft range while it directed the bomb's subsequent movements. In short order, this weapon sank a British battleship and killed two hundred sailors on the cruiser *USS Savannah*.

[25] Biography of Philip W. Anderson, Nobel Foundation, 1977.
[26] Henry E. Guerlac, *Radar in World War II* (Tomash Publishers and American Institute of Physics, New York, 1987), p. 224.

NRL specialists rushed to the Mediterranean and intercepted the radio frequency used by the glide bombs. With this information, Anderson's group successfully developed a jamming antenna system.[27] An account of the performance of their system appeared in a seven-page report (with thirty-nine figures) titled *Antennas for Guided Missile Countermeasures on Destroyer Escorts*. Anderson is one of four authors of this 1944 document. It was his first technical paper.

Nineteen-year-old Phil Anderson grew up socially in Washington, D.C. He explored the city and wandered around its many museums. Every Sunday, he and his NRL buddies went to dinner at a downtown seafood restaurant. It was easy to meet women and Anderson had his share of dating success. The technical work was a mixed bag. At one point, he allowed a high-voltage discharge to melt a polystyrene circuit board he was holding. He proved so clumsy at manipulating delicately wired circuits that his boss switched him to soldering copper tubes together as part of the "microwave plumbing" they needed for their testing work.[28]

Anderson's lack of manual dexterity notwithstanding, his experience at NRL contributed significantly to his education. Early on, he attended a radar course that taught him more about the subject than all his Harvard electronics courses put together. Later, NRL brought in outside experts to help them understand the propagation of microwaves through the atmosphere. Anderson attended these meetings and heard his old Harvard professor, John Van Vleck, warn his hosts that an antenna system they proposed to build would not be effective because water vapor in the atmosphere would strongly absorb microwaves at the operating

---

[27] A. Hoyt Taylor, *Radio Reminiscences: A Half-Century* (US Naval Research Laboratory, Washington, DC, 1948), pp. 227–30. The NRL jamming antenna operated at a frequency between the AM and FM bands in Figure 3.2

[28] "Profiles of 4 Nobel Prize Winners", *New York Times*, October 12, 1977, p. 92. In an October 18, 1977 letter, Ernst Krause (Phil's boss at NRL during World War II) disputes the *New York Times* statement above as "an apocryphal story. I have no recollection of such clumsiness." He recalls "a bright physicist who was suddenly thrust into an engineering environment and did well at it."

frequency of the antenna. They ignored his advice and the experiment failed.

Van Vleck based his warning on a fundamental difference between classical physics and quantum physics. Classical physics describes the behavior of *macroscopic* objects like dust particles, human beings, and planets. Quantum physics describes the behavior of *microscopic* objects like electrons, nuclei, atoms, and molecules. Van Vleck's analysis exploited the fact that classical objects absorb electromagnetic waves at all frequencies, whereas quantum objects like water molecules absorb electromagnetic waves only if the wave frequency matches one of a set of discrete frequencies which are unique to the molecule. These frequencies amount to a "fingerprint" for the molecule in question.

Anderson understood the relevant physics because an NRL colleague had loaned him a just-published book—*The Mathematics of Physics and Chemistry* by Henry Margenau and George Murphy—which clearly explained the basics of quantum mechanics. Phil read it cover-to-cover and loved it. He must have nodded vigorously when reading the statement in the book's Preface that "emphasis on extreme rigor often engenders sterility; the successful pioneer depends more on brilliant hunches than on the results of existence theorems."[29]

Anderson spoke to Van Vleck during his visit. The war was winding down and they discussed the possibility that he would return to Harvard to earn a PhD. This was not an obvious choice. The Manhattan Project was still a closely guarded secret, but nuclear physics was widely understood to be the exciting new field of the future. Based on the quality of the nuclear physics research done at several universities before the war, an aspiring young physicist might easily choose to study at Columbia, Berkeley, Chicago, Princeton, or Rochester.

---

[29] Henry Margenau and George Mosely Murphy, *The Mathematics of Physics and Chemistry* (D. Van Nostrand and Co., Inc., New York, 1943), p. iii.

Van Vleck understood the importance of nuclear physics, but it was not his field of research and, as the current Chair of Harvard's physics department, he felt no particular obligation to direct Phil to other institutions. For his part, Anderson felt that Harvard's faculty had robbed him of a modern physics education (through its Electronic Physics curriculum) and he resolved he would extract it from them as a graduate student. He was not interested in other schools.

World War II ended in August 1945. Anderson was lucky. No one who served from his extended family or from his high school class was killed or badly injured. Stepping back into civilian life was easy. Van Vleck arranged for a quick acceptance to Harvard's graduate physics program and he was back on campus in time for the fall term to begin.

4

First Fruits

Graduate school at Harvard was a wonderful experience for Anderson. He was more mature, he was in familiar surroundings, and the GI bill paid his tuition, room, and board.[1] Making up for lost time, he took a heavy load of classes for five straight semesters. Highlights included a laboratory course in atomic physics, a year of quantum mechanics taught by Wendell Furry, and a summer course of classical mechanics taught by visiting professor Samuel Goudsmit.[2]

Thanks to pent-up supply from the war years, a large and extremely talented group of students went through many of these classes with Phil. His undergraduate rival Tom Kuhn (Figure 4.1) was among them and the two renewed their competition by trying to best each other solving the homework problems assigned in a Theory of Functions course.[3]

Another graduate school competitor was Walter Kohn, a refugee from Vienna by way of the University of Toronto who later won a Nobel Prize for his work in theoretical solid-state physics.[4]

---

[1] The GI Bill was the popular name for the Serviceman's Readjustment Act of 1944.

[2] Samuel A. Goudsmit (1902–1978) earned his PhD at the University of Leiden. He was a co-proposer of the idea of electron spin and a specialist in theoretical atomic spectroscopy. Goudsmit served for many years as a Senior Scientist at Brookhaven National Laboratory and editor-in-chief of the *Physical Review*, the American physics journal of record.

[3] Interview of PWA by Lillian Hoddeson, May 10, 1988, Niels Bohr Library & Archives, American Institute of Physics, College Park, MD.

[4] Kohn won one-half of the 1988 Nobel Prize for Chemistry. See A. Zangwill, "The Education of Walter Kohn and the Creation of Density Functional Theory," *Archive for History of Exact Sciences* **68**, 775–848 (2014).

Anderson took two classes with Kohn. One of these, "Group Theory with Applications to Quantum Theory," was taught by John Van Vleck, who began every class by picking a name from his class roll and asking that person a question about material discussed the previous class meeting. Kohn dreaded the days when Van Vleck called his name. However, as he related sixty-seven years after the events, "what irked me then, and still irks me now, is that Phil always knew the correct answer to the question!"[5]

Anderson and Kohn had different temperaments as students and this carried over into their styles as mature scientists. According to Phil, "I attack a problem as A, B, and then jump to Z. Walter goes ABCDE.... Z."[6] A physicist who published papers with both men agreed:

> Walter was always trying to formulate things in a mathematically precise way and then work out the consequences in a clear, clean, and elegant way. That is not at all what Phil would do. He would look for the central point and then use his insight and speed to go for the jugular.[7]

The two theorists remained friends for their entire careers. Nevertheless, close associates always sensed a rivalry between them for prominence in the solid-state physics community.

Anderson and Kohn also took a three-semester course taught by the newest member of the Harvard faculty, Julian Schwinger.[8]

---

[5] Oral presentation by Walter Kohn at "PWA at 90: A Lifetime of Emergence," a conference held at Princeton University, December 14–15, 2013.

[6] Remark by PWA quoted by William Brinkman. Interview of William Brinkman by the author, March 19, 2016.

[7] Interview of Maurice Rice by the author, July 15, 2015. The biographer David Cassidy [*Uncertainty: The Life and Science of Werner Heisenberg* (W.H. Freeman and Company, New York, 1992)] remarks similarly that his subject rejected the "cautious, traditional, and rational approach of most of his colleagues" [p. 124] because "his legendary intuition permitted him to leap to a bold solution without stumbling over intervening steps." [p. 291].

[8] The third semester of Schwinger's course does not appear on Phil's Harvard transcript. His vivid memory of the class suggests that he simply attended the class without registering for it.

Harvard had just outbid Berkeley and Columbia to secure the services of this *wunderkind* who had completed the research needed for a PhD even before Columbia University awarded him a bachelor's degree at the age of eighteen.[9] During the war, Schwinger worked at the MIT Radiation Laboratory. He famously worked only at night and scientists who left the Lab in the evening with mathematical problems written on their blackboards would find them solved the next morning in Schwinger's handwriting. Harvard promoted him to full professor after one year of service.

Walter Kohn gave a memorable description of Schwinger's teaching style:

> Attending one of his formal lectures was comparable to hearing a major concert by a very great composer, flawlessly performed by the composer himself... Old and new material were treated from fresh points of view and organized in magnificent overall structures. The delivery was magisterial, even, carefully worded, irresistible like a mighty river. He commanded the attention of his audience entirely by the content and form of the material, and by his personal mastery of it, without a touch of dramatization.[10]

Anderson never developed a lecturing style even vaguely similar to Schwinger's. However, he did pay close attention in class and some techniques he learned in Schwinger's lectures became important to his thesis research.

Outside of class, Anderson enjoyed the friendship of about a dozen Harvard graduate students he met at VANSERG dining hall.[11] They were a diverse group in age and academic interests—mathematics, physics, chemistry, anthropology, history, economics, and English.

---

[9] Jagdish Mehra and Kimball A. Milton, *Climbing the Mountain: The Scientific Biography of Julian Schwinger* (Oxford University Press, Oxford, 2003).

[10] Walter Kohn, "Tribute to Julian Schwinger," in *Julian Schwinger: the physicist, the teacher, and the man*, edited by Y. Jack Ng (World Scientific, Singapore, 1996), p. 62.

[11] VANSERG is an acronym for the original occupants of the building: Veterans Administration, Naval Science, Electronic Research, Graduate dining hall.

The VANSERG pals almost never engaged in deep political or philosophical discussion.[12] Instead, group members met every night to play bridge, read science fiction, solve math and word puzzles, compose doggerel, and sing together around a battered piano.[13]

Mathematics doctoral student Chandler Davis shared a passion for science fiction with Anderson and later became notorious as an outspoken Communist. He remembers that Phil was very comfortable in the group and exuded "a supreme confidence that made us take his opinions very seriously."[14] Bridge partner David Robinson recalled that:

> Phil announced one day that he had "proved Murphy's Law." Namely, why buttered toast always lands with the buttered side down when it falls off of a kitchen table. Using physics, he had studied the rotation of a piece of bread after being pushed off a flat surface from various heights.[15]

Another mathematics student, Tom Lehrer, was the acknowledged instigator of the group (Figure 4.1). Lehrer was a clever wordsmith who wrote songs the gang could sing while he played the piano. Anderson joined the singing enthusiastically and Lehrer remembers him for his joviality and good sense of humor.[16] Lehrer began performing at parties and his fame grew as his repertoire grew.[17] Some of his songs reflected his love of math and science like *The New Math* (fads in education), *Lobachevsky* (academic plagiarism) and *The Elements* (the periodic table set to music).

[12] Author correspondence with David Z. Robinson, April 8, 2015.

[13] Biography of Philip W. Anderson, Nobel Foundation, 1977.

[14] Author correspondence with Chandler Davis, April 8, 2015.

[15] Interview of David Z. Robinson by the author, April 6, 2015. Fifty years later, Anderson's analysis was confirmed, extended, and published: Robert A.J. Matthews, "Tumbling toast, Murphy's Law and the fundamental constants," *European Journal of Physics* **16**, 172–6 (1995).

[16] Interview of Tom Lehrer by the author, April 10, 2015.

[17] Ben Smith and Anita Bedejo, *Looking for Tom Lehrer, Comedy's Mysterious Genius*, Buzzfeed, April 9, 2014. Accessed July 23, 2017.

Figure 4.1 Classmates of Anderson at Harvard. Left: Thomas Kuhn. Source: Harvard University Archives. Right: Tom Lehrer. Source: *Rolling Stone* magazine.

Other Lehrer songs displayed a prescient social consciousness, such as *We Will All Go Together When We Go* (nuclear annihilation), *The Old Dope Peddler* (drug addiction), and *National Brotherhood Week* (race relations). In 1953, he recorded a long-playing record, *The Songs of Tom Lehrer*, and sold it by mail order. One year later, 10,000 records had sold and he was performing his songs around the country and then internationally.

The playful stimulation provided by the VANSERG crowd was a welcome relief from the pressure of his day job to excel academically in the classroom. Any inadequacy he may have felt about his undergraduate Electronic Physics degree disappeared when he earned nearly straight A's as a graduate student.

Choosing a Thesis Advisor

In his fourth semester, Anderson turned his attention to securing a faculty member to oversee his thesis research. The choice of a thesis supervisor is an important decision in the life of any graduate student. The research agenda, creative style, and personal contacts of a potential supervisor often set the course for a student for years to come. Phil's experience at NRL taught him he probably could not manage creative experimental work; he was better suited to theoretical physics. At Harvard in 1946, the only possible advisors in that area were Wendell Furry, Julian Schwinger, and John Van Vleck.[18]

Anderson liked Furry personally and he listened carefully as Furry explained his interests in elementary particle (meson) theory.[19] As an undergraduate, physics had excited Anderson because it explained how the world worked. Meson theory seemed pretty remote from that. Furry sensed Phil's hesitation and asked him if he was enthused about any of the research topics he had outlined. Phil's answered "not really" and that settled the matter.[20]

Schwinger was the superstar among the theorists and, with only a few exceptions, all the other theory-minded students in Anderson's class wanted to work with him (twelve of them did).[21] At this point, Anderson first displayed a strain of contrarianism that would appear repeatedly in his career. If all the fish were swimming downstream, Anderson would doggedly (some would say perversely) swim upstream.

Phil convinced himself that nuclear physics was not as exciting as all his classmates thought. He also decided that the large group of students hoping to work for Schwinger were overly enamored with formal mathematical methods. Years later, Anderson characterized

[18] The theorist Edwin Kemble was available, in principle, but he left research after World War II to help spearhead a history of physics effort promoted by Harvard's president, James Conant. Alexi Assmus, "Edwin C. Kemble 1889–1984," *Biographical Memoir of the National Academy of Sciences* (National Academy Press, Washington, DC, 1999).

[19] Meson theory was the thesis topic of the only PhD student Furry supervised in the years immediately after World War II. James N. Snyder, *Stimulated Decay of Mesons*, PhD Thesis, Physics Department, Harvard University, 1949.

[20] Interview of PWA by Lillian Hoddeson on May 10 1988, Niels Bohr Library & Archives, American Institute of Physics, College Park, MD.

[21] Jagdish Mehra and Kimball A. Milton, *Climbing the Mountain: The Scientific Biography of Julian Schwinger* (Oxford University Press, Oxford, UK), p. 153.

Figure 4.2 Anderson's PhD thesis supervisor, John Hasbrouck Van Vleck, circa 1930, a few years before he moved permanently to Harvard. Source: University of Wisconsin Archives.

the group as "competitive, intellectually snobbish, even somewhat sycophantic."[22] Given his superb grades, these classmates did not intimidate him intellectually. His uncharitable portrayal then seems more like a retrospective rationalization for his unwillingness to work for Schwinger on a problem that he suspected would vary only slightly from (and require methods of solution nearly identical to) what a dozen other Schwinger students were doing.

Fortunately for Anderson, the last physicist at Harvard capable of suggesting and supervising a theoretical research project was John Hasbrouck Van Vleck, a man Phil liked and respected (Figure 4.2). Van Vleck was a tenth generation American (of Dutch descent) and a third generation professor.[23] Thanks to a wealthy uncle, his childhood and adolescence combined gracious

---

[22] P.W. Anderson, "BCS and Me," in *More and Different: Notes from a Thoughtful Curmudgeon* (World Scientific, Hackensack, NJ, 2011), p. 5.

[23] Brebis Bleaney, "John Hasbrouck Van Vleck, 1899–1980," *Biographical Memoirs of the Royal Society* **28**, 627–65 (1982).

living with a passionate involvement with books, art, and travel. He went to college at the University of Wisconsin and then, under the direction of Edwin Kemble at Harvard, he used the brand-new quantum theory to earn his PhD in 1922.[24]

Van (as all his friends and colleagues called him) began his career at the University of Minnesota where he did significant work on the theory of the absorption and emission of electromagnetic waves by atoms. He moved to the University of Wisconsin in 1928 and there began what was to be his life's work: the application of quantum mechanics to electric and magnetic phenomena in matter.[25] His 1933 monograph, *The Theory of Electric and Magnetic Susceptibilities*, was the first book to systematically apply quantum mechanics to a large swath of what would later be called solid-state physics.

In 1934, Van Vleck moved to Harvard where he retained his previous interests in atomic and molecular physics, but turned increasingly to the theory of magnetism in solids. In time, Van Vleck's exquisite taste in choosing problems, and his clarity as a thinker and as a writer, established him as the doyen of theoretical magnetism in the United States and beyond.

Van Vleck was a modest and cultivated man with a good sense of humor and a lively wit. After completing a series of eight lectures in Paris in the local language, he quipped "I had to go to the Riviera afterwards to recuperate; I don't know what the audience had to do."[26] He also had great charm and a well-deserved reputation for kindness. In 1946, a new graduate student arrived on campus two days before the start of the fall term and could not find a room to

[24] Frederick Hugh Fellows, *J.H. Van Vleck: The Early Life and Work of a Mathematical Physicist*, PhD Thesis, University of Minnesota, March 1985.

[25] Charles Midwinter and Michel Janssen, "Kuhn Losses Regained: Van Vleck from Spectra to Susceptibilities," in *Research and Pedagogy: A History of Early Quantum Physics Through its Textbooks*, M. Badino and J. Navarro eds. (Edition Open Access, Berlin, 2013), pp. 137–205.

[26] Brebis Bleaney, "John Hasbrouck Van Vleck, 1899–1980," *Biographical Memoirs of the Royal Society* **28**, 627–65 (1982).

rent. Van Vleck offered to let him stay in the finished basement of his house, asking only that the student shovel the snow around his property when necessary.[27] Well-known as a fastidious dresser, Van Vleck was also an enthusiastic fan of college football and a person whose encyclopedic knowledge of American and European train schedules was legendary among his acquaintances.[28]

Van Vleck wanted to understand the absorption of microwave radiation by small molecules. This interest followed naturally from his wartime interest in the absorption of microwaves by water vapor.[29] Before the war, physicists had studied molecules by analyzing how they absorbed visible light and its longer wavelength cousin, infrared radiation.[30] After the war, it was natural to repurpose radar technology to study how molecules absorbed microwaves, a radiation with even longer wavelength.[31]

Quickest off the mark in exploiting microwaves for physics research were universities (Columbia, Duke, Harvard, MIT, and Oxford) and industrial laboratories (Bell Laboratories, General Electric, RCA, and Westinghouse). The existence of a wartime radar program at all these places ensured that the necessary strong sources of microwave radiation were already on hand. Some of this research aimed at measuring the properties of the atomic nucleus. At Harvard, the focus was on the molecules themselves.

[27] Author correspondence with Hellmut Juretschke, May 30, 2018.

[28] Interview of Martin Blume by the author, August 5, 2015.

[29] J.H. Van Vleck, "The Absorption of Microwaves by Oxygen," *Physical Review* **71**, 413–24 (1947); "The Absorption of Microwaves by Unsaturated Water Vapor," *Physical Review* **71**, 425–33 (1947).

[30] Nathan Ginzburg, "History of Far-Infrared Research: II. The Grating Era, 1925–1960," *Journal of the Optical Society of America* **67**, 865–71 (1977).

[31] Paul Forman, "Swords into ploughshares: breaking new ground with radar hardware and technique in physical research after World War II," *Reviews of Modern Physics* **67**, 397–455 (1995). See also, Walter Gordy, "Early Events and Some Later Developments in Microwave Spectroscopy," *Journal of Molecular Structure* **97**, 17–32 (1983).

Anderson visited the laboratories of the experimenters among his graduate student friends and watched as they worked with equipment familiar to him from his Naval Research Laboratory days. The glowing curves that appeared on their oscilloscope screens during absorption made the invisible microwaves more real to him than the descriptions offered by contemporary *Scientific American* articles which focused on their applications to wireless communication and the rapid heating of food.[32]

The student experimenters were excited about their research and this helped Anderson decide to ask Van Vleck to supervise him. The forty-eight year old theorist agreed and only Harvard's oral qualifying exam stood in the way of their working together. A small committee (including Van) administered the exam and Phil breezed through it until a question came up about the motion of a spinning top. Unaccountably, he drew a blank, despite the fact that he had done well in Van's graduate course where that topic was discussed. Anderson left the room dispirited and the committee debated his prospects as a theoretical physicist. Van was hesitant, but he convinced his colleagues to give his new student a chance. They passed him, albeit narrowly.[33]

Thesis Problem Posed

In the fall of 1946, Van proposed a thesis problem to Anderson drawn from the field of *spectroscopy*, the study of the interaction of electromagnetic radiation with matter. All physicists learn about this subject from the hero's story of Niels Bohr, the Danish theorist who earned a permanent place in the history of physics with his ground-breaking quantum theory of the hydrogen atom. For years, experimenters had painstakingly measured the frequencies

[32] Harland Manchester, "Microwaves on the Way," *Scientific American* **174**(1), 28–35 (1946); Vin Zeluff, "Demobilized Microwaves," *Scientific American* **176**(6), 252–5 (1947).

[33] Interview of PWA by Lillian Hoddeson, May 10, 1988, Niels Bohr Library & Archives, American Institute of Physics, College Park, MD.

of the electromagnetic waves emitted by this atom. Bohr's theory brushed aside the teachings of classical physics yet correctly predicted every one of hydrogen's very precisely known fingerprint frequencies. The quantum era began in earnest with Bohr's insights.

Van piqued Anderson's curiosity in molecular spectroscopy by showing him just-published data obtained from a sample of ammonia (NH_3) gas. The experimenters had exposed the molecules to a source of microwaves and measured the relative amounts of electromagnetic wave energy absorbed over a range (spectrum) of frequencies. The absorption signal (black curve in Figure 4.3) exhibited a collection of narrow and partly overlapping peaks, each centered on one of the *spectral lines*—the fingerprint frequencies—of the ammonia molecule (vertical red bars). A complete theory for this experiment would predict both the fingerprint frequencies and the heights and widths of the peaks.

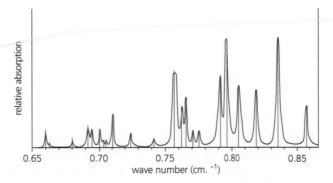

Figure 4.3 The absorption spectrum for an ammonia molecule in the microwave portion of the electromagnetic spectrum. The "wave number" on the horizontal axis is proportional to the frequency of the absorbed radiation. The vertical axis is the relative amount of radiation absorbed at each frequency. The black curve is experimental data. The vertical red lines indicate the fingerprint frequencies at the centers of the absorption peaks. Figure adapted from B. Bleaney and R.P. Penrose, "Ammonia spectrum in the 1 cm wavelength region," Nature 157, 339–40 (1946).

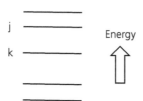

Figure 4.4 Ladder representation of the allowed states of quantum system arranged by total energy. Absorption of a photon with frequency f excites the system from a quantum state labelled k to a quantum state labelled j if $E_j - E_k = hf$.

Anderson knew that the mere existence of fingerprint frequencies confirmed three key elements of Bohr's theory. First, all microscopic systems possess only a discrete set of allowed states. Second, an electromagnetic wave with frequency f consists of many *photons*, each of which carries an energy hf, where h is a number called Planck's constant. Third, quantum systems absorb one photon at a time with a corresponding transfer of the photon's total energy hf to the system.

The rungs of the ladder shown in Figure 4.4 represent the allowed states of a quantum system arranged according to the energies of the states. Bohr's radical suggestion was that a system initially in an allowed state with energy E_k absorbs a photon with energy hf and jumps up to an allowed state with energy E_j. No energy is gained or lost overall (energy is conserved) as long as $E_j - E_k = hf$. Using this formula, different choices for the initial and final states produce the different fingerprint frequencies.

When Anderson began his work, physicists understood that all the fingerprint frequencies in Figure 4.3 corresponded to *inversions*. These are periodic motions of the pyramid-shaped NH_3 molecule where the nitrogen atom at the pyramid's apex oscillates back and forth across the plane formed by the triangle of three hydrogen atoms (see Figure 4.5).[34] Think of an open umbrella repeatedly turning itself inside out and then righting itself again. There

---

[34] Technically, the heavy nitrogen atom remains at rest and the three hydrogen atoms execute the inversion.

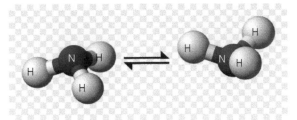

Figure 4.5 Cartoon of an ammonia (NH_3) molecule undergoing inversion.

are many different inversion frequencies because the ammonia molecule can *rotate* in different ways while it inverts. In quantum language, the nitrogen atom inverts by hopping back and forth at the frequencies $(E_j - E_k)/h$ where the energies E_j and E_k label different allowed states of rotation.

It happens that the energy of an ammonia molecule is largest just when the nitrogen atom passes through the plane formed by its three hydrogen molecules. Moreover, the photon absorbed during inversion does <u>not</u> quite supply enough energy to the molecule for the nitrogen atom to reach that energy. Inversion occurs nevertheless by virtue of a mind-bending quantum mechanical process called *tunneling*.

Tunneling exploits the wave-like nature of a quantum particle by permitting the particle to explore regions of space that are absolutely forbidden to it classically. This situation arises several times in the chapters to come. Here, the suspension of Newton's laws for the nitrogen atom sanctioned by tunneling permits it to pass *through* the energy barrier presented by the hydrogen atoms even though it lacks the energy to surmount it. Classically, the nitrogen atom would simply bounce off the barrier plane.

In January 1953, John Van Vleck discussed the inversion of the ammonia molecule as the centerpiece of a talk he gave when his term ended as president of the American Physical Society. He discussed the history of the phenomenon, showed photographs of the latest microwave technology used for contemporary experiments, quoted the English author Jonathan Swift when he needed a *bon mot*,

Figure 4.6 Cartoon used by John Van Vleck to illustrate quantum mechanical tunneling. Upper panel: A sufficiently high barrier protects a physicist from a classical lion. Lower panel: No barrier protects a physicist from a quantum mechanical lion capable of tunneling. Reproduced from J.H. Van Vleck, "Two Barrier Phenomena," *Physics Today* 6(6), 5–11 (1953) with the permission of the American Institute of Physics.

and used Figure 4.6 to give a qualitative explanation of quantum mechanical tunneling. The man behind the barrier in this cartoon is safe from a classical lion but not from a quantum lion.

With inversion events understood to be the origin of the microwave absorption peaks in Figure 4.3, physicists turned their attention to the *shapes* of the peaks. The pre-quantum American physicist Albert Michelson had proposed that collisions between the absorbing species increased the widths of spectral lines.[35] Accordingly, Van Vleck challenged his student Anderson to develop a quantum mechanical theory for microwave absorption which took full account of intermolecular collisions. This was not a deep question of theoretical physics, but it was far from trivial and unquestionably a hot topic: six experimenters and five

[35] Albert A. Michelson, "On the Broadening of Spectral Lines," *Astrophysical Journal* **2**, 251–63 (1895).

theorists published papers about the line widths of the ammonia inversion spectrum during the time it took Anderson to complete his thesis.[36]

Physicists were keenly interested in spectral line widths at this time because there was virtually no other way to learn about the forces that acted between molecules. This motivated Anderson to begin his research by studying everything the scientific world knew about these forces. Months passed and his progress was slow. It slowed even more in May 1947 when he went home to Urbana for a short vacation. No one could have predicted that he would meet and court the woman with whom he would spend the rest of his life.

Joyce Gothwaite

The 21-year old Joyce Charlotte Gothwaite dreamed of a better life. Her mother worked as the office manager at a paint company. Her father's undiagnosed dyslexia and mild alcoholism impaired his ability to hold a job for very long. As a result, Joyce and her parents lived in the two-bedroom apartment of her grandparents in the Hyde Park neighborhood of Chicago. This home had been forced on the older Gothwaites after they and 600,000 other shareholders saw much of their life savings disappear with the collapse of the utilities empire of Samuel Insull.[37] Joyce slept on a sofa in the living room.

Joyce's saviors were her cousin, aunt, and uncle. The latter was a paleontologist for the Illinois Geological Survey.[38] Spending time with them made it clear that a good education was her best hope to escape the life her parents led. She hung out with the children of University of Chicago faculty members, excelled in

[36] James E. Wollrab, *A Bibliography of Microwave (Rotational) Spectroscopy*, US Army Missile Command, Report RD-TM-65–14, August 4, 1965.

[37] John F. Wasik, *The Merchant of Power: Sam Insull, Thomas Edison, and the Creation of the Modern Metropolis* (St. Martin's Press, New York, 2006).

[38] Interview of Susan Anderson by the author, March 3, 2016.

Figure 4.7 Joyce Gothwaite, circa 1945–1946, a year or two before she met her future husband, Phil Anderson. Source: Susan Anderson.

high school, and earned a BA in English after only two years at the University of Illinois.[39] A coincidence put Joyce together with Phil Anderson. Her best friend and Illinois sorority sister, Martha Swain, was the sister of Phil's high school friend Hank Swain. After graduating from college in 1945, Joyce stayed in Urbana and worked as a secretary. Her longer-term plan was to move to New York City and look for a job there. Several previous summers working for the Chicago Chamber of Commerce had generated letters of reference attesting to her efficiency, intelligence, and independence.

Sometime in early June 1947, Hank brought Joyce around to the Anderson home to meet Phil. They hit it off immediately. He found her smart, funny, and attractive (Figure 4.7). Her piercing

[39] Joyce's parents opposed her decision to attend college. They wanted her to get a job and bring money into the family.

sapphire eyes were particularly striking. It helped that her liberal politics coincided with his. Their courtship was unconventional: a dinner date followed by an entire evening of talk, repeated again the next day. That was all it took; they were a couple from that moment on.

A few weeks later, Phil returned to Harvard and Joyce left for interviews in New York City. She found a job with the Coca Cola Company, lived in a borrowed apartment on the Upper West Side of Manhattan, and began a training program. But, after a few weekends of Phil hitchhiking back and forth between Boston and New York City, the couple decided that separation was intolerable. Joyce made the difficult decision to quit her job and join Phil in Boston. She took a low-paying position in Harvard's bursar office and the couple married on July 31, 1947.

A month later, Phil's parents drove from Illinois to Massachusetts to spend time with friends. The newlyweds visited with them for a few days and then borrowed their car for a honeymoon trip of touring and camping around Cape Cod. Joyce drove back to Urbana with her father- and mother-in-law and lived in Phil's old room while she completed an MA degree at the University of Illinois she had begun earlier. Most importantly, she earned several hundred much-needed dollars by teaching freshman English to GI Bill veterans. She returned to Boston in February 1948 and joined Phil in the half-house with bathroom privileges he had rented for $90 dollars a month in the town of Belmont, three miles northwest of Harvard.

Susan Anderson, the only child of Joyce and Philip Anderson, was born April 22, 1948 at the Boston Lying-In Hospital.[40] For the next ten months, the family lived on Joyce's savings from Urbana and the $70 dollars per month left over after paying the rent out of Phil's GI Bill and National Scholarship allowances. They lived as frugally as possible. A later family passion for walking and hiking

---

[40] Interview of Susan Anderson by the author, March 3, 2016. Today, the Boston Lying-In Hospital is part of Brigham and Women's Hospital.

began as the cost-free activity of bundling Susan in a rucksack and tramping along the stone walls that bordered the farms of Outer Boston. Whatever its deprivations, their austere life in Belmont focused Phil's mind wonderfully on the need to finish his thesis.

Thesis Problem Solved

It is typical of a young scientist to thrash around for a while figuring out how to make a useful contribution to a mature field. An idea that seems promising one day turns to ashes the next day. It is easy to get stuck and not know how to get unstuck. For reasons like this, Anderson periodically went to Van Vleck's office for discussions. These chats taught him to think and act like a physicist.

Good grades in physics courses and qualifying exams do not give an apprentice physicist automatic membership into the professional guild of physicists. One must learn *how* to do research. This includes learning to read scientific papers with a critical eye, to ask the right questions, to formulate a plan of action, to deal with frustration and adversity, to extract the physics in an intelligible manner, to bring a research project to conclusion, and to communicate the results to other physicists. Van Vleck provided advice and counsel on all these matters.

Anderson set himself apart by not discussing his thesis work with other graduate students. This included Van Vleck's two other physics graduate students, Thomas Kuhn and Arianna Wright. He distanced himself similarly from chemistry graduate student Robert Karplus, who was working with Julian Schwinger on precisely Phil's thesis topic—the effect of molecular collisions on the widths of absorption lines.[41] Anderson knew that Karplus and Schwinger were using a completely different approach to the theory, so it seems unlikely that he feared they would scoop him. A fair speculation is that his decision not to speak to others about

[41] Robert Karplus and Julian Schwinger, "A Note on Saturation in Microwave Spectroscopy," *Physical Review* **73**, 1020 (1948).

his thesis work reflected a need to prove he could solve the problem without input from his classmates. In later life, this attitude morphed into an aversion to working on any physics problem where another theorist had made the first important contribution.

That being said, here at the beginning of his career as a theoretical physicist, Anderson read and internalized all the previous work on his problem. The text of his thesis praises one approximate quantum theory of collision-induced line broadening and then criticizes more mathematically rigorous theories that came later because "little is to be gained in return for the great difficulty of these treatments." This comment again reflects Phil's lifelong dislike of theoretical work aimed at mathematical completeness without sharpening the physics or adding predictions relevant to experiments. By contrast, he generously acknowledged and exploited insightful theoretical work he found in two 1942 PhD theses devoted to his problem.[42]

The most original aspect of Anderson's research was his invention of a method to take explicit account of situations where a collision causes a molecule to make a transition from one of its allowed quantum states to another. In his thesis, Anderson ascribed his methodology to monographs written by the quantum pioneers Paul Dirac and Wolfgang Pauli.[43] A later recollection more accurately remarks that:

> I borrowed the methods I used in my thesis from Schwinger's course...They were not as sophisticated as quantum electrodynamics, but they were very sophisticated in terms of really using [scattering theory,] the full operators of tensor algebra, and representation group theory.[44]

[42] The two theses Anderson praised were the work of Einar Lindholm (University College Stockholm) and Henry Foley (University of Michigan).

[43] P.A.M. Dirac, *Quantum Mechanics*, 3rd edition (Oxford University Press, London, 1947); Wolfgang Pauli, *Handbuch der Physik*, 2nd edition, Volume 2, Part 1 (J.W. Edwards, Ann Arbor, MI, 1946).

[44] Interview of PWA by Alexei Kojevnikov, March 30, 1999, Niels Bohr Library & Archives, AIP, College Park, MD.

Anderson applied his theory to ammonia and several other small molecules. He discussed intermolecular forces and justified the approximations he used to account for state-changing collisions. This justification step is often ignored by young theoreticians and the reader detects the influence of Van Vleck at various points. Indeed, one of Van's earlier PhD students recalled that:

> Van was low key as a supervisor of graduate students. His guidance was deft, shrewd, but unobtrusive, so that you ended by believing the ideas were your own, only realizing gradually how your thinking had been nudged subtly in fruitful directions by his deceptively simple and apparently naive questions. Although Van had complete mastery of the most complex mathematical manipulations, his questions continually led you back to the basic physics.[45]

Anderson deemed his numerical results to be in "gratifying" agreement (within 10 percent) with the latest absorption line widths measured for fifteen lines of the ammonia spectrum.[46] It did not bother him that two of his calculated widths did not agree with the data. The fledgling physicist was confident he had included all the relevant physics and had made no mistakes in his calculations. Therefore, he concluded that:

> It is wrong to suggest errors in experiments with such small theoretical basis, but in view of the fact that all three [of our] formulas agree much better with each other than with the experiment,... one might suggest some kind of experimental error.[47]

Years later, he learned that the experimental values of the two widths at issue were indeed incorrect.[48]

[45] 'Remarks by Harvey Brooks' appended to Brebis Bleaney, "John Hasbrouck Van Vleck, 1899–1980," *Biographical Memoirs of the Royal Society* **28**, 627–65 (1982).

[46] B. Bleaney and R.P. Penrose, "Collision Broadening of the Inversion Spectrum of Ammonia: III. The Collision Cross-sections for Self-broadening and for Mixtures with Non-polar Gases," *Proceedings of the Physical Society of London* **60**, 540–9 (1948).

[47] P.W. Anderson, *The Theory of Pressure Broadening of Spectral Lines in the Microwave and Infrared Regions*, PhD thesis, Harvard University, 1949, p. 221.

[48] Interview of PWA by P. Chandra, Coleman, and S. Sondhi, October 15, October 29, and November 5, 1999, Niels Bohr Library & Archives, American Institute of Physics, College Park, MD.

Van was not happy with the first draft of his student's PhD thesis. "You write English as though it was German," he said, and "your notation is too compact."[49] Anderson dutifully increased the number of equations and, for the first of many times in future years, he enlisted Joyce to edit his writing. It took them four drafts to produce a document acceptable to Van Vleck. Readers of Anderson's mature work can attest that he never completely cured himself of the infelicities Van criticized.

Anderson defended his PhD thesis, *The Theory of Pressure Broadening of Spectral Lines in the Microwave and Infrared Regions*, on January 19, 1949. Figure 4.8 reproduces one of its pages. It is a sophisticated document for a novice researcher. One expert noted that "elegance is present in abundance, but simplicity certainly appears to be lacking."[50] Henry Margenau, the co-author of the book Anderson read at NRL to learn quantum mechanics, had a long-standing interest in the theory of spectra. In 1953, he and a PhD student published a paper whose only purpose was to reproduce Anderson's results using other theoretical methods.[51]

Anderson published his results in the September 1, 1949 issue of the *Physical Review*, the principal American physics journal at the time.[52] A few months later, a university researcher published experimental data confirming his prediction for the temperature dependence of the collisional broadening effect.[53] As late as 1981, a survey article deemed Anderson's theory still the most complete available.[54]

[49] Jeremy Bernstein, *Three Degrees Above Zero, Bell Labs in the Information Age* (Charles Scribner's & Sons, New York, 1984), p. 127.

[50] R.G. Breene, Jr., *The Shift and Shape of Spectral Lines* (Pergamon Press, New York, 1961), Section 8.19.

[51] Stanley Bloom and Henry Margenau, "Quantum Theory of Spectral Line Broadening," *Physical Review* **90**, 791–4 (1953).

[52] P.W. Anderson, "Pressure Broadening in the Microwave and Infrared Regions," *Physical Review* **76**, 647–61 (1949).

[53] Raydeen Howard and William V. Smith, "Temperature Dependence of Microwave Linewidths," *Physical Review* **77**, 840–1 (1950).

[54] G. Peach, "Theory of the pressure broadening and shift of spectral lines," *Advances in Physics* **30**, 367–474 (1981).

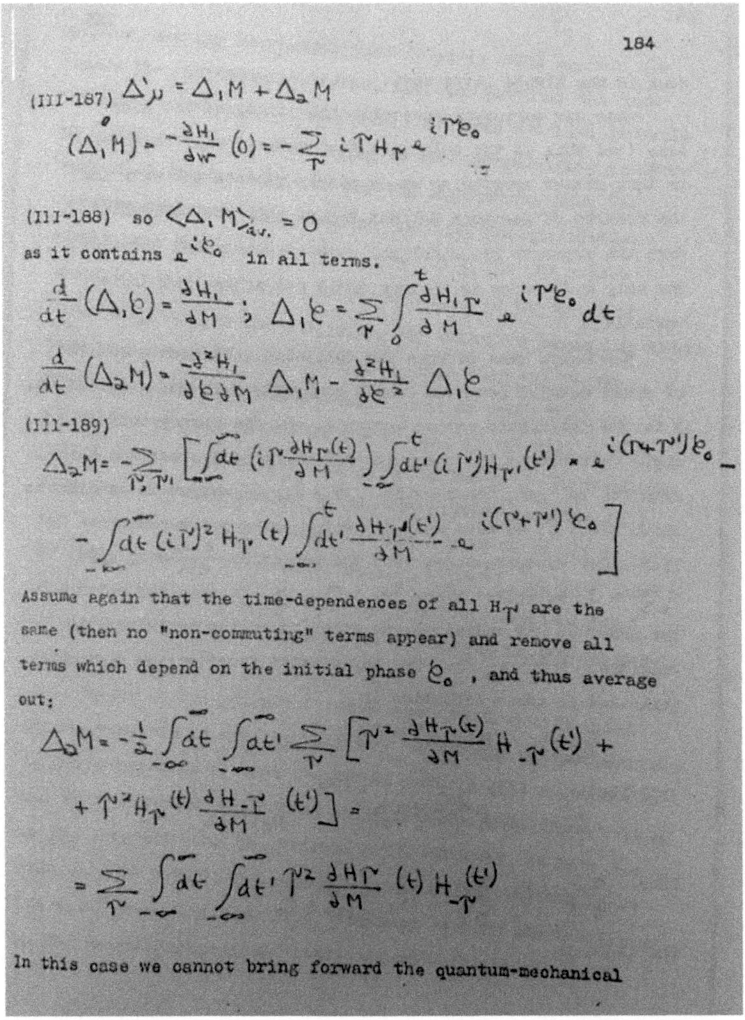
184

(III-187) $\Delta'_{\nu} = \Delta_1 M + \Delta_2 M$

$$(\Delta_1 M) = -\frac{\partial H_1}{\partial w}(0) = -\sum_{\tau} i\tau H_{\tau} e^{i\tau\varphi_0}$$

(III-188) so $\langle \Delta_1 M \rangle_{av.} = 0$

as it contains $e^{i\varphi_0}$ in all terms.

$$\frac{d}{dt}(\Delta_1 \varphi) = \frac{\partial H_1}{\partial M}\,;\quad \Delta_1\varphi = \sum_{\tau}\int_0^t \frac{\partial H_{1\tau}}{\partial M} e^{i\tau\varphi_0}\,dt$$

$$\frac{d}{dt}(\Delta_2 M) = \frac{-\partial^2 H_1}{\partial\varphi\,\partial M}\Delta_1 M - \frac{\partial^2 H_1}{\partial\varphi^2}\Delta_1\varphi$$

(III-189)

$$\Delta_2 M = -\sum_{\tau,\tau'}\Big[\int_{-\infty}^{\infty} dt\,\Big(i\tau\frac{\partial H_{\tau}(t)}{\partial M}\Big)\int_{-\infty}^{t} dt'\,(i\tau')H_{\tau'}(t')\times e^{i(\tau+\tau')\varphi_0} - \int_{-\infty}^{\infty} dt\,(i\tau')^2 H_{\tau'}(t)\int_{-\infty}^{t} dt'\,\frac{\partial H_{\tau}(t')}{\partial M} e^{i(\tau+\tau')\varphi_0}\Big]$$

Assume again that the time-dependences of all $H_{\tau'}$ are the same (then no "non-commuting" terms appear) and remove all terms which depend on the initial phase φ_0, and thus average out:

$$\Delta_2 M = -\frac{1}{2}\int_{-\infty}^{\infty} dt\int_{-\infty}^{\infty} dt'\sum_{\tau}\Big[\tau^2\frac{\partial H_{\tau}(t)}{\partial M}H_{-\tau}(t') + \tau^2 H_{\tau}(t)\frac{\partial H_{-\tau}}{\partial M}(t')\Big] =$$

$$= \sum_{\tau}\int_{-\infty}^{\infty} dt\int_{-\infty}^{\infty} dt'\,\tau^2\frac{\partial H_{\tau}}{\partial M}(t)\,H_{-\tau}(t')$$

In this case we cannot bring forward the quantum-mechanical

Figure 4.8 A page from Phil Anderson's 1949 PhD thesis. Source: Philip W. Anderson

Scientists show respect for the work of a peer by citing that person's published articles in their own articles. Between 1949 and 2016, other researchers cited Anderson's thesis work 1360 times.[55]

[55] Google Scholar Search, July 26, 2017.

This is a large number, but far from unheard of. Much more striking is the fact that 300 of these citations appear in papers published between 2000 and 2016. It is the rare PhD thesis that garners twenty citations a year almost seventy years after its publication.

Phil was excited to finish his thesis. However, the apartment in Belmont was cramped and the three Andersons lived a very frugal existence. It was time to find a real job.

5

A Solid Beginning

Most professional physicists begin their working careers with a postdoctoral fellowship. This is a one- or two-year paying job with no responsibilities beyond conducting research and publishing scholarly papers. The newly minted PhD Phil Anderson did not seek such a position.[1] He worried that a postdoc's salary was too small to support his family. He also believed, rightly or wrongly, that the purveyors of postdoctoral fellowships had a prejudice against married applicants. When recruiters made their rounds at Harvard, he focused exclusively on those bearing assistant professor jobs at colleges and universities, or research staff jobs at government and industrial laboratories.

Job Hunting

A serious impediment to Anderson's job search was that the United States was in recession for much of 1948–1949.[2] In addition, for academic jobs, he was competing with postdocs trained in nuclear physics looking for permanent positions. The relative popularity of nuclear physics compared to his own field of molecular spectroscopy can be judged from the articles published in the *Physical Review* in 1949. Out of 1050 articles, 630

---

[1] Interview of PWA by P. Chandra, P. Coleman, and S. Sondhi, October 15, October 29, and November 5, 1999, Niels Bohr Library & Archives, American Institute of Physics, College Park, MD.

[2] Benjamin Caplan, "A Case Study: the 1948–1949 Recession" in *Policies to Combat Recession*, (National Bureau of Economic Research, Washington, 1956), pp. 27–58.

addressed some aspect of nuclear physics while only 80 dealt with spectroscopy. In the event, the only academic offer he received was for an assistant professorship at Washington State College, an institution with no graduate physics program.[3]

An interview trip to Brookhaven National Laboratory did not go well. His contact there was Samuel Goudsmit, a Dutch-American who had been the guest instructor for a course Anderson took at Harvard. Goudsmit was a well-known figure in the physics world.[4] He had immigrated to the United States at age twenty-three, but only after co-authoring a famous paper that predicted the spin of the electron. He worked at the MIT Radiation Laboratory during World War II and then served as the scientific head of the secret *Alsos* mission, a US government mission to determine the progress German scientists had made toward the development of an atomic bomb.

Sam was friendly at the interview and asked Anderson to discuss some of the questions left open from his thesis. The new PhD replied that *no* questions remained open. This answer surprised and annoyed Goudsmit. He did not believe anyone could write a doctoral thesis that did not leave open questions. Phil did not get a job offer, but he did maintain a good relationship with Goudsmit. This was a good outcome because, only a year later, Goudsmit became editor of *Physical Review* and thus was on the receiving end of many future communications from Anderson the author and Anderson the manuscript referee.[5]

A better job result for Anderson came after a visit to the Westinghouse Electric Corporation. The interviewer was

[3] P.W. Anderson, *More and Different: Notes from a Thoughtful Curmudgeon* (World Scientific, Hackensack, NJ, 2011), p. 9; The State College of Washington Catalog, February 1950, p. 422.

[4] Benjamin Bederson, "Samuel Abraham Goudsmit (1902–1978)," *Biographical Memoir of the National Academy of Sciences* (National Academy of Sciences, Washington, DC, 2008).

[5] The editors of technical journals typically ask one or more "referees" to read submitted manuscripts to help them decide whether or not to publish.

Theodore Holstein, a theorist who was the "reigning royalty" of the company's research laboratory.[6] Phil was excited because Holstein had published papers on molecular spectroscopy and had shown a real understanding of his thesis work.

Westinghouse offered Anderson a research staff position, but not the one he wanted. His supervisor would not be Holstein, but another person whose responsibility was to reverse engineer a shipment of transistors received from Bell Labs. The invention of the transistor by three Bell Labs scientists was less than six months old, but even Anderson knew that the inventors had already gone public with an analysis of the physical principles that under-pinned the device.[7] He was not sanguine that Westinghouse could compete with Bell.

The starting salaries offered by Washington State and Westinghouse were identical, $5400 per year.[8] He and Joyce were unsure what to do, so they made a list of pros and cons for each opportunity.[9] In the end, they chose to follow Phil's father into academia. His parents bought him a car and he prepared to drive his family the 2800 miles in the dead of winter from Boston, Massachusetts to Pullman, Washington.

John Van Vleck had not involved himself in Anderson's job search. When he finally grew curious and learned of his student's plans, he asked if Washington State was his first choice. The answer was "No, I want to go to Bell Labs."[10] Anderson had admired AT&T since his wartime experience with their equipment, and the

[6] Walter Kohn, "Density Functional Theory of Excited States," in *Condensed Matter Physics: The Theodore D. Holstein Symposium*, edited by Raymond L. Orbach (Springer-Verlag, New York, 1987), pp. 67–73.

[7] See footnote 4 of J. Bardeen and W.H. Brattain, "Physical Principles Involved in Transistor Action," *Physical Review* **75**, 1208–25 (1949).

[8] Adjusted for inflation, a salary of $5400 in 1949 had the same buying power as a salary of $56,000 in 2017.

[9] Interview of David Z. Robinson by the author, April 6, 2016.

[10] Interview of PWA by Alexei Kojevnikov, March 30, 1999, Niels Bohr Library & Archives, American Institute of Physics, College Park, MD.

transistor was an exciting development. Unfortunately, the Bell recruiter at Harvard had shown no interest in him.

Van Vleck swung into action and used a previously scheduled consulting trip to plead Anderson's case in person to William Shockley, the co-head of the Bell Labs Solid-State Physics group. Phil duly received an invitation to interview at the main Bell Labs facility in suburban Murray Hill, New Jersey. The visit went well, in part because Shockley did *not* ask Anderson to solve a tricky logic puzzle as he often did with job candidates.[11]

A few days later, Shockley offered Anderson a one-year post-doctoral position working directly with him. Phil demurred and tried to convince the senior scientist to offer him a permanent staff position. Abruptly, Shockley did exactly that. Anderson was elated, but he did not know that junior Bell Labs staff members at the time worked on annual contracts. Management could still let him go after one year, just like a postdoc.

Phil Anderson was now a member of the Bell Labs Solid-State Physics group. However, his only training in the subject was a graduate course he had audited at Harvard where John Van Vleck used Frederick Seitz's *Modern Theory of Solids* (1940) as the textbook. It did not help that Van's teaching style was unexciting and Seitz's writing style was dry.[12] Perhaps that is why Anderson found the subject "diffuse and boring."[13] If so, why had Bell Labs devoted so many resources to it? The answer to that question requires the answer to two others: what was the mission and identity of Bell Labs and what was the field of solid-state physics at the time?

[11] William Poundstone, *How Would You Move Mount Fuji?* (Little, Brown and Co., New York, 2004), pp. 3–5, 31–32.

[12] Author correspondence with Hellmut Juretschke, May 30, 2018.

[13] Interview of PWA by Lillian Hoddeson, May 10, 1988, Niels Bohr Library & Archives, American Institute of Physics, College Park, MD.

What Was Bell Labs?

For fifty years in the middle of the twentieth century, Bell Telephone Laboratories was arguably the greatest research and development organization in the world.[14] A short list of Bell Labs innovations and inventions provides evidence: radio astronomy (1933), speech synthesis (1936), the transistor (1947), cellular communications (1947), information theory (1948), solar cells (1954), the laser (1958), digital transmission (1962), communication satellites (1962), the UNIX operating system (1969), charge-coupled devices (1969), and the digital signal processor (1979).[15] Through 2017, sixteen scientists had received a share of a Nobel Prize for work they did at Bell Labs.[16]

At its peak in 1982, Bell Labs had a budget of $2 billion, received its 20,000th patent, and employed 25,000 people (3000 with PhD degrees and 5000 with MS degrees) at twenty-one locations in New Jersey and elsewhere.[17] The Labs allocated 10% of its budget to basic research, the monies coming from an assessment of the revenues of the 22 Bell operating companies (which provided local telephone service) and the AT&T Long Lines division (which provided long-distance telephone service).

The research budget of Bell Labs was unusually stable compared to other industrial laboratories because AT&T's status as a regulated monopoly guaranteed predictable revenues for its

[14] Jon Gertner, *The Idea Factory: Bell Labs and the Great Age of American Innovation* (Penguin Press, New York, 2012), p. 1; Francis Bello, "The World's Greatest Industrial Laboratory," *Fortune*, November 1958, pp. 148–57.

[15] Charge-coupled devices are the enabling technology for digital photography.

[16] The Bell Labs Nobel laureates are Clinton Davisson (1937), John Bardeen (1956, 1972), Walter Brattain (1956), William Shockley (1956), Charles Townes (1964), Philip Anderson (1977), Arno Penzias (1978), Robert Wilson (1978), Arthur Schawlow (1981), Douglas Osheroff (1996), Steven Chu (1997), Horst Störmer (1998), Daniel Tsui (1998), Willard Boyle (2009), George Smith (2009), and Eric Betzig (2014).

[17] US Congress, Office of Technology Assessment, *Information Technology and R&D: Critical Trends and Issues*, OTA-CIT-268 (US Government Printing Office, Washington, DC, 1985), Chapter 4.

subsidiaries.[18] This ended on January 1, 1982, when an anti-trust agreement with the federal government compelled AT&T to divest itself of its operating companies. The parent company shrank to one-quarter of its previous size and a significant fraction of Bell Labs moved to a new organization charged to provide technical support for the divested operating companies.

The pre-divestiture funding arrangement gave Bell Labs scientists access to state-of-the-art equipment and the freedom to focus on long-range objectives without the need to teach, attend committee meetings, or apply for external research grants as their academic counterparts had to do. Uniquely among industrial laboratories, it was not necessary for research projects at Bell Labs to produce short-term benefits to the company as long as it was possible to articulate some potential future benefit. All these advantages had tangible consequences. Bell Labs recruited the best people from the best institutions, a high percentage of them came, and they challenged each other to excel on a daily basis.

Bell Telephone Laboratories was born on January 1, 1925 from a merger of the engineering departments of the American Telephone and Telegraph (AT&T) Company and the Western Electric Company, a wholly owned subsidiary of AT&T. Most of the 2000 Bell Labs staff members worked on product development. However, about 300 scientists conducted basic and applied research in physical and organic chemistry, metallurgy, magnetism, electrical conduction, radiation, electronics, acoustics, phonetics, optics, mathematics, mechanics, physiology, psychology, and meteorology.

Walter Gifford, the president of AT&T at the time, boasted that Bell Labs could "carry on scientific research on a scale that is probably not equaled by any organization in the country, or in the world."[19] Of course, the research done was not 100% curiosity-driven.

[18] The 1921 Willis-Graham Act of the US Congress created the AT&T regulated monopoly.
[19] Quoted in Jimmy Soni and Rob Goodman, *A Mind at Play: How Claude Shannon Invented the Information Age* (Simon & Schuster, New York, 2017), p. 66.

It rather had a dual nature, being fundamental from the point of view of the researchers while at the same time supported by the company for its possible future applications.

In the early 1930s, the prescient Bell Labs managers Oliver Buckley and Mervin Kelly (both trained physicists) recognized it was important to gain a fundamental understanding of the solid materials from which so much of their equipment was fabricated.[20] It was equally far-sighted of them to realize that acquiring expertise in the new subject of quantum mechanics was an important step needed to achieve that goal. In practice, that did not happen until 1936, when a six-year hiring freeze necessitated by the Great Depression ended and Mervin Kelly became the Bell Labs Director of Research.[21]

The first person Kelly hired was a fresh PhD from MIT named William Shockley. Shockley was an expert in solid-state physics and Kelly directed him to establish a weekly study group so a select group of Bell Labs scientists could teach one another the quantum theory of solids. Concurrently, Kelly tasked Shockley to start thinking about how the company might replace the expensive and bulky vacuum tube amplifiers that were so important to the operation of their telephone network.[22] A class of solids called semiconductors seemed promising in this regard and research in that area proceeded vigorously at Bell Labs until America's entry into World War II redirected most of its efforts to military projects.

---

[20] Lillian Hartmann Hoddeson, "The Entry of the Quantum Theory of Solids into the Bell Telephone Laboratories, 1925–1940: A Case Study of the Industrial Application of Fundamental Science," *Minerva* **18** (3): 422–47 (1980).

[21] Jon Gertner, *The Idea Factory: Bell Labs and the Great Age of American Innovation* (Penguin Press, New York, 2012), p. 36. Kelly earned his PhD in 1918 at the University of Chicago.

[22] Lillian Hoddeson, "Innovation and Basic Research in the Industrial Laboratory: The Repeater, Transistor and Bell Telephone System," in *Between Science and Society*, edited by Andries Sarlemijn and Peter Kroes (Elsevier, New York, 1990), pp. 181–214.

Figure 5.1 Mervin Kelly committed Bell Telephone Laboratories to solid-state physics during his tenure as Director of Research (1936–1944), Executive Vice President (1944–1951), and President (1951–1959) of the organization. Source: American Institute of Physics Emilio Segrè Visual Archive.

The war in Europe ended in June 1945 and Kelly announced a reorganization of the Research Division of the Labs. Taking a cue from the large interdisciplinary research teams that had functioned so well during the war at MIT and Los Alamos, he created a Solid-State Physics group co-headed by Shockley and the chemist Stanley Morgan. The authorization Kelly (Figure 5.1) wrote for the group was very clear:

> Employing the new theoretical methods of solid-state quantum physics and the corresponding advances in experimental techniques, a unified approach to all of our solid-state problems offers great promise. Hence, all of the research activity in the area of solids is now being consolidated in order to achieve a unified

approach to [our] theoretical and experimental work [in] the solid-state area.[23]

Kelly's mandate distinguished experimental work from theoretical work because physicists tended to specialize in one or the other of these tasks. Experimental solid-state physicists maintained laboratories filled with specialized equipment designed to measure the properties and behavior of different kinds of matter. Theoretical solid-state physicists worked with pen and paper (and later computers) with the aim of constructing mathematical descriptions of solid-state behavior.

Bell Labs rapidly became the center for solid-state physics research in the United States, if not the world. They dominated by the quality of their work-product and also by the sheer number of physicists they employed. In 1960, Bell Labs employed 650 physicists, the vast majority engaged in some kind of solid-state physics.[24] The research laboratories of Westinghouse Electric, General Electric, and RCA each employed about half that number. No other industrial laboratory reported as many as 200 physicists on staff.

What is Solid-State Physics?

According to one long-time practitioner:

> Solid-state physics is a haven for physicists who are drawn to physics by a desire to understand the world around them on a personal level. It is the part of physics where research programs exist to answer questions raised by the simplest of observations. Why is iron magnetic? Why does a transparent crystal of sodium chloride turn colored when exposed to ultraviolet light? Why does a pencil stroke conduct electricity? Each line of experiment designed to answer one of these easily posed questions tends to

---

[23] Quoted in Michael Riordan and Lillian Hoddeson, *Crystal Fire: The Birth of the Information Age* (W.W. Norton, New York, 1997), p. 116.

[24] John H. Gribbin and Sue Singer Krogfus, *Industrial Research Laboratories of the United States*, 11th edition (National Academy of Sciences, Washington, DC, 1960).

produce data that in turn poses questions at a deeper level. The field is an endless frontier."[25]

Modern solid-state physics began with the 1912 discovery by Max von Laue of the *diffraction* of x-rays by solids. By analogy with what scientists understood about the diffraction of visible light, von Laue deduced that the arrangement of atoms in most solids is a *crystal*.[26] A crystal is a repeated stacking of a single three-dimensional building block called a *unit cell*, each of which contains one or a few atoms arranged in exactly the same way. Thus, either of the shaded unit cells shown in Figure 5.2 for a calcium fluoride crystal plays the same role for a crystal as a ceramic tile does for a bathroom floor. The collection of points that lie at the center of every unit cell is called the *lattice* of the crystal.

The information provided by x-ray diffraction—the position and identity of all the atoms in the crystal—is exactly the input required by the Schrödinger equation, which is the central equation of quantum mechanics. In principle, the solution of this equation provides all the information needed to calculate the physical properties of any crystal. Not all solids are crystals (the arrangement of atoms is not orderly in window glass), but the study of crystals provides the foundation for the study of non-crystals.

In practice, experiments often reveal unexpected phenomena that challenge theory to provide an explanation. This, then, is the

---

[25] John J. Hopfield, "Whatever Happened to Solid-state Physics?" *Annual Reviews of Condensed Matter Physics* **5**, 1–13 (2014).

[26] *Diffraction* is a phenomenon characteristic of all waves, including electromagnetic waves. A typical diffraction effect observed and explained in the seventeenth century involves a beam of light made to strike a grating (e.g., a glass plate with a large number of equally spaced parallel grooves etched onto its surface). The effect of diffraction is to split the original beam into several beams, each of which exits in a different direction. This happens only if the wavelength of the light is comparable to the distance between the grooves. A simple formula relates the groove separation to the wavelength of the light and the angles between the exiting beam directions and the original beam direction. The same diffractive effect occurs when x-rays strike a crystal because the parallel rows of atoms in the crystal play the role of the grooves.

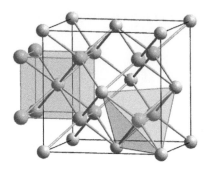

Figure 5.2 A calcium fluoride crystal. The spheres denote calcium (green) and fluorine (grey) atoms. The two shaded regions are alternatives for the unit cell of this crystal.

broad task of solid-state physics: to use theory, experiment, and computation to construct a coherent understanding of solid matter, both qualitatively and quantitatively.

The theoretical part of the solid-state physics program is daunting for a simple reason pointed out in 1929 by quantum mechanics pioneer Paul Dirac:

> The underlying physical laws necessary for the mathematical theory of a large part of physics and the whole of chemistry are thus completely known [i.e., the Schrödinger equation], and the difficulty is only that the exact application of these laws leads to equations much too complicated to be solvable.[27]

This state of affairs motivated many physicists to bypass the "completely known" and move on to the new frontier of nuclear physics. Until the Bell Labs reorganization in 1945, there was no identifiable group of people looking at solids as an exclusive endeavor.[28] Indeed, the term "solid-state physics" was unknown

[27] P.A.M. Dirac, "Quantum Mechanics of Many-Electron Systems," *Proceedings of the Royal Society of London. Series A* **123**, 714–33 (1929).

[28] Spencer Weart, "The Solid Community," in *Out of the Crystal Maze: Chapters from the History of Solid-state Physics*, edited by Lillian Hoddeson, Ernest Braun,

to Niels Bohr when Anderson's classmate Walter Kohn began a fellowship at Bohr's institute in 1951.[29]

Slowly, through the early 1950s, a community of physicists committed to studying the physics of solids began to grow around the world.[30] That growth did not begin from a single embryo; there was no solid-state analog of Ernest Rutherford's dramatic discovery of the atomic nucleus.[31] Instead, the discipline of solid-state physics was stitched together from a group of pre-existing technical specialties focused on material types (metals, alloys, compounds, dielectrics) and measurement techniques (electrical resistance, magnetic susceptibility, heat capacity, x-ray diffraction, spectroscopy).

The quilt-like construction of solid-state physics almost guaranteed that it would suffer from a lack of deep unifying principles, a defect not shared by atomic physics or nuclear physics. This permitted nuclear physicists (and later particle physicists) to claim the intellectual high ground and assert the "fundamental" nature of their endeavors compared to anything that solid-state physicists were doing.

F. Wheeler Loomis, the Saturday Hiker and Head of the Physics Department at the University of Illinois spoke to this issue in a 1949 talk on "The Future of Physics" delivered during his term as

---

Jürgen Teichmann, and Spencer Weart (Oxford University Press, New York, 1992) pp. 617–69.

[29] Walter Kohn, Nobel biography, 1998.

[30] For the United States, see Joseph D. Martin, *Solid-state Insurrection: How the Science of Substance Made American Physics Matter* (University of Pittsburgh Press, Pittsburgh, PA, 2018). For England, Europe, and Russia, see *Out of the Crystal Maze: Chapters from the History of Solid-state Physics*, edited by Lillian Hoddeson, Ernest Braun, Jürgen Teichmann, and Spencer Weart (Oxford University Press, New York, 1992). For Japan, see Katsuki Atsushi, "History of Solid-state Physics in Japan," *Historia Scientiarum. International Journal of the History of Science Society in Japan* 7 (2) 107–24 (1997).

[31] See, e.g., Emilio Segrè, *From X-rays to Quarks: Modern Physicists and Their Discoveries* (W.H. Freeman, New York, 1980), Chapter 6.

President of the American Physical Society.[32] He asked, "What parts of present day physics will no longer be physics?" His answer was "radar, electronics, acoustics, crystallography, metallurgy, and solid-state physics."[33]

It would take years for the solid-state physics community to find a vocal and persuasive champion who would, first, clearly identify and make explicit the underlying principles that illustrated the intellectual coherence of the subject, and second, publicly battle the particle physicists on the issue of fundamentality. That champion turned out to be Phil Anderson, but his efforts in that direction lay twenty years in the future.

William Shockley

Phil Anderson joined Bell Labs as a solid-state neophyte. For someone in his position, there was probably no better place to be. Not only did everyone who was anyone in the solid-state community pass through the Labs in those days, there was probably no greater concentration of solid-state physics talent working anywhere else in the world. At least five outstanding theorists were available to him daily: William Shockley, Gregory Wannier, Charles Kittel, Conyers Herring, and John Bardeen. In one way or another, each of these men played an important role in helping Anderson complete the transition from apprentice to professional that Van Vleck had begun.

William Shockley was the senior theoretical physicist at Bell Labs (Figure 5.3). He was responsible for hiring Anderson and he served as his first mentor. A direct descendant of a crewman and a passenger on the Pilgrim ship *Mayflower*, Shockley majored in physics at Caltech and wrote a PhD thesis at MIT on the

---

[32] Loomis was a Saturday Hiker who wrote a letter of recommendation for Phil Anderson when he applied to Harvard College (see Chapter 1).

[33] F. Wheeler Loomis, "The Future of Physics—APS, 1949." F. Wheeler Loomis Papers, 1920–1976, University of Illinois Archives, Record Series Number 11/10/22, Box 3.

Figure 5.3 Phil's first mentor at Bell Labs was William Shockley. Source: American Institute of Physics Emilio Segrè Visual Archive.

computation of the allowed energies of electrons in a sodium chloride crystal using an approximate solution of the Schrödinger equation.[34]

After the war, Shockley focused his efforts on finding a solid-state replacement for the vacuum tube amplifier. This effort bore fruit spectacularly in December 1947 when two members of his team, Walter Brattain and John Bardeen, demonstrated to Bell Labs management their invention of the semiconductor-based *point-contact transistor*. However, rather than be happy for them, Shockley was angry he had not made the breakthrough himself.

[34] William Shockley, "Electronic Energy Bands in Sodium Chloride," *Physical Review* **50**, 754–9 (1933).

This drove him to violate an unwritten Bell Labs rule that forbade managers from competing with their supervisees.[35] Secretly, he worked day and night until he had conceived an alternative to the point-contact transistor. Shockley believed (correctly, as it turned out) that his *junction transistor* design was superior to the Brattain–Bardeen design.[36]

When Anderson started at Bell Labs on February 1, 1949, Shockley's experimental team was two months from producing the first working junction transistor. Shockley was completing his soon-to-be-classic monograph, *Electrons and Holes in Semiconductors with Applications to Transistor Electronics* (1950). He had also so completely alienated Brattain and Bardeen that they both threatened to quit rather than remain in his group.

Brattain took up other aspects of experimental semiconductor research. Bardeen returned to an old interest—the search for a theory to explain the phenomenon of superconductivity. Anderson, the new hire, was ignorant of the bad blood between Shockley and Brattain and Bardeen. So ignorant, in fact, that when Phil and Joyce invited the Shockleys and the Bardeens to their apartment for dinner, the hosts were perplexed by the obvious tension in the air.

Anderson was hired at Bell Labs because Shockley needed a theorist to work out the consequences of his ideas about a class of solids called *ferroelectrics*. These materials were interesting because of their ability to convert electrical signals into acoustic signals and vice versa. Their potential was so great that Shockley arranged for Bell Labs to hire an expert in the synthesis of new ferroelectrics to complement the theoretical expertise Anderson

[35] P.W. Anderson, "Giant Who Started the Silicon Age." Review of Joel N. Shurkin, *Broken Genius: The Rise and Fall of William Shockley* (Macmillan, London, 2006), *Times Higher Education Supplement*, June 16, 2006.

[36] Michael Riordan and Lillian Hoddeson, *Crystal Fire: the Birth of the Information Age* (W.W. Norton, New York, 1997), Chapter 8; Joel Shurkin, *Broken Genius: The Rise and Fall of William Shockley, Creator of the Electronic Age* (Macmillan, London, 2006), Chapter 6.

was supposed to develop. The person hired was Bernd Theodor Matthias, a strong-willed German who ultimately discovered hundreds of new ferroelectric materials.[37] Anderson began chatting regularly with Matthias and later chapters will discuss Matthias' role in his life as a friend, scientific collaborator, and philosophical foil.

Anderson focused his initial energy on learning as much as he could from Shockley. The need was immediate because his self-study bible, Seitz's *The Modern Theory of Solids*, did not discuss ferro-electrics. Among other things, he learned that all solids respond to an external electric field by producing an electric field of their own. This self-generated electric field disappears when the exter-nal electric field disappears.

Shockley explained to Anderson that the self-generated elec-tric field of a *ferroelectric* solid was special because it appears spon-taneously, at low temperature, *even if there is no external electric field present at all*. Shockley asked Phil to test his ideas about the origin of the electric field by solving the Schrödinger equation approxi-mately for a particular ferroelectric crystal called barium titanate ($BaTiO_3$).

It annoyed Anderson that Shockley seemed to treat him like a postdoc, rather than as a staff member as Shockley had agreed. On the other hand, he was the youngest person in the group and it was best for his job security if he did Shockley's bidding, at least at first. He set himself to the task and soon convinced himself that Shockley's ideas about ferroelectrics were probably correct. He came to this conclusion using qualitative reasoning and simple numerical estimates.

Anderson resisted doing the numerical work his boss wanted because he did not trust the approximations that would be required

[37] T.H. Geballe and J.K. Hulm, "Bernd Theodor Matthias 1918–1990," *A Biographical Memoir of the National Academy of Sciences* (National Academies Press, Washington, DC, 1996).

to carry out the calculations.[38] He would always disdain numerical work when he could figure out the answer to a problem more simply. In this respect, the paper he eventually wrote about barium titanate is a model of many of his future publications.

Despite his low opinion of the calculations Shockley asked him to perform, Anderson had a high opinion of Shockley as a scientist. As he later recalled,

> [Bill Shockley] was one of the most brilliant men I have ever met, and that includes all the great theoretical physicists of my time...His scientific work was characterized by quickness and clarity. I have often said that he could take the first steps to solve a problem faster than anyone I know.[39]

Shockley's brilliance came with a mixed bag of other characteristics. He showed great kindness when he insisted that the Andersons stay at his home for a few days when they first arrived in New Jersey. However, he was unsparingly competitive when he took Anderson on a rock-climbing expedition. He also had a distinct lack of empathy. When Anderson tried to argue that detailed calculations for barium titanate were unwarranted, Shockley responded by urging his superiors to terminate his protégé's employment. A distillation of this behavior came near the end of his life when he damaged his reputation badly by making unsupported statements about race, human intelligence, and eugenics.[40]

---

[38] It is not clear if Anderson had access to the one general purpose computer in service at Murray Hill at that time. See Russ Cox, "Computing History at Bell Labs." Posted April 9, 2008 at https://research.swtch.com/bell-labs. Accessed April 3, 2020. Shockley hedged his bets by convincing John Slater (his PhD supervisor at MIT and a paid consultant at Bell Labs) to work on the barium titanate problem. Slater's quantitative results [J.C. Slater, "The Lorentz Correction for Barium Titanate," *Physical Review* **78**, 748–61 (1950)] were accepted until more accurate work by others vindicated Anderson's claim that the approximations used by Slater were too drastic.

[39] P.W. Anderson, "BCS and Me," in *More and Different: Notes from a Thoughtful Curmudgeon* (World Scientific, Hackensack, NJ, 2011), p. 12.

[40] Joel Shurkin, *Broken Genius: The Rise and Fall of William Shockley, Creator of the Electronic Age* (Macmillan, London, 2006), Chapter 10.

At the time, Anderson knew nothing of Shockley's attempts to fire him. Even if he had, it is unlikely he would have thought about leaving. He had a 9–5 job doing pure research in a subject he had learned to appreciate. Unlike other industrial labs, there was no dress code and, barely four months into the job, the Labs paid to send him to a meeting of the American Physical Society.[41] In light of the advantages, Anderson figured he could manage his relationship with Bill Shockley. As it turned out, that relationship was not to last very long.

[41] P.W. Anderson, "Physics at Bell Labs, 1949–1984," in *More and Different: Notes from a Thoughtful Curmudgeon* (World Scientific, Hackensack, NJ, 2011); Jon Gertner, *The Idea Factory: Bell Labs and the Great Age of American Innovation* (Penguin Press, New York, 2012), p. 151.

6

Breaking Symmetry

The first half of the 1950s marked the beginning of Anderson's ascent in the world of theoretical physics (Figure 6.1). New mentors at Bell Labs directed him to problems in the theory of magnetism and he attacked these with great success. The high point of this activity was his discovery of the phenomenon of spontaneous symmetry breaking, although that name had not yet been coined. We begin, however, with a political situation that is revealing of his character because it forced him to respond to a basic threat to civil liberty. He defied the wishes of Bell Labs management, broke ranks with most of his colleagues, and thereby earned their respect as a person willing to act on his principles.

The Red Scare

For ten years after World War II, the Red Scare was a nearly hysterical belief by some Americans that Communism posed a serious threat to their way of life.[1] This attitude was not completely unwarranted as there were legitimate reasons to be concerned about the global ambitions of the Soviet Union. In 1948–1949 alone, Joseph Stalin orchestrated a Communist coup in Czechoslovakia, blocked access to the city of Berlin, and ordered the detonation of an atomic bomb.

President Harry Truman responded by creating a "loyalty-security" program designed to identify and dismiss employees of

[1] Richard M. Fried, *Nightmare in Red: The McCarthy Era in Perspective* (Oxford University Press, Oxford, 1990).

Figure 6.1 Phil Anderson at Bell Labs in 1950. Source: Susan Anderson.

the federal government deemed to be subversive. Several universities and large companies followed suit. The University of California, Berkeley fired thirty-one faculty members who refused to sign an oath declaring that they were not members of the Communist Party.[2]

The US House of Representatives Un-American Activities Committee (HUAC) was quick to issue subpoenas and demand that witnesses supply the names of persons they knew or suspected to be sympathetic to Communism. Joseph McCarthy, an

[2] J.D. Jackson, "Panofsky Agonistes: The 1950 Loyalty Oath at Berkeley," *Physics Today* **62** (1), 41–7 (2009).

opportunistic and bombastic senator from Wisconsin, used the Permanent Subcommittee on Investigations of the US Senate to press unsubstantiated claims of widespread Communist Soviet penetration in the government, academia, Hollywood, and other American institutions.

Anderson knew two people deeply involved in this matter.[3] One was his favorite Harvard professor, Wendell Furry, who had flirted with Communism in the 1930s. Testifying before McCarthy's committee, Furry waived his Fifth Amendment right not to incriminate himself and freely discussed his interest in the Communist party. McCarthy threatened to cite Furry for contempt of Congress if he refused to name other Communists.[4] McCarthy backed down but it was apparent to many that the incident left Furry badly scarred.[5]

More disturbing to Anderson was the experience of his graduate dining hall buddy Chandler Davis. Unbeknownst to his friends, Davis had been a member of the Communist Party his entire time at Harvard. When subpoenaed by HUAC in 1954, Davis invoked his First Amendment right to free speech and refused to answer any questions about his personal political beliefs. Congress cited him for contempt and he eventually served six months in a federal prison.

Anderson's turn came one day in 1962 when bright yellow notices appeared on the walls of every corridor at Bell Labs. The notices read:

> The Secretary of Defense has designated this facility as a "defense facility"... It is unlawful for any member of a Communist organization... to seek, accept, or hold employment at any defense facility, or to conceal or fail to disclose the fact that he is a member of such an organization.[6]

[3] Ellen W. Schrecker, *No Ivory Tower. McCarthyism & The Universities* (Oxford University Press, New York, 1986), p. 117.

[4] Harold Taylor, "The Dismissal of Fifth Amendment Professors," *Annals of the American Academy of Political and Social Science* **300**, 79–86 (1955).

[5] Jeremy Bernstein, *The Life it Brings, One Physicist's Beginnings* (Ticknor & Fields, New York, 1987), pp. 71–7.

[6] P.W. Anderson supplied the author with an original of this notice.

Bell Labs management followed up by asking all their employees to fill out and sign a security questionnaire. Phil had no sympathy for the Communist Party, but he did not believe his employer had the right to extract information about his personal political views. He was one of a handful of scientists who refused to sign the questionnaire. People watched and waited, but the non-signers suffered no consequences, even when Anderson and a few accomplices removed the yellow notices from the walls.

This was the first of several instances when Anderson protested the abridgement of civil liberties in the pursuit of security. For example, he consistently declined to participate in programs that required a security clearance because he regarded the required background checking as a violation of his right to privacy. This included an invitation to join JASON, a group of prominent scientists who consulted with the US government on defense matters.[7] The fact that JASON included members from across the political spectrum was irrelevant to Anderson's fundamental objection to clearances.

In 1984, the American Association for the Advancement of Science (AAAS) asked 100 leading American scientists if there was "a basic conflict between the principle of open scientific communication and national security." The executive summary submitted by the AAAS to hearings of the Judiciary Committee of the US House of Representatives concerned with *Civil Liberties and the National Security State* quoted three scientists. Phil Anderson was one of them:

> It is a demonstrable fact that security classification and secrecy impedes scientific and technical progress very effectively... Secrecy tends to cloak inefficiency, ignorance, and corruption more often than it hides genuine secrets.[8]

---

[7] See Anne Finkbeiner, *The Jasons. The Secret History of Science's Postwar Elite* (Viking Press, New York, 2006). Winners of the Nobel Prize in Physics who have been members of JASON include Hans Bethe, Murray Gell-Mann, Donald Glaser, Burton Richter, Charles Townes, and Steven Weinberg.

[8] "1984: Civil Liberties and the National Security State," Hearings of the Judiciary Committee of the House of Representatives, Serial No. 103, p. 741.

This strong and colorfully worded statement is typical of Anderson when his purpose is to persuade. In fact, he had no evidence to support the "demonstrable fact" he claimed, either personally or from experiences related to him by others. He simply believed it to be true.

New Mentors

Six months into his new job, Anderson's disaffection with William Shockley's authoritarian manner had grown to the point where he began to look elsewhere for mentorship. John Bardeen was friendly and helpful, but he was focused single-mindedly on superconductivity—a topic which did not interest anyone else at the Labs at the time. The most natural choices for mentors were Gregory Wannier, Conyers Herring, and Charles Kittel, all about ten years his senior (Figure 6.2). With their help, he became an expert in the theory of magnetism and grappled for the first time with symmetry as a deep issue in physics.

Wannier was a mathematically sophisticated Swiss physicist who developed a reputation as an expert in the theory of phase transformations in the late 1930s.[9] He spent the war years in the United States and worked in both academia and private industry before arriving at Bell Labs at the same time as Anderson. They shared an office and quickly became friends. Just as Bill Shockley had instructed Anderson in the basics of solid-state physics, Wannier taught Phil about phase transitions and some subtle aspects of the spectrum of electron energy states in crystals.

Conyers Herring learned theoretical solid-state physics from the two giants of that subject in the 1930s: Eugene Wigner at Princeton and John Slater at MIT. He worked on underwater acoustics during World War II and joined Bell Labs in 1946. William Shockley placed him in a research group focused on vacuum tubes, but Herring always functioned as a general-purpose

---

[9] P.W. Anderson, "Gregory Wannier," *Physics Today* **37** (5), 100–101 (1984).

Figure 6.2 Phil's mentors at Bell Labs. Gregory Wannier (left), Conyers Herring (middle), and Charles Kittel (right). Source: American Institute of Physics Emilio Segrè Visual Archive.

theorist.[10] Anderson admired his older colleague's intellect and was happy to be tutored by him in statistical mechanics, the branch of physics which treats the thermal properties of matter from a microscopic point of view.[11] Herring readily shared with Anderson his near-encyclopedic knowledge of the physics literature. This included the Russian literature, which he read and translated himself.

Long before searchable databases, Herring used 3" × 5" index cards to summarize information he deemed important from the many journal articles he read. He filed the cards topically in a black suitcase he called his "brain box." At the time of his death in 2009, Herring's brain box contained 15,000 handwritten cards containing over 100,000 citations to the scientific literature.[12]

[10] Interview of Addison White by Lillian Hoddeson, September 30, 1976, Niels Bohr Library & Archives, American Institute of Physics, College Park, MD. In 1954, the 39-year-old Herring was singled out by *Fortune* magazine as one of the "Ten Top Young Scientists in US Industry": Francis Bello, "The Young Scientists," *Fortune* **49** (6), pp. 142–8, 172–8 (1954).

[11] Phil quickly realized that the statistical mechanics course he took his last semester at Harvard from Wendell Furry was actually a course in the kinetic theory of gases, a subject which uses classical physics and statistical ideas to compute macroscopic thermal properties.

[12] Philip W. Anderson, Theodore H. Geballe, and Walter A. Harrison, "W. Conyers Herring, 1914–2009" (National Academy of Sciences, Washington,

Charlie Kittel trained in nuclear physics at the University of Wisconsin. His wartime research at the Naval Ordnance Laboratory transformed him into a solid-state physicist with an expertise in magnetism. Kittel served as the "house theorist" for the experimental magnetism groups during his years at Bell Labs (1947–1951). He was very successful in that role, as befits someone who was a consummate phenomenologist. That is, someone deeply concerned with the results of experiments and their interpretation rather than with the construction of formal theories for their own sake.[13] This philosophy resonated strongly with Anderson and he eagerly became a student of magnetism with Kittel as his personal instructor.

Ferromagnetism

Magnetism was a concern of Bells Labs from the very beginning.[14] Magnets were essential to the switches and relays used to route telephone calls and to the transformers that appeared everywhere in the company's communication technology. A priority was the search for novel magnetic materials that could operate without much energy loss from heating. This need spawned an extensive research program devoted to the synthesis and analysis of non-metallic magnets. After World War II, there was a sustained effort to understand the microscopic origin of magnetic phenomena of all sorts.

Anderson's first foray into magnetism concerned a phenomenon called *antiferromagnetism*. This is an exotic topic that is best approached after first gaining an appreciation of the much more common phenomenon of *ferromagnetism*—also called permanent

DC, 2010). The contents of the Herring Brain Box can be viewed at http://large. stanford.edu/herring/brain/. Accessed April 3, 2020.

[13] Interview of Morrel Cohen by Lillian Hoddeson, June 5 1981, Niels Bohr Library & Archives, American Institute of Physics, College Park, MD.

[14] S. Millman, *A History of Engineering and Science in the Bell System 1925–1980* (Bell Telephone Laboratories, Murray Hill, NJ, 1983), Chapter 1.

magnetism. Refrigerator magnets, horseshoe magnets, and bar
magnets are all made from ferromagnetic materials.

In 1600, the English physician and early scientist William Gilbert
dismissed years of speculation about the magnetic mineral lode-
stone, including its supposed ability to free women from witch-
craft and put demons to flight. He then described his own careful
and reproducible experiments with the substance. Two-and-one-
half centuries later, James Clerk Maxwell "mathematized" the
experimental observations of Michael Faraday and analyzed permanent magnetism in his *Treatise on Electricity and Magnetism*, one of
the crown jewels of nineteenth century theoretical physics.

According to Faraday, a bar magnet produces an invisible magnetic field B at every point in space. He used this field to understand the forces normally associated with magnetism.
The magnetic field "lines" drawn in Figure 6.3(a) encode the

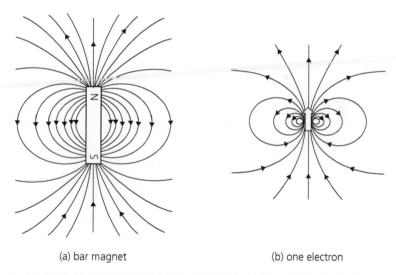

(a) bar magnet   (b) one electron

Figure 6.3 The magnetic field B of (a) a bar magnet and (b) a single electron, each drawn as a set of smooth curves. Each curve is actually a
dense collection of arrows drawn head-to-tail. Only a few arrowheads
are drawn to reduce clutter. The vertical arrow at the center of panel (b)
represents the spin vector S of the electron.

magnitude and direction of this field at every point in space.[15] This identifies B as a *vector* quantity, which we indicate using bold-face letters in the text and arrows in the diagrams.

A bar magnet is a ferromagnet because it creates a *macroscopic* B entirely on its own. More precisely, a ferromagnet self-generates a macroscopic magnetic field when its temperature falls below a material-dependent value called the Curie temperature, T_C. Iron is the most common ferromagnet, where $T_C = 1044\,\mathrm{K}$ or $1420\,°\mathrm{F}$.[16] Other elemental ferromagnets include nickel, cobalt, gadolinium, and dysprosium.

Ferromagnetism arises from a peculiar property of the electron called *spin*. For our purposes, this is a shorthand way of saying that every electron produces a *microscopic* magnetic field like the one shown in Figure 6.3(b). Spin is a vector quantity and rotating the spin vector S at the center of Figure 6.3(b) rotates the entire magnetic field pattern to follow it. Comparing the two panels of Figure 6.3 shows why physicists often think of electrons as microscopically sized bar magnets.

Solids are made from atoms and atoms are made from electrons. This raises the question: why aren't all solids ferromagnetic? In other words, why don't the microscopic magnetic fields produced by all the electrons in every solid combine to produce the macroscopic magnetic field characteristic of a ferromagnet? The answer begins at the level of a single atom.

The quantum pioneer Niels Bohr taught that every electron of an atom occupies one of several planetary-type orbits centered on the atomic nucleus. Each orbit accommodates exactly two electrons (with oppositely pointing spin vectors) and groups of orbits with similar size are called *shells*. Experiment shows that an atom produces

---

[15] At any point in space in Figure 6.3, the direction of B is parallel to the magnetic field line that passes through that point and the magnitude (strength) of B is proportional to the density of magnetic field lines at that point.

[16] Scientists use the Kelvin temperature scale rather than the Fahrenheit temperature scale. The formula $\mathrm{K} = 273 + (5/9)(°\mathrm{F} - 32)$ converts one to the other. Room temperature is $68\,°\mathrm{F}$ or $293\,\mathrm{K}$. The lowest possible temperature is $0\,\mathrm{K}$.

a net magnetic field only if at least one of its shells is <u>not</u> completely filled with electrons. A filled shell contains only filled orbits, so half the electron spin vectors point in one direction and the other half point in the opposite direction. These "up" and "down" spin vectors cancel in pairs to give zero net spin and zero net magnetic field. Conversely, an atom with an unfilled shell is magnetic; it has a non-zero spin S and produces a microscopic magnetic field.

In 1928, the soon-to-be Nobel laureate Werner Heisenberg exploited his experience as the inventor of quantum mechanics and proposed a microscopic model to explain the occurrence of ferromagnetism for a crystal composed of identical magnetic atoms.[17] The twenty-six-year-old *wunderkind* had been a full professor at the University of Leipzig for only six months. He squeezed his ferromagnetism research in between lecturing to undergraduates and challenging the members of his research group to ping pong matches.[18]

Heisenberg's model is easiest to understand for the case when one magnetic atom occupies each site of a crystal lattice. His first great insight was to replace the atom at the kth lattice site ($k = 1,2,3,\ldots$) by its atomic spin vector, S_k. For visual clarity, Figure 6.4 illustrates this

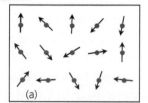 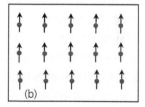

Figure 6.4 A model ferromagnetic crystal where the arrow at each lattice site (red dots) denotes the spin vector of the magnetic atom at that site: (a) the disordered non-magnetic state at high temperature; (b) the ordered magnetic state at low temperature.

[17] W. Heisenberg, "On the Theory of Ferromagnetism" (in German), *Zeitschrift für Physik* **49**, 619–36 (1928).

[18] Nevill Mott and Rudolf Peierls, "Werner Heisenberg, 5 December 1901—1 February 1976," *Biographical Memoirs of Fellows of the Royal Society* **23**, 212–51 (1977).

for a square lattice in two dimensions. However, for technical reasons, our discussion applies only to three-dimensional crystals.[19]

Heisenberg's prescription may seem like an unacceptable over-simplification. It is nevertheless typical of the best models in physics because it identifies and retains only the essential features of the real physical system and discards the rest. Our task is to show that this model reproduces the known macroscopic behavior of a ferromagnet.

At high temperature, all materials are non-magnetic and thus do not produce a macroscopic magnetic field. In Heisenberg's model, thermal energy causes the direction of each spin vector to fluctuate randomly resulting in a *disordered* state like the snapshot shown in Figure 6.4(a). This is consistent with a fundamental principle first stated by the nineteenth-century American physicist Josiah Willard Gibbs: *all physical systems at high temperature favor configurations with maximum disorder.*[20] The crystal in Figure 6.4(a) is indeed non-magnetic because the randomly oriented microscopic magnetic fields produced by the randomly oriented spins cancel one another to give zero net macroscopic field.

Another fundamental principle enunciated by Gibbs is that *all physical systems at zero temperature adopt the lowest energy configuration available to them.* For a ferromagnet, this configuration happens to be the *ordered* state shown in Figure 6.4(b) where all the spins point in the same direction. Here, the microscopic magnetic fields produced by the individual S_k reinforce (rather than cancel) and the result is the self-generated, macroscopic magnetic field characteristic of a conventional permanent magnet.

Heisenberg's second great insight was to realize that the cooperative behavior seen in Figure 6.4(b) is a consequence of a phenomenon called *exchange*. Exchange is not a new force in Nature. It is simply a shorthand language used to describe the

---

[19] A mathematical theorem precludes the existence of magnetism in the Heisenberg model for one- and two-dimensional crystals.

[20] Gibbs is credited with an explicit statement of this principle in Arnold Sommerfeld, *Thermodynamics and Statistical Mechanics* (Academic Press, New York, 1956), p. 48.

results obtained when one applies quantum mechanics to study the repulsion of two identically charged particles, both with spin (like two electrons). Exchange dictates that the lowest electric energy of a pair of neighboring electrons occurs when their spins point in the same direction. Heisenberg reasoned that if every unpaired spin experienced an energy-lowering with only its nearest-neighbor spins, the energetic advantage of parallelism would spread from spin to spin throughout the crystal.

To make Heisenberg's ideas quantitative, we approximate the energy of a pair of spins at the nearest-neighbor lattice sites i and k by the scalar product $-J S_i \cdot S_k$, a quantity whose numerical value varies with the angle between the spin vectors S_i and S_k.[21] The constant J characterizes the energy of interaction between the two spins. When $J > 0$, the lowest energy $(-JS^2)$ occurs when the spins i and k are parallel and the highest energy $(+JS^2)$ occurs when these spins are antiparallel, i.e., when they point in opposite directions. The total energy of the ferromagnet in Heisenberg's model is the sum of the contributions from all pairs of nearest-neighbor spins.

The exchange interaction with $J > 0$ guarantees that the Heisenberg model produces the ordered state of Figure 6.4(b) at zero temperature. Therefore, Gibbs' two principles (high temperature favors disorder and low temperature favors low energy) imply that a ferromagnet reversibly switches between its non-magnetic state [Figure 6.4(a)] and its magnetic state [Figure 6.4(b)] as the temperature cycles above and below the temperature T_C, respectively. The magnetic phase transition exhibited by the Heisenberg model was the sort of thing that fascinated Anderson's colleague and mentor, Gregory Wannier.

Antiferromagnetism

In the early summer of 1949, Charlie Kittel wandered into Phil's office and casually remarked that he wanted to learn more about

---

[21] If S_i and S_k both have magnitude S, the scalar product $S_i \cdot S_k = S^2 \cos\theta$ where θ is the angle between the vectors S_i and S_k.

a new-fangled phenomenon called antiferromagnetism.[22] It was not exactly new-fangled. Louis Néel in France and Lev Landau in Russia had independently predicted its existence and properties in the early 1930s. According to them, an antiferromagnet behaved in certain ways like a ferromagnet, but it never produced a macroscopic magnetic field of its own.

Antiferromagnetism is a cooperative magnetic phenomenon like ferromagnetism. The difference is that the energy of an anti-ferromagnet is lowest when neighboring spins are *antiparallel* rather than parallel.[23] The metal chromium is an antiferromag-net, as are many metal oxides, including common rust (iron oxide). Remarkably, perhaps, it is possible to describe this phe-nomenon using the Heisenberg model. The only modification needed is to change the algebraic sign of the exchange constant J from positive to negative in the exchange energy, $-J S_i \cdot S_k$.

Figure 6.5 shows the ordered configuration of spins that minimizes the Heisenberg model energy for an antiferromagnet. Gibbs' fundamental principles guarantee that this *Néel state* appears at low temperature while the disordered phase [Figure 6.4(a)] appears at high temperature. The transition between the two occurs at a material-dependent temperature, T_N.

Overall, the Heisenberg model is capable of reproducing the most basic *qualitative* facts about ferromagnetism and antiferro-magnetism. However, it does not purport to give a *quantitative* account of any particular magnet. This is typical of many models in physics where one gives up a truly realistic description in favor of mathematical simplicity. In this case, the Heisenberg simplifica-tion to limit the exchange interaction to nearest-neighbors spins

---

[22] P.W. Anderson, "Some Memories of Developments in the Theory of Magnetism," *Journal of Applied Physics* **50**, 7281–4 (1979).

[23] A historical survey of antiferromagnetism is Stephen T. Keith and Pierre Quédec, Chapter 6, "Magnetism and Magnetic Materials" in *Out of the Crystal Maze: Chapters from the History of Solid-state Physics*, edited by Lillian Hoddeson, Ernest Braun, Jürgen Teichmann, and Spencer Weart (Oxford University Press, New York, 1992).

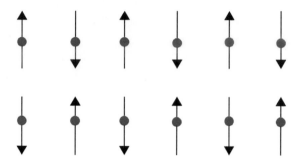

Figure 6.5 The spin arrangement of the ordered Néel phase for the Heisenberg model of an antiferromagnet. The nearest-neighbor pair energy $-J S_i \cdot S_k$ is lowest (most negative) for this configuration of spins if the exchange constant $J < 0$.

only may not be strictly true for a real antiferromagnet. This was among the issues Anderson pondered as he studied the experimental literature of magnetism.

Two months after Kittel's casual remark, Anderson's study of antiferromagnetism blossomed into a full-blown research project. The catalyst was a visit to Bell Labs by the experimenter (and future Nobel laureate) Clifford Shull. Shull described a new technique called *neutron diffraction*, which made it possible to deduce the spin orientation of every unpaired spin in the ordered state of a magnetic crystal.[24]

Shull presented experimental data for the antiferromagnetic electrical insulator manganese oxide (MnO). The physics he discussed merits a digression because it reappears and plays an important role at least twice more in Anderson's subsequent career. Unfortunately, the crystal structure of MnO is much more complicated than Figure 6.5, not least because it contains both magnetic and non-magnetic ions. The cartoon in Figure 6.6

---

[24] Because particles behave like waves in quantum mechanics, a crystalline solid diffracts a beam of neutrons just like it diffracts a beam of x-rays. However, because a neutron has a spin like an electron, it feels the magnetic field produced by all the unpaired electrons in the crystal. This makes neutron diffraction sensitive to the orientations of the atomic spins.

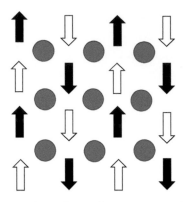

Figure 6.6 A cartoon of one plane of atoms in a MnO crystal. The spins represent Mn ions. An O ion (blue circle) lays midway between every pair of next-nearest neighbor Mn spins. The latter have the same color (black or white), lie along diagonals in the diagram, and are oriented antiparallel to each other in the magnetically ordered state shown.

is a better representation of one plane of atoms in this three-dimensional crystal.[25]

The blue dots in Figure 6.6 stand for non-magnetic oxygen ions. The black and white arrows stand for the spins of the magnetic Mn ions. Choose any Mn ion and focus on the four nearest-neighbor ions which lay above it, below it, to its right, and to its left. The color of all four is the same but their spin directions are not all the same. This was Shull's first message: *there is no consistent magnetic order between any Mn ion and its four nearest-neighbor Mn ions.*

Now focus on the four Mn ions (along the diagonals) that are the *next*-nearest-neighbors to the chosen Mn ion. The color of all four is again the same, but this time so are their spin directions: they <u>all</u> point antiparallel to the chosen ion. This was Shull's second message: *there is consistent antiferromagnetic order between every Mn ion and its four next-nearest-neighbor Mn ions.*

Antiparallelism between next-nearest neighbors is peculiar because the distance between them is too large for direct

[25] C.G. Shull, W.A. Strauser, and E.O. Wollam, "Neutron Diffraction by Paramagnetic and Antiferromagnetic Substances," *Physical Review* **83**, 333–45 (1951).

antiferromagnetic exchange to be relevant. Worse, there is always a non-magnetic oxygen ion interposed midway between every pair of next-nearest neighbor Mn spins. This was the puzzle Shull presented to his audience: *what physics forces every Mn spin in MnO to orient itself antiparallel to each of its next-nearest-neighbor Mn spins?*

For the first of many times in his long career as a theoretical physicist, Anderson elected to *follow the data.* He exchanged letters with Shull to better understand the experimental details. A consultation with Kittel led him to study some old work by the Dutch theorist Hendrik Kramers.[26] Slowly, a solution to Shull's puzzle came together in his mind.

In work unrelated to antiferromagnetism, Kramers had established that an interaction existed between two magnetic ions separated by a non-magnetic atom. He called this interaction *superexchange.* The formula Kramers derived for superexchange was difficult to evaluate and his paper attracted almost no attention.[27] But it was exactly what Anderson needed. First, he worked out the detailed physics of superexchange for the crystal structure adopted by most transition metal oxides. Then, he recast Kramers' formal analysis in the language of Heisenberg's spin vectors. Finally, he applied his version of superexchange to these oxides. For MnO, he discovered that superexchange yielded the lowest energy when the spin configuration was precisely the one revealed by Shull's neutron scattering data.

The paper Anderson wrote detailing his calculations circulated as a preprint.[28] His manuscript impressed enough people that he was invited to present his results at a big national meeting—the 1950 March Meeting of the American Physical Society

[26] Letter from PWA to C.G. Shull, September 2, 1949; Letter from C.G. Shull to PWA, October 13, 1949; Letter from PWA to C.G. Shull, February 20, 1950. AT&T Bell Laboratories Archives, 5 Reinman Road, Warren, New Jersey. H.A. Kramers, "The Interaction Between Magnetic Atoms in a Paramagnetic Crystal" (in French), *Physica* 1, 182 (1934).

[27] Kramers' paper received only seven citations in the fifteen years between its publication and Anderson's use of it. Google Scholar search September 14, 2017.

[28] A preprint is a pre-publication copy of a scientific paper. Before the Internet, physicists maintained a list of colleagues to whom they mailed

Invited talks at such meetings are the coin of the realm in the physics business. For that reason, Anderson knew that his prospects to remain a staff physicist at Bell Labs were very good, despite Bill Shockley's unhappiness that his protégé was working on antiferromagnetism rather than ferroelectricity.

Bell Labs rewarded Anderson with a salary raise. He and Joyce used the extra money to buy a 100-year-old farmhouse on one-and-one-half acres of land. The price was right because the previous owners—a group of followers of the early twentieth century mystic George Gurdjieff—had left the structure in serious disrepair. Room-by-room renovations followed and, after a few years, they installed a 30,000-gallon swimming pool using a contractor only to dig the necessary hole. They lived at the farm house for over a decade and made a point of inviting Lab friends and summer visitors over to swim and enjoy Phil's signature martinis.

Symmetry Lost

Anderson was now hooked on antiferromagnetism. Over the next two years, he wrote three more papers on the subject, one of which contained the germ of an idea that would later appear as part of his grand synthesis of fundamental ideas in condensed matter physics. The idea was "broken symmetry," although neither that phrase, nor the generality of the concept, would crystallize in his mind for a number of years to come (see Chapter 9).

Anderson's motivation was simple: virtually nothing was known about the ground (lowest energy) state of an antiferromagnet in three dimensions if one treated the spins as truly quantum mechanical objects rather than as classical vectors. Previous

preprints, usually just after they submitted the original manuscript to a journal for consideration for publication. The published version is P.W. Anderson,
"Antiferromagnetism. Theory of Superexchange Interaction," *Physical Review* **79**,
350–6 (1950).

approximate theory by John Van Vleck had predicted the ordered Néel spin arrangement shown in Figure 6.5.[29]

Anderson attacked this problem using a much more sophisticated approximation that allowed the spins to jiggle back and forth in a manner required by the laws of quantum mechanics.[30] He confirmed Van Vleck's prediction for the spin arrangement and then went far beyond, obtaining convincing results for the total energy of that state. Today, his results are textbook material.[31]

At this point, many authors might have declared victory and moved on to another problem. However, even at this early stage of his scientific career, Anderson devoted the longest section of his paper to a set of issues that would only grow in importance as his scientific taste and talent matured. The issues were the symmetry of the system of interest, the effect that symmetry had on the behavior of the system, and the implications that symmetry had for experiments.

Physicists say that a system possesses a *symmetry* if some feature of it remains the same before and after some action or transformation has been applied to it. For example, the smooth solid ball in Figure 6.7(a)

(a)   (b)

Figure 6.7 Geometrical symmetry: (a) a sphere exhibits continuous rotational symmetry in three dimensions; (b) a snowflake exhibits discrete rotational symmetry in two dimensions.

[29] J. H. Van Vleck, "On the Theory of Antiferromagnetism," *Journal of Chemical Physics* **9**, 85–90 (1941).

[30] P.W. Anderson, "An Approximate Quantum Theory of the Antiferromagnetic Ground State," *Physical Review* **86**, 694–701 (1952). This paper extends to antiferromagnetism a method used for ferromagnetism in Martin J. Klein and Robert S. Smith, "A Note on the Classical Spin-Wave Theory of Heller and Kramers," *Physical Review* **80**, 1111 (1950).

[31] See, e.g., K. Yosida, *Theory of Magnetism* (Springer-Verlag, Berlin, 1996), Section 9.2.

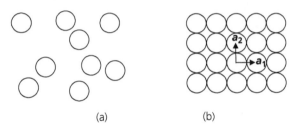

(a)   (b)

Figure 6.8 Symmetry breaking: (a) The high-temperature gas phase exhibits continuous translational symmetry; (b) the low-temperature crystalline phase loses or "breaks" this symmetry of the gas phase and exhibits only discrete translational symmetry, i.e., the crystal looks the same only after a rigid translation of all the atoms by integer multiples of the lattice vectors a_1 and a_2.

exhibits continuous *rotational symmetry* in three dimensions because it appears exactly the same before and after rotating it by any amount around any axis that passes through it center. The snowflake-shaped object in Figure 6.7(b) exhibits six-fold *discrete rotational symmetry* in two dimensions because it appears exactly the same only if it is rotated by discrete angular increments of sixty degrees around an axis perpendicular to the page which passes through its center.

The symmetry associated with the gas of atoms shown in Figure 6.8(a) is more subtle. This high-temperature, positionally disordered system possesses *continuous translational symmetry* because a rigid translation of all the atoms by the same amount in any direction leaves the gas completely unchanged—in a very specific sense. Namely, the atomic positions before and after such a rigid translation are equally valid configurations of the gas. However, because each atom in a gas moves randomly, it may take a very long time before the gas happens to adopt the arrangement of atoms that we have just called the "translated" configuration.

Figure 6.8(b) shows the same group of atoms at lower temperature after the gas has condensed into a crystal. This system has lost or "broken" the continuous translational symmetry of the gas phase. Instead, it possesses a less robust *discrete translational symmetry*. This means it only looks exactly the same before and after a rigid translation of all of its atoms by an integer multiple of either one or both of the two lattice vectors drawn as black arrows. A rigid translation by any other amount produces a shifted configuration of atoms that could never appear in a later snapshot of this crystal because its atoms are immobile.

The Heisenberg exchange energy possesses *continuous spin rotational symmetry* because the numerical value of $-JS_i \cdot S_k$ (which depends only on the angle between the vectors S_i and S_k) does not change when all the spin vectors of the model crystal rotate together as a single unit with respect to any fixed axis.

The high-temperature, disordered phase of the Heisenberg model [Figure 6.4a)] also exhibits continuous spin rotational symmetry also. Similar to the gas in Figure 6.8 (a), this figure is merely a snapshot of a physical situation where thermal energy causes all the spins to rotate wildly. Other snapshots at different times show the spins pointing in different directions. All such snapshots are equally likely, so the rigid rotation in question always leaves the system in a valid high-temperature configuration. Hence, the disordered phase is symmetric with respect to spin rotational symmetry.

The low-temperature ordered phase of the Heisenberg model shown in Figure 6.4(b) has lost or broken the continuous spin symmetry of Figure 6.4(a) because all of its spins point along one specific direction in space. A rigid rotation of all these spins leaves the Heisenberg energy unchanged but it does *not* leave any *broken symmetry* spin configuration unchanged. A rotation transforms each into a physically distinct spin configuration where all the spins point in a different direction.

The brilliant Russian theorist Lev Landau was the first person to appreciate the deep connection between symmetry loss and the development of ordered phases at low temperature.[32] Fifteen

[32] L. D. Landau, "On the Theory of Phase Transformation I & II" (in Russian), *Zhurnal Éksperimental'noĭ i Teoreticheskoĭ Fiziki* **11**, 26–46 (1937); ibid. **11**, 545 (1937). English translation in *Collected Papers of L.D.* Landau, edited by D. Ter Haar (Gordon and Breach, New York, 1965), pp. 193–216. See also, V.L. Pokrovsky, "Landau and Modern Physics," *Physics—Uspekhi* **52**(11), 1169–76 (2009).

years later, Anderson pondered a subtle difference between the ferromagnet ($J > 0$) and the antiferromagnet ($J < 0$). For the quantum *ferromagnet*, every broken symmetry spin pattern obtained from Figure 6.4(b) by a rotation of all the spins is a perfectly acceptable ground state. Surprisingly perhaps, this is not true for the quantum *antiferromagnet* where <u>none</u> of the broken symmetry solutions obtained by a rigid rotation of the Néel spin pattern in Figure 6.5 is an acceptable ground state.

In 1955, the British theorist Walter Marshall proved that the ground state of a quantum antiferromagnet is what physicists call a *singlet*. This means that the magnet treats all spin directions democratically and never picks out one particular direction for the spins to point.[33] It is interesting, then, that Anderson's 1952 paper on the subject confidently asserts that "one knows the ground state to be a singlet."[34] At the time, the truth of this statement was known only for the dimensional case studied by the quantum pioneer Han Bethe in 1931.[35]

Anderson credits his colleague Conyers Herring with the suggestion that he create a singlet solution by averaging the spin pattern in Figure 6.5 over all possible directions the spins could point. However, because it costs no energy to rotate all the spins collectively, Phil realized (in a flash of deep insight) that *the quantum antiferromagnet possessed a mechanism to perform the averaging itself*:

> The ground state of the antiferromagnet cannot…show in any way a preference for one direction over another. Therefore, while [nearest-neighbor] spins will certainly point in opposite directions…the two sets will rotate [together] in the course of time.

Anderson estimated it would take *ten years* for all the spins to complete one full rotation of the sort just suggested. As he drolly

[33] W. Marshall, "Antiferromagnetism," *Proceedings of the Royal Society of London. Series A, Mathematical and Physical Sciences*, **232**, 48–68 (1955).
[34] P.W. Anderson, "An Approximate Quantum Theory of the Antiferromagnetic Ground State," *Physical Review* **86**, 694–701 (1952). The quoted statement appears in the abstract.
[35] H.A. Bethe, "On the theory of metals. 1. Eigenvalues and eigenfunctions for the linear atomic chain" (in German) *Zeitschrift für Physik* **71**, 205–26 (1931).

remarked, "the tendency of the [spins] to rotate around is a weak one." This explained why Shull's experiments revealed a broken-symmetry state for MnO. Ten years later, physicists would refer to this peculiar behavior as *spontaneous symmetry breaking* and the gla-cially slow rotation responsible for restoring spin direction democracy would be regarded as an example of a *Goldstone mode* (see Chapter 10).

At the time, only a handful of theorists noticed Anderson's work on antiferromagnetism. The quantum pioneer Wolfgang Pauli was one of these and, at an invitation-only meeting of top-echelon physicists from around the world, he wondered out loud whether Anderson's approximate theoretical treatment was accurate enough to be trusted.[36] Otherwise, it was not until the mid-1960s that Anderson's conclusions about symmetry breaking attracted the attention of the greater physics community.[37] His analysis of the quantum antiferromagnet was truly ahead of its time.

Comings and Goings

A major disruption of the Bell Labs Solid-State Physics group occurred in late 1951 when John Bardeen and Charlie Kittel accepted professorships at the University of Illinois and the University of California, Berkeley, respectively. Both had been fed up with William Shockley's arrogance and therefore susceptible to offers from academia. Illinois attracted Bardeen by offering him a professorship shared between its physics and electrical

[36] Pauli's comments about Anderson's work appear in the discussion remarks following L. Néel, "Antiferromagnetism and Metamagnetism" (in French) in *Les Electrons dans les Metaux*, Proceedings of the 10th Solvay Conference on Physics, edited by R. Stoops (Institut International de Physique Solvay, Brussels, 1955).

[37] G.W. Pratt, "Necessity and Experimental Consistency of Antiferromagnetic Ground State without Long-Range Order," *Physical Review* **122**, 489–90 (1961); H. Stern, "Broken Symmetry, Sum Rules, and Collective Modes in Many-Body Systems," *Physical Review* **147**, 94–101 (1965). See also, Robert Brout, *Phase Transitions* (W.A. Benjamin, Inc., New York, 1965).

engineering departments. That way, he could pursue research in semiconductors and superconductivity with equal ease.[38]

Kittel's move to California illustrates how national politics influenced a seemingly unrelated issue like the future of solid-state physics in the United States. In the summer of 1950, Berkeley's physics department was in disarray because the loyalty oath controversy born of the Red Scare had led to the departure of all four of its theoretical physicists.[39] Solid-state physics did not exist at Berkeley at the time.[40] Nevertheless, in a bit of a panic, members of their senior faculty approached Kittel and asked him to spend the 1950 fall semester on their campus as a trial run.[41] He did so, organized a solid-state seminar series, and started seven PhD students on research projects before returning to Bell Labs at the end of the semester.

The seven students kept the seminar going and worked on their own until Kittel returned permanently a year later. If Kittel had not returned, there is no question all seven would have completed PhDs in nuclear or particle physics. Instead, by the end of 1953, they had fanned out and begun their careers as leaders of the next generation of solid-state theorists.[42] Kittel's *Introduction to Solid-State Physics* appeared that year and it served as the basic text for teaching the subject for years to come.

Ironically, Bell Labs replaced Bardeen and Kittel with two former Berkeley physicists: Peter Wolff, a recent PhD, and Harold

---

[38]   For Bardeen, see Lillian Hoddeson and Vicki Daitch, *True Genius. The Life and Science of John Bardeen* (Joseph Henry Press, Washington, DC, 2002), Chapter 9.

[39]   J.D. Jackson, "Panofsky Agonists: The 1950 Loyalty Oath at Berkeley," *Physics Today* **62**, 41–7 (2009).

[40]   In the Chemistry Department at UC Berkeley, William Giauque maintained a world-class laboratory devoted to studying the behavior of matter (particularly molecular crystals) at very low temperature. He won the 1949 Nobel Prize for Chemistry for his discovery of the powerful refrigeration technique called adiabatic demagnetization.

[41]   Morrell Cohen, "Berkeley Days. How Great Events Shaped Our Careers." April 20–22, 2007. Available at http://abrahamsfest.rutgers.edu/Recollections. html. Accessed August 7, 2019.

[42]   The seven students were Elihu Abrahams, Morrel Cohen, Harvey Kaplan, Fred Keffer, Jack Tessman, John Weymouth, and Yako Yafet. Weymouth turned mostly to experiment in later years.

Lewis, a refugee of the loyalty-oath controversy. The fact that Bell Labs hired a scientist with unorthodox political views was consistent with the laissez-faire attitude they had shown at the time of the security questionnaire flap.

Anderson was unhappy to see his colleagues Kittel and Bardeen leave. He liked them personally and he enjoyed learning from them. However, he was still a relatively new employee and it was important for him to continue to solidify his position. This led him to combine his new expertise in theoretical magnetism with his old expertise in molecular spectroscopy. The result was a calculation relevant to solid-state *magnetic resonance* spectroscopy. In physics, two oscillating systems in communication with each other are said to be "in resonance" when their frequencies match. Magnetic resonance spectroscopy exploits the fact that a spin in a static magnetic field can absorb energy from a time-oscillating magnetic field if the oscillation frequency is chosen properly.[43]

Anderson adapted his previous work on the *broadening* of molecular absorption lines due to collisions to calculate the *narrowing* of electron spin resonance absorption lines due to quantum mechanical exchange.[44] He sent preprints of his results to various people, one of whom was Ryogo Kubo, a solid-state theorist at the University of Chicago on leave from the University of Tokyo. Kubo and Anderson had met at a meeting of the American Physical Society and Phil's antiferromagnetism paper had stimulated Kubo to work on the same problem.

A Journey to Japan

Kubo had considerable influence back home in Japan. As a result, Anderson soon received an invitation from the University of

---

[43] This technique is called electron paramagnetic resonance (EPR) if an electron spin is involved. It is called nuclear magnetic resonance (NMR) if a proton spin is involved. The latter is the basis of magnetic resonance imaging (MRI) in diagnostic medicine.

[44] P.W. Anderson and P.R. Weiss, "Exchange Narrowing in Paramagnetic Resonance," *Reviews of Modern Physics* 25, 269–76 (1952).

Tokyo to spend half a year there, courtesy of funds provided by the Fulbright Scholar Program.[45] After some negotiations, Bell Labs agreed to send Anderson to a September 1953 International Conference of Theoretical Physics in Japan and then allow him to remain there on a six-month unpaid leave of absence to work in Kubo's research group. Therefore, one week before the conference began, 30-year-old Phil, Joyce, and 5-year-old Susan Anderson boarded an ocean liner in Seattle, Washington and sailed for Japan.[46]

The conference attracted one thousand physicists. Anderson was an active participant and he made comments after six different talks, all involving some aspect of magnetism.[47] He networked extensively and met a number of people who would become life-long scientific and personal friends. The most important of these was the solid-state theorist Nevill Mott, who would soon be named the sixth Cavendish Professor at the University of Cambridge.[48] Twenty-four years later, Anderson shared the Nobel Prize for Physics with Mott and John Van Vleck. By an odd coincidence, Anderson also met Per-Olov Löwdin, the theorist who chaired the subcommittee of the Swedish Academy of Science that recommended to the full Academy that Anderson, Mott, and Van Vleck share the Nobel Prize for Physics.

When the conference ended, the Andersons settled into a small house they had rented a few blocks from Kubo's house in the Komagome district of Tokyo. Evidence of the Allied firebombing of the city during the war was visible in other districts where no tree stood that was more than eight years old. Life for the family

[45] A United States government program, which, since 1947, has provided financial support for scholars and professionals to study and conduct research in other countries.

[46] Interview of Susan Anderson by the author, March 3, 2016.

[47] Proceedings of the International Conference of Theoretical Physics, Tokyo and Kyoto, Japan, September 1953 (Science Council of Japan, Tokyo, 1954), pp. 699, 713, 740, 750, 801, and 836.

[48] The Cavendish Professor is also Head of the Physics Department at the University of Cambridge.

was difficult but rewarding.[49] Almost no one around them spoke English and the deprivations of postwar Japan were severe. The winter of 1953–1954 was unusually cold and their stove burned a soft coal that left grimy soot everywhere. It was difficult to keep the house clean and sanitary. When the stove failed, Joyce burned scientific reprints in a sooty fireplace to keep herself and Susan warm.[50]

At the University, Anderson and Kubo had adjacent offices and they talked every day at length. They discussed methods to calculate the response of a many-particle system to an external probe, and Anderson always felt proud of the famous "Kubo formula" his host published two years later.[51] He also honored Kubo by publishing a long paper on the widths of spectral lines in the *Journal of the Physical Society of Japan*.[52]

As a visitor with the status of a professor, Anderson taught a twenty-week course on the theory of magnetism. The students in the class were PhD candidates, postdocs, and professors, including several future leaders of the Japanese solid-state physics community. Few of the class members understood spoken English, but Phil's lecture notes were transcribed, reproduced, and made available in a document called "The Little Red Book."

Anderson admits on the first page of the 152-page Little Red Book that his lectures would not present a comprehensive account of the theory of magnetism. Instead, he focused on topics of recent interest that were familiar to him. His notes show particular respect for two authors: his former PhD supervisor John Van Vleck and

[49] Biographical notes written by PWA for the 25th anniversary of the graduation of the Class of 1943 from Harvard College. Courtesy of Philip W. Anderson.

[50] Interview of David Z. Robinson by the author, April 6, 2016.

[51] P.W. Anderson, "Scientific and Personal Reminiscences of Ryogo Kubo" in *More and Different: Notes from a Thoughtful Curmudgeon* (World Scientific, Hackensack, NJ, 2011), pp. 62–7.

[52] P.W. Anderson, "A Mathematical Model of the Narrowing of Spectral Lines by Exchange or Motion," *Journal of the Physical Society of Japan* **9**, 316–39 (1954).

John Slater, the doyen of solid-state physics at MIT who had been training theoreticians in that area for twenty years.[53]

Anderson approvingly quotes Slater's opinion that the Heisenberg model is suitable for a magnetic insulator like MnO but not for a magnetic metal like iron. For that case, Slater endorsed solving the Schrödinger equation directly to decide whether the metal was ferromagnetic, antiferromagnetic, or non-magnetic.[54] Unfortunately, as Slater pointed out, it was not technically possible to do this at the time using existing computers.

Anderson's support of Slater's views was consistent with his earlier refusal to perform the detailed calculations demanded by Bill Shockley. That problem would similarly have required solving the Schrödinger equation with a level of accuracy that was not yet available. Later, as computer power increased, Anderson came to believe that Slater "became obsessed with the electronic computer and with the idea that all of the important theoretical problems of solid-state physics could be reduced to mechanical computation."[55] It was an idea that Anderson fought against his entire career.

Phil and Joyce did not confuse the Japanese people with their pre-war government and they made a conscious attempt to learn Japanese culture. They enrolled Susan in Japanese language classes, went to the public baths, and played *pachinko* in one of the many saloons devoted to that pinball-machine-type entertainment. They toured all around the country, following an itinerary developed by Joyce. She had taken the time to educate herself about Japanese architecture and the couple stayed in traditional

---

[53] John C. Slater, *Solid-State and Molecular Theory: A Scientific Biography* (Wiley-Interscience, New York, 1975).

[54] J.C. Slater, "Ferromagnetism and the Band Theory," *Reviews of Modern Physics* **25**, 199–210 (1953).

[55] P.W. Anderson, "Nevill Mott, John Slater, and the 'Magnetic State': Winning the Prize and Losing the PR Battle," in *More and Different: Notes From a Thoughtful Curmudgeon* (World Scientific, Hackensack, NJ, 2011), p. 126.

Figure 6.9 Japanese newspaper photograph of Phil, Susan, and Joyce Anderson on a train from Tokyo to Sendai in 1952. Source: Susan Anderson.

inns as often as possible. American tourists were still very rare in 1953 and the very photogenic Susan Anderson appeared on the front page of local newspapers wherever they went.[56]

The couple learned firsthand about the respect ordinary Japanese citizens had for scholarship and learning. At a train station near Kyoto, Joyce happened to be carrying their touring maps in a large official envelope Phil had taken from the Yukawa Institute for Theoretical Physics (established in 1952 to honor the 1949 Nobel laureate Hideki Yukawa). Two men in work clothes approached Joyce on the platform and pointed excitedly at the envelope. They were impressed that these foreigners had some connection to Yukawa, whom they knew to be a very great scientist.

Kubo and Anderson developed a warm personal relationship during this period. In later years, they corresponded by letter and Kubo visited Anderson whenever he could fit a visit to Bell Labs into one of his business trips to America. In 1983, Kubo made a

---

[56] Interview of Susan Anderson by the author, March 3, 2016.

point of flying to Murray Hill to attend a small scientific symposium organized to honor Anderson on his 60th birthday.[57] Thirty years earlier, Anderson had planted a Japanese quince (an attractive flowering shrub) in the garden of Kubo's home as a token of his friendship and appreciation.[58]

The Andersons returned to the United States in the spring of 1954 with an admiration for Japanese culture, art, and architecture. For thirty years, Phil and a group of similarly minded enthusiasts at Bell Labs played the board game Go every day after lunch. At his best, he ranked as a 3-*dan*, which is a good advanced player.[59]

Anderson's distinct preference for Go compared to, say, chess, is revealing.[60] Pattern recognition is important to both games, but Go tends to favor strategic skills while chess favors tactical skills. Throughout his career, Anderson mostly avoided step-by-step logical argument and approached what he called "scientific anomalies" very much like a puzzle-solver. He collected pieces—both theoretical and experimental—and then fitted them together until he saw and understood the bigger picture.

The most lasting effect on Anderson of his visit to Japan was his sense of his own place in the world of physics. Before the trip, Anderson saw himself as a junior person. He was still a bit tentative and he felt guilty about abandoning the problem Bill Shockley had given him. Then, at the International Conference for Theoretical Physics, he discovered he could talk comfortably with physicists of the first rank. The extremely positive reaction of his Tokyo hosts to his seminars and his lecture course boosted his ego more. He realized on the three-day trip flying home that he was no longer a neophyte solid-state physicist. He felt secure in his abilities, secure in his scientific taste, and confident he could strike out independently at Bell Labs.

[57] Private communication with Hidetoshi Fukuyama, June 14, 2016.
[58] Private communication with Hiroto Kono, March 25, 2020.
[59] Interview of PWA by Kaylee Ding, summer 2015.
[60] Author correspondence with John J. Hopfield, October 13, 2015.

7

Disorderly Conduct

Phil Anderson's research achievements in the second half of the 1950s catapulted him to the top rank of theoretical physics in America. He published papers on the subjects of magnetism, ferroelectricity, semiconductors, superconductivity, liquid helium, disordered solids, electrical conductivity, and the newly invented maser. These papers are notable for their depth, but it is their topical range which truly amazes.

In a typical five-year period, the vast majority of physicists focus their research on at most a small handful of different topics. By contrast, Anderson's output between 1955 and 1960 includes at least one paper in eight of the nine areas of solid-state physics that were most popular at the time.[1] He avoided only the most popular topic: the calculation of electron energy levels in particular crystalline solids. This decision, once again, set himself apart from many of his colleagues.

Anderson wrote two papers during this period which together led the Nobel selection committee to award him a one-third share of the 1977 Prize for Physics. The first, on the motion of electrons in disordered crystals, is the subject of this chapter and the next. The second, on magnetism in metals, is the subject of Chapter 11. Our focus in both cases is how he came to study these problems, how he achieved his solutions, how the physics

[1] As judged by the relative number of papers published in *Physical Review*, the nine most popular areas of solid state physics between 1955 and 1960 were electron energy levels in solids, magnetic resonance, properties of insulators and semiconductors, magnetism, superfluidity, electrical conductivity, properties of metals, defects in solids, and superconductivity.

community received them, and how their influence grew to merit recognition by the Swedish Academy of Sciences.

Van Vleck's Legacy

Anderson's successes in the late 1950s reflect the fruits of a theoretical legacy left to him by his PhD mentor, John Van Vleck: follow the data, address fundamental principles, and use simple theoretical models.

Van Vleck taught Anderson to respect experimental data and the people who obtained it. Sixty years after the event, Phil vividly recalled standing at a physics conference as experimenters swarmed around Van Vleck. Each knew that the Harvard luminary would listen carefully to the details of their experiment and help them understand the meaning of their results. Van Vleck communicated to Anderson the value in having early access to new experimental data. But it was even more important to follow the data rather than impose a set of pre-existing theoretical prejudices on information brought to him by experimentalists.

Anderson internalized this message completely. Talking to experimenters was enjoyable and he was eager to learn the technical details of their work. He took the time to understand their motivations and laboratory strategies and he relished grappling with the raw data himself. One theoretical collaborator observed him more than once use a photocopy machine to magnify an experimental curve, trace the curve on wax paper, and replot it on graph paper so he could examine the bumps and wiggles in the data in great detail.[2] Later in life, Anderson went so far as to characterize himself as "six-tenths theorist and four-tenths experimentalist," despite never having performed an experiment himself.[3]

[2] Private communication with Vangal Muthukumar, August 1, 2020.
[3] Interview of PWA by Alexei Kojevnikov, March 30, 1999, Niels Bohr Library & Archives, American Institute of Physics, College Park, MD.

Van Vleck was a purist who viewed physics as an organic whole devoted to fundamental principles. He was interested in atoms, molecules, and solids only in so far as they provided test systems for these principles. Consistent with this perspective, he opposed the creation of divisions within the American Physical Society (including the Division of Solid-State Physics) because he believed doing so would dilute and balkanize the Society by welcoming chemists, metallurgists, and others not fully committed to the physics enterprise.[4]

Anderson carried on Van's allegiance to fundamental principles in two ways. After his Shockley-imposed experience with the ferroelectric barium titanate, he never again tried to analyze the detailed behavior of any one particular material system. Over and over in his long career, he sought out the fundamental principles that lay behind the most important behavior and aimed to understand the consequences of those principles for entire classes of materials. Later, he took up fundamentality as a philosophical issue when he challenged the claims of the particle physicists that only their activities deserved to be called fundamental.

John Van Vleck's research career amply illustrated to Anderson the value of theoretical model-making. An insightful discussion of the then-novel Heisenberg model for a magnetic solid appears in Van's 1932 book *Theory of Electric and Magnetic Susceptibilities*. In his research papers, Van Vleck repeatedly used what he called "effective" energy expressions to study the behavior of magnetic ions in crystalline environments.

Anderson has described model-making in physics as "discarding almost all of the apparently relevant features of reality in order to create a model which has two almost incompatible features: enough *simplicity* to be solvable, or at least understandable, [with] enough *complexity* left to…mimic the actual behavior."[5] Future

---

[4] Joseph D. Martin, *Solid-State Insurrection. How the Science of Substance Made American Physics Matter* (University of Pittsburgh Press, Pittsburgh, PA, 2018), Chapter 3.

[5] P.W. Anderson, in "BCS and Me," in *More and Different: Notes from a Thoughtful Curmudgeon* (World Scientific, Hackensack, NJ, 2011), p. 38.

Figure 7.1   Four lithograph plates from Pablo Picasso, *Bull* (1945–1946). Source: Arts Right Society.

events would demonstrate that one of Anderson's greatest strengths as a theoretical physicist was precisely his ability to strip away the details from a complicated problem and expose its essence.

Figure 7.1 is a pictorial representation of this process, executed by the artist Pablo Picasso for the case of a bull.[6] Here, one may actually prefer the model bull to the real bull for aesthetic reasons, and Anderson would later invoke aesthetics in connection with model-making in physics.[7] However, on the occasion of his Nobel Prize address, he focused on the practical benefits:

> Very often such a model throws more light on the real workings of Nature than any number of [more realistic] calculations…Even when correct, [such calculations] often contain so much detail as to conceal rather than to reveal reality. It can be a disadvantage rather than an advantage to compute so accurately…After all, the perfect computation simply reproduces Nature, it does not explain Her.[8]

Localized and Delocalized

The interior walls of inexpensive apartments famously do a poor job of blocking out sounds coming from adjoining apartments. However, according to a theoretical discovery made by Phil

[6] I thank Mohit Randeria and Nandini Trivedi for bringing Picasso's *Bull* to my attention.

[7] See Chapter 13 and A. Zhang and A. Zangwill, "Four Facts Everyone Ought to Know about Science: The Two-Culture Concerns of Philip W. Anderson," *Physics in Perspective* **20**, 342–69 (2018).

[8] P.W. Anderson, "Local Moments and Localized States," *Reviews of Modern Physics* **50**, 191–201 (1978).

Anderson in 1958, if human hearing occurred at much higher frequencies than it actually does, the very same walls could bring these sound waves to a complete standstill and every apartment would be soundproof.

The phenomenon Anderson discovered—now called *wave localization*—is a possible fate for waves of all kinds: sound waves, water waves, seismic waves, electromagnetic waves, elastic waves, brain waves, etc. The essential requirement is that the arrested wave propagation occurs in a medium that is *disordered* in some way. Our example is imaginable because apartment walls are commonly made from sheetrock, a building material composed of a random packing of tiny gypsum particles.[9]

Rather than sound or light waves, Anderson was interested in the waves that quantum mechanics associates with microscopic particles. These waves dictate how an electron hops from atom to atom in a solid and therefore how that solid conducts electricity. It is impossible to overstate the importance of this subject to contemporary society. The operation of every cell phone, every computer—indeed, every microelectronic device one can think of—depends critically on the flow of electrons through the microscopically thin layers of metallic and semiconductor materials used to fabricate integrated circuits.

Disorder is important to our story because it is responsible for electrical resistance, the friction-like phenomenon that heats up a conducting wire and slows down its current-carrying electrons. Connecting a crystal to a battery exposes its electrons to an electric force that accelerates them and initiates current flow. Collisions between electrons and sites of disorder retard this motion and thus limit the magnitude of the current. By "sites of

[9] Anderson's theory predicts that the arrest of wave propagation through a medium is possible only if the wavelength of the wave is close to the size of (and the distance between) the randomly arranged particles which constitute the medium. For an experimental demonstration of this phenomenon for ultrasound, see Hefei Hu, A. Strybulevych, J.H. Page, S.E. Skipetrov, and B.A. Van Tiggelen, "Localization of Ultrasound in a Three-Dimensional Elastic Network," *Nature Physics* **4**, 945–8 (2008).

disorder," we mean (1) atoms displaced from their normal lattice position by small-amplitude vibrational motion; (2) missing atoms (vacancies), or (3) foreign atoms (impurities). All three contribute to electrical resistance, as does anything else which disrupts the repeated stacking of identical unit cells which constitutes a perfect crystal.

A puzzling experimental observation (discussed below) by one of Anderson's colleagues motivated him to ask a novel question: was it possible for disorder to completely halt the motion of an electron rather than merely retard it? The language he used to answer this question exploited a distinction of some importance: the difference between a *localized* electron and a *delocalized* electron.

Classical mechanics treats an electron exactly like a golf ball. Given some initial data, Newton's laws of motion predict a unique trajectory. Quantum mechanics radically denies that a trajectory even exists for a microscopic object like an electron. In its place, quantum theory provides only the *probability* that an electron will be observed at this or that point in space. In detail, the theory asks practitioners to solve the Schrödinger equation to find a quantity called the *wave function* $\psi(x,t)$ for the electron. The squared magnitude of this function, $|\psi(x,t)|^2$, is proportional to the probability of finding the electron at a point x in space at a time t.

Despite its name, a graph of the wave function does not always resemble a wave. An example is the top panel of Figure 7.2, which shows ψ for an electron in an isolated atom. $|\psi|^2$ is large near the nucleus but becomes small (although never exactly zero) a short distance away.[10] Quantum mechanics does not locate the position of the electron exactly, but in this case, ψ guarantees that the electron is *localized* in the immediate vicinity of the nucleus.

The lower panel of Figure 7.2 shows ψ for an electron in a perfect crystal composed of a periodic arrangement of identical atoms. This wave function is *delocalized* because the tails of the

---

[10] ψ in the upper panel of Figure 7.2 is nearly zero at a distance of 1–2 Ångstrom (one ten-billionth of a meter) from the nucleus. This is roughly the size of an atom.

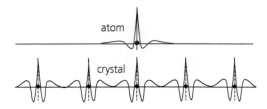

Figure 7.2   Upper panel: The wave functio ψ for an electron in an iso-lated atom is *localized* in space. Lower panel: The wave functio ψ for an electron in a perfect crystal is *delocalized* in space. The solid dots indicate atomic nuclei.

wave functions from adjacent atoms overlap and link together to form a single wave function which extends over the entire length of the crystal. An electron described by this ψ can contribute to an electric current because it can "hop" from one atom to the next.

What happens if disorder is present at some of the lattice sites in Figure 7.2? Similar to the classical collision argument sketched above, a quantum treatment of electric resistance amounts to a calculation of the electric current. This means there is an implicit assumption that the electron wave functions remain delocalized when disorder is present. Can this assumption be justified?

Imagine a situation where there is disorder at only one lattice site of an otherwise perfect crystal. The Schrödinger equation can be solved for this case and one finds delocalized wave functions similar to those of the perfect crystal. In addition, however, a new *localized* wave function appears just at the site of the disorder. This localize ψ often looks very much like the single-atom wave function shown in the upper panel of Figure 7.2.

The case where disorder occurs at many random sites throughout the crystal is more realistic. It is also more difficult to analyze because randomness introduces significant mathematical challenges to solving the Schrödinger equation. Nevertheless, in the 1950s, the consensus among experts was that the tails of wave functions localized near different sites of disorder—even if they were widely separated in space—would simply overlap to form

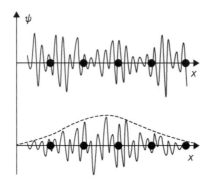

Figure 7.3 Typical wave functions for a disordered crystal. The solid dots indicate atomic nuclei. Upper panel: a portion of a delocalized wave function that extends over the entire solid. Lower panel: a localized wave function that does not extend over the entire solid. The envelope of this function—the dashed line—approaches zero exponentially rapidly outside the region of localization.

some delocalized wave function.[11] At most, the randomness was expected to produce random variations in the amplitude of ψ as shown in the upper panel of Figure 7.3.

This belief of the experts was not based on special knowledge of solid-state physics. It was based on their understanding of quantum mechanical tunneling. As discussed in Chapter 4, tunneling exploits the wave-like nature of quantum particles and permits them to move essentially anywhere they like, including passing through barriers which would cause a classical particle to bounce back. The ability of an electron to tunnel from one disorder-induced localized state to another inevitably leads to a delocalized ψ and thus to the ability of a disordered solid to conduct electricity.

Despite this conventional wisdom, experimental evidence discussed below convinced Anderson that it was possible for disorder to shut off quantum tunneling entirely. If so, the electrons would be trapped, i.e., localized, in regions of space which did not extend from one end of the crystal to the other. Ultimately, Anderson

[11] Paul W. Henriksen, "Solid-state Research at Purdue," *Osiris* **3**, 237–60 (1987). Arthur Samuel Ginzbarg, *Electronic Structure of Mixed Crystals*, PhD Thesis, Purdue University, 1949 (unpublished).

proved to his satisfaction that sufficiently strong disorder *always* produces a localized ψ . However, the size of the region to which an electron becomes localized depends on the details. A localized wave function might extend over just one atom (like the atomic function in the upper panel of Figure 7.2) or it might extend over many atoms (like the function in the lower panel of Figure 7.3). Neither of these is capable of passing an electric current from end to end through a disordered crystal.

Anderson's discovery of a situation in quantum mechanics where all particle motion stops was literally unbelievable to more than few theoretical physicists.[12] It led some to ignore him and others to assume he was wrong in some subtle way. The rescue and promotion of his theory by a senior colleague is part of the story told in the next chapter.

George Feher

Anderson returned from Japan in the spring of 1954 and reprised his lectures on the theory of magnetism. Bill Shockley was on leave at Caltech and soon left Bell Labs to found the Shockley Semiconductor Laboratory—the first technology company located in what would become known as Silicon Valley.[13] This made it easier for the 30-year-old Anderson to settle comfortably into Charlie Kittel's old role as "house theorist" for the expanding group of experimentalists at the Labs who worked on magnetism and magnetic resonance.

Around this time, Anderson began making weekly visits to the laboratory of a new Bell Labs hire named George Feher.[14] Feher is important to our story because a puzzling result obtained in his

[12] Author interview with Roger Haydock, March 3, 2015.

[13] In 1957, Shockley's abusive style led eight of his engineers to leave and form Fairchild Semiconductor. This company, in turn, spawned dozens of other Silicon Valley technology companies. See Joel Shurkin, *Broken Genius: The Rise and Fall of William Shockley, Creator of the Electronic Age* (Macmillan, London, 2006), Chapter 7.

[14] George Feher, "My Road to Biophysics: Picking Flowers on the Way to Photosynthesis," *Annual Reviews of Biophysics and Biomolecular Structure* **31**, 1–44 (2002).

laboratory was the principal motivation for Anderson's work on disorder-induced localization. At the time, however, Phil was simply curious about Feher's plans and the sweltering heat of the New Jersey summer made Feher's air-conditioned laboratory a welcome refuge from his office, which had only a fan.

A native of Bratislava, Feher was expelled from public school at age fourteen after the Nazi annexation of parts of Czechoslovakia ignited anti-Jewish policies throughout the country. He fled to Palestine and earned a living repairing radios and working as a laboratory assistant. At the same time, he worked as an electronics expert for *Haganah*, a paramilitary organization dedicated to forcing the ruling British to create an independent Jewish state.

Feher left Palestine before the establishment of the State of Israel to attend college at the University of California, Berkeley. He stayed there for graduate school and earned his PhD in experimental solid-state physics.[15] Bell Labs snapped him up as soon as he graduated. From his first day at the Labs, Feher wanted to build a state-of-the-art laboratory to perform electron and nuclear spin resonance experiments at very low temperature.[16] The specialized equipment he needed to do this was available as part of the cost Bell Labs management paid to retain the services of their ferroelectricity expert Bernd Matthias.

Matthias had taken a leave of absence at the University of Chicago where he became interested in discovering new superconductors. To lure him back, Bell Labs purchased the equipment Matthias needed to perform measurements at the low temperatures where ordinary metals become superconductors. This made

[15] Feher's PhD supervisor was Arthur Kip. Kip joined the Berkeley physics department in 1951 (almost simultaneously with Charles Kittel) and established a laboratory to use microwave resonance methods to study the behavior of electrons in metals and semiconductors.

[16] George Feher, "The Development of ENDOR and Other Reminiscences of the 1950s," in *Foundations of Modern EPR*, edited by Gareth R. Eaton, Sandra S. Eaton, and Kev M. Salikhov (World Scientific, Singapore, 1998), Chapter H.8.

it possible for other Bell Labs experimenters—like George Feher—to initiate programs in low temperature physics.[17]

Feher performed his first low temperature experiments on high quality crystals of silicon that had been subjected to *doping*, a process which significantly enhances their electrical properties.[18] Doping creates chemical disorder by intentionally replacing a small fraction of the atoms in a crystal by foreign, impurity atoms. By design, the doping process distributes the impurity atoms randomly through the volume of the host crystal. To appreciate the experiment Feher did, and its influence on Anderson, we pause to discuss the concept of a semiconductor and the importance of doping for a semiconductor like silicon.

Silicon and Doped Silicon

Today, crystalline silicon is the backbone of the electronics industry. However, when George Feher began taking data in 1955, the transistor was only seven years old and virtually all transistors were made from germanium. There were theoretical reasons to prefer silicon to germanium, but relatively little was known about the properties of solid silicon compared to solid germanium. For that reason, Bell Labs management was happy to support Feher's efforts to understand the behavior of electrons in silicon crystals.

---

[17] The essential pieces of equipment—both purchased from the AD Little Company—were a large Bitter electromagnet to produce magnetic fields up to 20 kilogauss and a Collins cryostat to produce liquid helium and cool samples to a temperature of 4.2 K. R.M. Bozorth, "Magnetization and Crystal Anisotropy of Single Crystals of Hexagonal Cobalt," *Physical Review* **96**, 311–16 (1954). P.W. Anderson, "Some Memories of Developments in the Theory of Magnetism," *Journal of Applied Physics* **50**, 7281–4 (1979).

[18] The Dutch word *doop* (a thick sauce) entered the English language in the late nineteenth century to refer to a viscous mixture of liquid opium and cocaine. The word soon migrated to horse racing and industrial processes with the meaning of a performance-enhancing additive. Scientists working with semiconductors adopted the term around 1949.

Figure 7.4 George Feher in the early 1960s. Source: American Institute of Physics Emilio Segrè Visual Archive.

Return now to Figure 4.4 and recall that each rung of the ladder shown there represents an allowed rotational energy state of the ammonia molecule. In a different regime of energy, a similar ladder represents the allowed quantum states of the *electrons*. Like the orbitals of an atom, each of these quantum states accommodates exactly two electrons with opposite spin. The occupation of the states by the electrons begins with the lowest energy state and proceeds upward in energy from rung to rung until every electron has a home.

In a crystal, the valence electrons detach from their parent nuclei and occupy quantum states which extend over the entire volume of the crystal. However, unlike the well-separated rungs of a ladder, these delocalized states are extremely closely spaced in energy and collectively form what is called an *energy band* of states. Distinct energy bands are separated by energy gaps

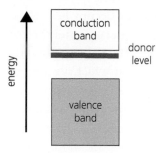

Figure 7.5 Energy band diagram for pure silicon and (by including the red donor level) doped silicon. An energy gap where no quantum states appear separates the valence band from the conduction band. Yellow indicates states occupied by two electrons with opposite spin per state beginning with the lowest energy state. Hence, the valence band is filled and the conduction band is empty.

where no allowed states occur. Figure 7.5 illustrates this for a typical semiconductor where electrons occupy all the *valence* band states and none of the *conduction* band states.[19]

You can win a bet at a bar not frequented by solid-state physicists by insisting that pure silicon—a semiconductor—is not useful for electronics. The explanation for this was given in 1931 by a 25-year old theoretician at the University of Cambridge named Alan Wilson. He pointed out that a battery supplies only a tiny amount of energy when it accelerates electrons to create an electric current. This tiny energy is just enough to promote electrons

[19] The origin of energy bands and energy gaps. Begin with an isolated "molecule" identical to the contents of one unit cell of a crystal of interest (see Figure 5.2). The energies of the allowed quantum states of this molecule form a ladder with well-separated rungs. Now, build the macroscopic crystal by stacking an enormous number of these unit cells together. At the end of this process, the widths of the energy rungs have increased (since each rung—or energy band—now includes many closely spaced allowed energy levels) with a consequent reduction in the size of the gaps between the rungs. The final crystal is a semiconductor if a small gap remains between the occupied rung with the highest energy (the valence band) and the unoccupied rung with the lowest energy (the conduction band). The final crystal is a metal if the rungs broaden so much that the highest occupied energy rung overlaps with the lowest energy unoccupied rung so there is no gap between them (see Figure 9.1).

from the filled states they occupy before the acceleration to the unfilled states they must occupy after the acceleration. However, according to Figure 7.5, the only empty states of a silicon crystal are in the conduction band, which is very distant (in energy) from the fully occupied states of the valence band. This means that a perfect silicon crystal cannot conduct electricity.

Doping replaces a fraction of the atoms in a semiconductor by foreign impurity atoms. In another flash of insight, Wilson explained how this process transforms a semiconductor from a virtually useless non-conductor into a conductor with exquisitely controllable electrical properties. Years later, when precision doping became easy to do, this fact made doped silicon the standard-bearer of the microelectronics revolution.

Here is Wilson's argument, specialized to George Feher's silicon crystals where the doping was done using phosphorus atoms. Compared to a silicon atom, a phosphorus atom has one extra proton and one extra electron. When a phosphorus atom replaces a silicon atom in a silicon crystal, the electric force of attraction between the extra, so-called *donor electron* and the extra proton ensures that this electron remains localized in the general vicinity of the impurity atom. Figure 7.6, taken from the book by William Shockley that Phil Anderson used to learn semiconductor physics, illustrates this where the plus sign indicates the charge of the extra proton and the density of small black dots is proportional to the amplitude of the donor electron wave function. Those dots show that the latter spreads out a fair distance from the impurity phosphorus atom.

The energy of a phosphorus donor electron in silicon lies quite close to the bottom of the silicon conduction band (horizontal red line in Figure 7.5). Close enough, in fact, that there is enough thermal energy at room temperature to excite the donor electron from its localized state near the impurity atom to a delocalized state in the conduction band. Once it resides in the nearly

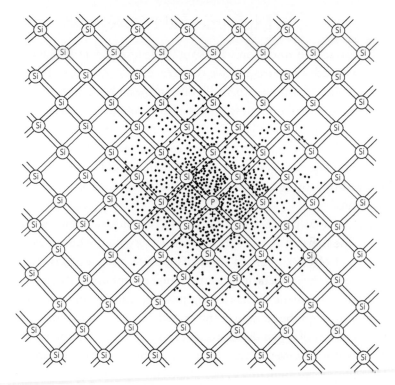

Figure 7.6 William Shockley's cartoon of a silicon crystal with one silicon atom replaced by a phosphorus atom. The density of small black dots is proportional the amplitude of the wave function of the extra (donor) electron of the P atom. Figure adapted from William Shockley, *Electrons and Holes in Semiconductors, with Applications to Transistor Electronics* (D. Van Nostrand Company, New York, 1950), p. 23.

empty conduction band, the donor electron can contribute to an electric current. This explains the appeal of doped silicon as a material for electronics. The number of impurity atoms controls the number of donor electrons and the number of donor electrons controls the conductivity of the crystal.

An Experimental Puzzle

Feher presented Anderson with some perplexing data. The issue was the width of the microwave resonance absorption lines Feher had obtained from his doped silicon samples. Phil had studied a similar issue for his doctoral thesis. However, the physical processes responsible for Feher's doped silicon line widths were entirely different from the processes responsible for the ammonia molecule line widths he had studied at Harvard. For silicon, the line widths provided information about the motion of the donor electrons.

At the low temperature where Feher's measurements were done, the thermal energy was too small to excite the donor electrons out of their localized states. Therefore, the electrons were localized in the immediate vicinity of their parent impurity atoms. The latter, by virtue of the doping process, were distributed randomly through the volume of the crystal. Figure 7.7 illustrates this for a portion of a two-dimensional crystal doped with one phosphorus atom for every six silicon atoms. Feher's crystals contained one phosphorus atom for every million silicon atoms.

Figure 7.7 A two-dimensional crystal of doped silicon where one phosphorus (blue) atom substitutes randomly for every six silicon (gray) atoms. George Feher's samples contained about one phosphorus atom for every million silicon atoms.

According to the consensus view, quantum mechanical tunneling facilitated communication between donor wave functions localized on different impurity atoms. As a result, a band of delocalized states was expected to form and all the donor electrons would occupy this band. Moreover, a donor electron tunneling from impurity atom to impurity atom in Figure 7.7 necessarily traced out an erratic path similar to that of a drunkard staggering around a parking lot. This is the hallmark of a kind of motion physicists call *diffusion*.

To Anderson and Feher's surprise, the experimental line widths did not conform to this scenario. Instead, they were most consistent with microwave absorption by *isolated* donor electrons confined to isolated phosphorus atoms. The data provided no evidence for tunneling and no evidence that an impurity band formed.

Feher's data also showed no evidence for *spin diffusion*. This was a process where the quantum mechanical exchange interaction causes pairs of antiparallel spins to interchange their spin orientations.[20] Although no spin ever moves, a sequence of such interchanges has the appearance of a spin with fixed orientation moving from site to site. The "motion" of this spin is also diffusive because, like the previous case of electron motion, the spatial arrangement of impurity sites is random.

Alan Portis, a young Berkeley physicist skilled in both theory and experiment, had calculated the effect of spin diffusion on resonance absorption measurements.[21] If Portis was correct, the effect of spin diffusion on Feher's data should have been glaringly obvious; but it was not. *It appeared that neither the donor electrons nor their spins moved at all.*

Apparently, there was something about random disorder which interfered with the quantum mechanics of both electron

[20] N. Bloembergen, "On the Interaction of Nuclear Spins in a Crystalline Lattice," *Physica* **15**, 386–426 (1949).

[21] A. M. Portis, "Spectral Diffusion in Magnetic Resonance," *Physical Review* **104**, 584–8 (1956).

tunneling and spin exchange. Both operated perfectly well when disorder was absent. The theoretical prejudice favoring delocalized donor electrons notwithstanding, Anderson resolved to *follow the data* and think hard about the possibility of disorder-induced localization of the donor electrons.

8

Law in Disorder

Hal Lewis had always noticed a vague unhappiness among his fellow theorists at Bell Labs. They were free to work on any problem they wished, but there was no clear path to decision-making roles because each was attached to a much larger group of experimenters. Issues of fairness resonated particularly strongly with Lewis after his resignation from the Berkeley physics faculty over the loyalty oath controversy. He was also a bit of a rabble-rouser and he pondered how he might improve the situation for his colleagues.

An opportunity arose when Lewis was recruited to join the physics faculty at the University of Wisconsin. Just before leaving, he fomented an "insurrection of the theorists" among Phil Anderson, Conyers Herring, Peter Wolff, and the recently arrived Melvin Lax. At Lewis' urging, they went to Bell Labs management and demanded that the Labs create a Theory Department for them. It would have all the perquisites of an academic department, including rotating leadership, the opportunity to hire postdoctoral fellows, travel opportunities, and an active summer visitor program. A bit to their surprise, the big Bell bosses agreed and the new arrangements established the character of the theory group at Bell Labs for decades to come.

The visitor program of the new Theory Department debuted in the summer of 1956, just when Anderson was trying to make sense of George Feher's data for doped silicon. Four visiting theorists took an active interest in the problem: Phil's former Harvard classmate Walter Kohn, the mathematical physicist Joaquin Luttinger, Elihu Abrahams, a recent graduate from Charlie Kittel's group at Berkeley, and David Pines, most recently a postdoc with John Bardeen at Illinois.

Kohn and Luttinger, who were both paid as consultants, had already collaborated to solve the Schrödinger equation for a single isolated phosphorus atom in an otherwise perfect silicon crystal. The excellent agreement between their calculated results and George Feher's data was strong evidence that the wave functions of the donor electrons in his samples were localized.[1]

For their part, Anderson, Abrahams, and Pines worked hard but were unable to rebut the conventional wisdom that quantum mechanical tunneling was sufficient to permit even widely spaced donor electrons to hop from one impurity site to the next. Because this conclusion flatly contradicted Feher's observations, they turned to an argument suggested for a related problem by Nevill Mott, the English theorist Anderson had met in Japan.

In 1949, Mott discussed crystals composed of atoms with partially filled electron shells.[2] These crystals are generally metals because an applied voltage induces electrons in each partially filled shell to hop into the partially filled shell of a neighboring atom. However, there is a classical, electric (Coulomb) force of repulsion between any two electrons which varies as $1/d^2$, where d is the distance between the electrons, i.e., the force *increases* as the distance between the electrons *decreases*. Mott suggested that there might be cases where the Coulomb energy cost for an electron to overcome this repulsive force and hop into a neighboring partially filled shell was so large as to cause the crystal to suppress the hopping process in the first place. This halts metallic conduction and practitioners now refer to such a crystal as a *Mott insulator*.

Returning to doped silicon, if tunneling *did* occur in Feher's samples, the most frequent event would be a donor electron

[1] W. Kohn and J.M. Luttinger, "Theory of Donor States in Silicon," *Physical Review* **98**, 915–22 (1955).

[2] N.F. Mott, "The Basis of the Electron Theory of Metals, with Special Reference to the Transition Metals," *Proceedings of the Physical Society (London)* A **62**, 416–22 (1949).

hopping from one phosphorus atom to a neighboring phosphorus atom whose own donor electron had not yet hopped away. However, this would not happen if Mott's argument applied and there was a prohibitive Coulomb energy cost to doubly occupying the localized donor state of the second phosphorus atom. The result would be no electron mobility, no impurity band formation, and no delocalization of the donor electrons.

By the end of the summer, all the departing theorists agreed that Mott's idea provided the only mechanism capable of keeping every donor electron "at home" with its parent phosphorus atom. Anderson was not so sure. Mott's theory did not invoke disorder and it did not suppress spin diffusion. His intuition told him that disorder was important and, to him, the absence of spin motion in Feher's samples was just as significant as the absence of particle motion. The localization puzzle was not yet solved.

A Theory of Localization

A first hint toward understanding the absence of diffusion in Feher's doped silicon samples appeared when Conyers Herring drew Anderson's attention to a preprint with the peculiar title, "Percolation Processes I. Crystals and Mazes."[3] The authors, mathematicians Simon Broadbent and John Hammersley, had studied how a fluid spreads through a collection of pores in a medium like a sponge.[4] To appreciate this "percolation process," return to Figure 7.7 and reinterpret the gray atoms as circular depressions scooped out of a flat beachfront and the blue atoms as similar depressions which have been filled with blue sand. Now, fill each of the gray depressions along the bottom row with liquid.

---

[3] S.R. Broadbent and J.M. Hammersley, "Percolation Processes I. Crystals and Mazes," *Mathematical Proceedings of the Cambridge Philosophical Society* **53**, 629–41 (1957).

[4] The original motivation was an interest by Broadbent to improve the performance of gas masks used by coal miners. Nicolas Bacaër, *A Short History of Mathematical Population Dynamics* (Springer-Verlag, London, 2011), Chapter 22.

If we reinterpret the thin grey lines in Figure 7.7 as narrow connecting channels, some of the liquid in each depression will flow out and spread into each of the four nearest-neighbor depressions that are also gray, i.e., not filled with blue sand.

Question: if the spreading continues according to this rule, will fluid ever appear in a gray depression along the top row? It is easy to see that the answer is yes for Figure 7.7 where the percentage of blocked depressions is only 15%. However, if the total number of depressions is very large, it is possible to prove that the answer is *no* if the percentage of filled depressions (randomly arranged) rises above 41%.[5] In that case, the fluid becomes trapped in disconnected pools and there is no percolation path across the entire sample.

The classical problem of a fluid spreading through a randomly blocked medium differs in important ways from the problem of quantum objects diffusing in a randomly disordered medium. Nevertheless, Anderson's intuition spoke to him in a way that perhaps happens to only the very best physicists. He challenged himself to prove that a sufficiently large amount of disorder would arrest quantum mechanical diffusion and induce localization. This required him to solve the Schrödinger equation to find the time-evolution of the wave function ψ of a typical spin in an imperfect crystal. This, in turn, required that he specify what physicists call the *Hamiltonian H* of the system. The Hamiltonian is a mathematical expression for the total energy.

It is important to appreciate that Anderson did *not* write down *H* for a silicon crystal doped randomly with impurity atoms. In fact, he did not write down *H* for any actual solid. Instead, he adopted the same strategy that Werner Heisenberg had used when he proposed his model for ferromagnetism: excise all the

---

[5] B.I. Shklovskii and A.L. Efros, *Electrical Properties of Doped Semiconductors* (Springer-Verlag, Berlin, 1984), p. 104. The critical fraction of closed pores below which percolation occurs depends on the geometry and dimensionality of the lattice.

details that distinguish one material from another and retain only the bare bones.

Anderson devised a *model Hamiltonian* for his problem composed of only two terms. One term used an energy V to specify the rate at which a quantum object (a particle or a spin) hops from site to site on a lattice. The other term labeled each site with an energy E (chosen at random from an interval of width W) and assigned that energy to the quantum object when it hopped to that site. This site energy is nothing but the "donor level" in Figure 7.5. Therefore, Anderson built disorder into his model using random variations of the donor level energy rather than using random spatial locations for the impurity atoms.

It took several months for Anderson to analyze his model thoroughly and use it to prove what Feher's data told him must be true: *the wave function of an electron localizes if the ratio V / W is small enough.* That is, localization occurs if the energy interval from which the sites energies are randomly chosen (the source of the disorder here) is sufficiently large compared to the hopping energy.

What is now called "Anderson localization" is a consequence of the facts of life for wave functions.[6] First, it is easiest for neighboring localized wave functions to link together to form a single delocalized wave function if the electron energy on every site is the same. This is the situation in the lower panel of Figure 7.2 where the spread in site energies W is zero. However, this linking ability becomes progressively harder as the electron energy on neighboring sites become progressively different. Eventually, W gets so large that the electron energies at neighboring sites differ so much that their localized wave functions fail to link at all. The corresponding total wave function consists of disconnected pieces, not unlike the disconnected pools of water in the percolation problem.

---

[6] A different (but ultimately equivalent) argument based on the phenomenon of wave interference is better suited to understand the Anderson localization of classical waves like light and sound.

On July 10, 1957, Anderson outlined his theory to the Bell Labs theory group. The hand-written script of his presentation gives some insight into his state of mind at the time:

> Most of what I have to say today will concern a theorem I have proved, at least to my satisfaction.... I really went through with it all not in order to solve any explicit physical problem but because I became fascinated with the model and the methodology.... However, for the sake of conventionality, I will say a few words about the direct, physical motivation for my work, which I consider relatively unimportant.[7]

The last sentence is unusual because it minimizes the role played by Feher's experiment. It is rare in Anderson's vast output to find experiment discounted in this way. A fair speculation is that he was, at that moment, overly proud of the mathematics he used to prove his "theorem," some of which was quite new to him.

What sort of analysis did Anderson carry out to deduce the existence of disorder-induced localization? The technical tools he used came from many sources. The most important one was borrowed from Walter Kohn and Quin Luttinger, who had spent the summer of 1956 constructing a quantum theory of electrical resistance.[8] They began with the delocalized wave functions of a perfect crystal and studied how weak disorder changes their character. Their method was a well-known technique of quantum mechanics called *perturbation theory*.

Physicists use perturbation theory for situations where it is too difficult to solve the Schrödinger equation exactly to find the wave functions and quantum state energies associated with a Hamiltonian H. However, one can often split H into two pieces

---

[7] Script of a seminar talk on localization, PWA file, AT&T Archives, Warren Township, New Jersey. For much of his career, Phil wrote out the text of what he planned to say at hour-long talks. However, he never read from these scripts at the time of the presentation. Phil submitted his work to the *Physical Review* three months after this talk.

[8] W. Kohn and J.M. Luttinger, "Quantum Theory of Electrical Transport Phenomena," *Physical Review* **108**, 590–611 (1957).

and write $H = H_0 + H_1$ where H_0 is a portion of the total Hamiltonian where the Schrödinger equation <u>can</u> be solved exactly to find the wave functions and quantum state energies. Perturbation theory is a theoretical procedure which systematically corrects the wave functions and energies of H_0 to account for the effect of the "perturbation" H_1. In principle, the theory provides an infinite number of correction terms collected into what is called a *perturbation series*. In practice, the first few terms of the series are usually the largest numerically and it is uncommon to proceed beyond them.

For their problem, Kohn and Luttinger chose H_0 to describe a crystal with *delocalized* wave functions. They then introduced a perturbation H_1 to introduce the effect of *weak disorder*. Retaining only the first correction term and averaging over all possible realizations of the disorder, they reproduced the simple collision theory of resistance sketched in the last chapter.

For his problem, Anderson chose H_0 to describe a crystal with *localized* wave functions. He then introduced a perturbation H_1 to introduce the effect of *weak hopping*. Retaining only the first correction term and averaging over all possible values of the random site energy E, he found that a particle initially localized at one impurity site always became delocalized, just as the conventional wisdom predicted. This was true no matter how strong he made the disorder parameter in his model Hamiltonian formula.

A mild depression settled over Phil. He had worked harder than at any time in his career, but he was no closer to understanding Feher's experimental data than when he began. What was he doing wrong? A eureka moment occurred when he realized that his error was averaging over all the possible values of the random donor level energy. Instead, he had to avoid such averages and study the *probability* that the hopping perturbation H_1 would or would not cause an initially localized particle to hop arbitrarily far away from its starting point. The mathematics he needed to do this was not part of his training, but it was available in the literature.

After much effort, a theorem revealed itself. If the spread in random site energies W was small enough compared to the hopping energy V (large values of V/W), an initially localized wave function delocalized and spread out across the entire lattice. However, similar to the critical fraction of closed pores in the percolation problem, Anderson's analysis yielded a critical value of the hopping to disorder ratio, $(V/W)_c$. If the spread in random site energies W was large enough that $V/W < (V/W)_c$, an initially localized wave function remained localized for all time. By the time he wrote up his results for publication, Anderson realized that disorder-induced localization was probably responsible for the absence of both spin diffusion *and* electron diffusion in George Feher's lightly doped silicon samples. There was no need to invoke Nevill Mott's Coulomb repulsion mechanism at all.

The Fortunes of Localization

The immediate reaction to Anderson's theory was not auspicious. After his presentation to his Bell Labs colleagues, Conyers Herring said nothing, Quin Luttinger thought his theorem was obvious, and Walter Kohn thought it was wrong.[9] Phil was sure he was not wrong, so he concluded that his result was at least subtle. To his chagrin, he discovered that physicists outside the Bell Labs sphere of influence ignored his work in droves.

Leo Kadanoff—a theorist who appears later in our story—was a graduate student at Harvard when he asked his mentors if there was anything interesting about Anderson's localization result and was told "no."[10] Even Phil's old Harvard classmate Rolf Landauer, a theorist who was working on disordered solids himself at the time, didn't know what to make of Anderson's

---

[9] Interview of PWA by Alexei Kojevnikov, March 30, 1999, Niels Bohr Library & Archives, American Institute of Physics, College Park, MD.

[10] Interview of Leo P. Kadanoff by the author, November 8, 2014.

theorem.[11] This view was widely held and Anderson's published paper earned only 34 citations in the first ten years after its appearance.[12] It did not help that the paper was regarded as "very complicated and difficult to read," and "exceedingly difficult."[13]

The outstanding exception to the rule of disinterest was Nevill Mott. Mott thought very highly of Anderson's work and he rapidly became a champion of Anderson localization as a plausible alternative to his own mechanism of localization based on Coulomb repulsion.[14] Because Mott framed the subject in the language of electrical conductivity, essentially all subsequent writers discussed localization in the context of electron motion rather than spin motion.

Nevill Mott had the good fortune to be an undergraduate at Cambridge University during the birth of quantum mechanics (1924–1927).[15] Reading voraciously and working by himself, he was the first to solve the Schrödinger equation for a two-particle collision problem.[16] He began PhD work under the supervision of Ralph Fowler, who was the only professor of theoretical physics at Cambridge at the time. However, because Fowler was often away, Mott again found himself working alone. He published several mathematically sophisticated papers on the quantum theory of collision processes before leaving Cambridge without a degree

---

[11] Rolf Landauer, "A Personal View of Early History," in *Coulomb and Interference Effects in Small Electronic Structures*, edited by D.C. Glattli, M. Sanquer, and J. Trân Thanh Vân (Editions Frontières, Gif-sur-Yvette, France, 1994), pp. 1–24.

[12] Google Scholar search performed December 24, 2017.

[13] These comments appear, respectively, in E.N. Economou and Morrel H. Cohen, "Existence of Mobility Edges in Anderson's Model for Random Lattices," *Physical Review B* 5, 2931–48 (1972) and J.M. Ziman, "Localization of Electrons in Ordered and Disordered Systems II: Bound Bands," *Journal of Physics C* 2, 1230–47 (1969).

[14] Commentary by Mott in *Sir Nevill Mott, 65 Years in Physics*, edited by N.F. Mott and A.S. Alexandrov (World Scientific, Singapore, 1995), p.259.

[15] Nevill Mott, *Sir Nevill Mott, A Life in Science* (Taylor & Francis, London, 1986).

[16] Mott used quantum mechanics to derive Ernest Rutherford's formula that describes the collision between two charged particles.

Figure 8.1 Nevill Mott. Source: American Institute of Physics Emilio Segrè Visual Archive.

in 1933 to become Professor of Theoretical Physics at the University of Bristol.

Mott's interest in both the "pure" and applied aspects of physics led him to <u>not</u> specialize in the burgeoning field of nuclear physics. Instead, he spent the next twenty years at Bristol building his department into the leading center of solid-state physics in Great Britain. His influential books, *The Theory of the Properties of Metals and Alloys* (1936) with Harry Jones and *Electronic Processes in Ionic Crystals* (1940) with Ronald Gurney give an indication of his breadth of interests. Constant contact with experimenters became a hallmark of his method and his papers abandoned mathematical sophistication in favor of what would become his signature style of creativity, physical arguments, and simple mathematics.

By 1970, Mott's tireless advocacy of Anderson's work had caused citations to his localization paper to jump to more than 100 per year and remain that way for a decade.[17] Crucial to this success were two concepts Mott proposed in the late 1960s. First, if the disorder was not great enough to localize *all* the quantum states in a solid, Mott suggested there was an energy E_C such that all the quantum states with energies $E < E_C$ were localized and all the quantum states with energies $E > E_C$ were delocalized. Therefore, in addition to his own Coulomb energy-driven mechanism, it was possible to imagine a disorder-driven mechanism to drive a system through what he called a *metal-insulator transition*.[18] Mott's second idea was that the electrical conductivity would jump abruptly from zero to a "minimum metallic conductivity" as one caused the energy of the fasted electron to exceed E_C.

These predictions motivated a great many physicists to apply their expertise and insights to the problem of wave function localization. One person who set his research group to work on the topic was John Ziman, a distinguished theoretical physicist at the University of Bristol in England.[19] Ziman was suspicious of localization and he convinced David Thouless, a brilliant Scot and professor of mathematical physics at the University of Birmingham, to look closely at Anderson's paper. Forty years later, Thouless recalled that:

> I was overwhelmed by the strength of the arguments in Anderson's 1958 paper and offered no comfort to Ziman. There was little to revise ... so the paper I wrote was basically a review of it, which I hoped would be more accessible than the original

[17] N.F. Mott, "Electrons in Disordered Structures," *Advances in Physics* **16**, 49–144 (1967); N.F. Mott, "Conduction in Non-Crystalline Systems I. Localized Electronic States in Disordered Systems," *Philosophical Magazine* **17**, 1259–68 (1968).

[18] N.F. Mott, "Metal-Insulator Transition," *Review of Modern Physics* **40**, 677–83 (1968).

[19] By an odd coincidence, Anderson's early 1950s papers on the quantum antiferromagnet scooped very similar work Ziman was doing for his PhD thesis at Oxford University.

paper appeared to be.... Most of us found features of Anderson's theory that were unfamiliar and uncomfortable, because they challenged assumptions that we had taken for granted. Within a few years, it moved from being a misunderstood set of ideas to being one of the core components of our understanding of dis-ordered condensed matter.[20]

The "unfamiliar and uncomfortable" feelings about localization mentioned by Thouless received an interesting expression in Russia. An early review article published there concludes with a two-page dialog between the article's *Author* and an imaginary *Theorist* who expresses skepticism about Anderson's theorem.[21] The *Theorist* admits to not spending much time studying the issue and his subsequent comments reveal that he is confused about how to deal with the randomness of disorder, just as Anderson was at first. The *Author* admits that the topic is subtle and tries to enlighten his interlocutor. In the end, he advises the *Theorist* to read the papers by Thouless and others if he wishes to lessen his confusion.

The Two-Person Career

The solid-state community's slow appreciation of Anderson localization had relatively little effect on his rise at Bell Labs. His overall achievements between, say, 1954 (when he returned from Japan) and 1967 (when he began spending half his time at Cambridge University), saw to that.[22] These achievements, in turn, owed quite a bit to the influence of his life-partner Joyce.

---

[20] D. Thouless, "Anderson Localization in the Seventies and Beyond," in *Fifty Years of Anderson Localization*, edited by Elihu Abrahams (World Scientific, New Jersey, 2010). The paper Thouless wrote at the time was D. J. Thouless, "Anderson's Theory of Localized States," *Journal of Physics C: Solid-state Physics* **3**, 1559–66 (1970).

[21] A.L. Éfros, "Electron Localization in Disordered Systems (the Anderson Transition)," *Soviet Physics Uspekhi* **21**(9), 746–60 (1978).

[22] See the first paragraph of Chapter 7.

Sociologists use the term "two-person career" to describe situations in mid-twentieth century America where the wives of men in middle- and upper-middle class occupations contributed substantially to the success of their husbands at work.[23] Besides their stereotypical roles as supporters, comforters, home managers, and child raisers, these women embraced enabling roles to help guarantee that their husbands reached their professional potentials.[24]

In Joyce's case, enabling meant providing a discipline and structure to Phil's life which prevented him from straying off into "feckless irrelevancies and foolishness."[25] Her contributions were particularly important when it came to organizing the social life expected of a rising star at Bell Labs. She took charge of their housing, friendship circles, clothing, manners, charitable work, etc. For example, the Andersons became well-known for an annual New Year's Eve party (fueled by an effective gin punch) to which a widening group of Bell Labs staff came, including management types.

Beginning with virtually no skill at entertaining, Joyce turned herself into a gourmet cook, a brilliant hostess, and a considerable asset to her husband.[26] In later years, her help became very concrete when she used her MA-level writing skills to edit the dozens of non-technical pieces Phil authored over the years. He, in turn, was totally devoted to her.

The Gang of Four

Anderson worked on disorder-induced localization only sporadically in the twenty years after his original paper appeared. For that

---

[23] Hanna Papanek, "Men, Women, and Work: Reflections on the Two-Person Career," *American Journal of Sociology* **78**, 852–72 (1973).

[24] Eliza K. Pavalko and Glen H. Elder, "Women Behind the Men: Variations in Wives' Support of Husbands' Careers," *Gender and Society* **7**, 548–67 (1993).

[25] Author correspondence with PWA.

[26] Author correspondence with James Philips, March 29, 2015.

reason, he was honestly surprised when the 1977 Nobel Committee honored him for his work on that topic. We will explore the possible thinking of the Nobel Committee in Chapter 11. What is certain is that winning a share of the Nobel Prize did *not* reignite Anderson's interest in localization. Nevertheless, less than a year after the award ceremony, he was the driving force behind a new and equally groundbreaking paper on that very topic. The impetus came from David Thouless and the occasion was a summer school for physicists in the French Alps.

Summer schools have long been venues where junior physicists improve their skills and learn about the latest developments in their field. Between 1928 and 1941 the Michigan Summer Symposium in Theoretical Physics helped elevate American physics to a very high level by bringing the world's greatest theorists together to lecture to senior graduate students, postdocs, and young faculty members.[27] In 1951, the young French physicist Cécile DeWitt-Morette founded the Les Houches School of Physics with the aim of helping rebuild the teaching and practice of physics in her country after World War II. She chose a setting designed to attract the best lecturers: a wooded hillside 3300 feet above sea level that overlooked the Chamonix valley and boasted a spectacular view of the Mont Blanc massif.[28]

For six weeks in the summer of 1978, Thouless, Anderson, and five other senior physicists lectured at Les Houches to 50 recent PhD graduates and postdocs. The accommodations were very modest and it rained half the time, but the level of enthusiasm was very high.[29] The School was devoted to disordered matter and Anderson gave a series of talks on non-crystalline solids. Unfortunately, many of his listeners found him incomprehensible—an outcome

[27] Samuel A. Goudsmit, "The Michigan Symposium in Theoretical Physics," *Michigan Alumnus Quarterly Review* **67** (Spring 1961), pp. 178–182.
[28] "The School High in the Alps," *Europhysics News*, May/June 1999, p. 68; History of the School, École de Physique des Houches, https://www.houches-school-physics.com/the-school/history/. Accessed December 28, 2017.
[29] Author correspondence with Robert Pelcovits, December 28, 2017.

consistent with the oracular lecturing style he had developed by this time.[30] As the distinguished Russian theorist Anatoly Larkin once remarked, "God speaks to us through Phil Anderson. The only mystery is why He chose a vessel that is so difficult to understand."[31]

By contrast, Thouless' lectures on percolation and localization were models of pedagogy. Ever since John Ziman prodded him ten years earlier, Thouless had devoted the majority of his effort to disordered systems.[32] He and his research group had devised several novel ways to study localization and his lectures elegantly summarized their achievements.

Anderson was very interested in Thouless' treatment of the electrical resistance R, or more precisely, the inverse resistance, $G = 1/R$, a quantity called the *conductance*. His key result was that the conductance exhibited certain "scaling" properties, which determined whether a quantum state was localized or delocalized. Scaling was an idea borrowed from the theory of phase transitions where the existence of a characteristic length scale or energy scale plays an important role. Anderson had developed a scaling theory for an entirely different problem (see Chapter 10) and it fired his imagination to learn that an argument of this kind might be useful to understand localization.

Anderson returned home from Les Houches and began talking about scaling and localization. One person listening was Don Licciardello, an ex-postdoc of David Thouless. Another listener was Elihu Abrahams, one of the visitors to Bell Labs who had labored with Phil to understand George Feher's data. T.V. Ramakrishnan, a former PhD student of Quin Luttinger, rounded

---

[30] Author correspondence with Julia Yeomans, April 3, 2019. Yeomans was a student moderator at the 1978 Les Houches Summer School. Her assignment was to poll the attendees and provide feedback to the speakers.

[31] This remark was made after an Anderson seminar at Argonne National Laboratory. Author interview with Richard Klemm, March 17, 2016.

[32] Thouless earned a share of the 2016 Nobel Prize for Physics for his work on topological issues in condensed matter physics.

out what later came to be called the Gang of Four.[33] At first, Ramakrishnan found it hard to make sense of what Anderson was saying. He consulted Luttinger and was advised that "Phil does not get things out beyond a certain point of clarity. But he is the most talented person in our field."[34]

It focused the minds of the four collaborators wonderfully when Abrahams discovered two papers in a German physics journal where the authors applied scaling ideas to some of Mott's ideas.[35] The Gang decided to cast their net more broadly. Where the Thouless group had proceeded numerically, Anderson urged his collaborators to try to develop an algebraic scaling theory for the disorder-driven transition from extended states to localized states.

The beauty and appeal of a scaling theory is that the microscopic details that distinguish one solid from another play no role. What matters are global issues like the choices of the scaling variables, the symmetries of the system, and its dimensionality. Of course, all real crystals are three-dimensional. However, a layered material like graphite (see Figure 8.2) *behaves* like a two-dimensional crystal because the bonding within each of its layers is much stronger than the bonding between the layers. Other, more exotic crystals are quasi-one-dimensional because they consist of parallel chains of atoms that bond only weakly to each other.

[33] The name "Gang of Four" played on then-current events. The real Gang was a political faction of four high ranking Chinese Communist Party officials (including the last wife of Mao Zedong) who were charged with treason immediately after Mao died in 1976.

[34] T.V. Ramakrishnan, "Anderson and Condensed Matter Physics," in *PWA90: A Lifetime of Emergence*, edited by P. Chandra, P. Coleman, G. Kotliar, P. Ong, D. L. Stein, and C. Yu (World Scientific, New Jersey, 2016).

[35] Franz J. Wegner, "Electrons in Disordered Systems. Scaling Near the Mobility Edge," *Zeitschrift für Physik B* **25**, 327–37 (1976); H.G. Schuster, "On a Relation Between the Mobility Edge Problem and an Isotropic XY Model," *Zeitschrift für Physik B* **31**, 99–104 (1978).

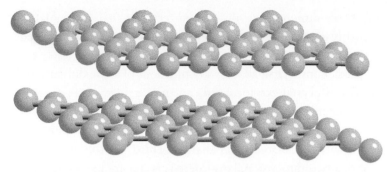

Figure 8.2 A ball-and-stick model of a graphite crystal. Each layer behaves nearly like a two-dimensional system because the chemical bonding between the layers is very weak.

Anderson challenged his collaborators to develop an equation to determine the conductance $G(L)$ as a function of the size L of the crystal.[36] This may seem like a pedestrian thing to do, but earlier research by others had demonstrated the importance of analyzing physical systems from the perspective of different length scales. Sometimes, the relevant scale was the size of the system. Other times, it was the length scale over which one viewed the system, as one might do with a microscope fitted with a variable field-of-view.

The Gang began with some reasonable guesses about the behavior of $G(L)$ for large and small values of G. Confirmation for some of the guesswork came when a classroom lecture Anderson gave on disordered systems led Ramakrishnan to remember a ten-year-old calculation of the electrical resistance of a metal as a function of the concentration of impurities.[37] In the end, they simply drew smooth curves for $G(L)$ for one-, two-,

---

[36] They considered a one-dimensional solid of length L, a two-dimensional solid of area L^2, and a three-dimensional crystal of volume L^3.

[37] T.V. Ramakrishnan, "One Subject, Two Lands: My Journey in Condensed Matter Physics," *Annual Reviews of Condensed Matter Physics* **7**, 1–10 (2016); J.S. Langer and T. Neal, "Breakdown of the Concentration Expansion for the Impurity Resistance of Metals," *Physical Review Letters* **16**, 984–6 (1966).

and three-dimensional disordered crystals by interpolating smoothly between the portions of those curves they had guessed.

The qualitative behavior of $G(L)$ allowed the Gang to make several predictions.[38] First, all the wave functions of a one-dimensional disordered crystal are localized, no matter how weak the disorder. Second, all the wave functions of a three-dimensional disordered crystal are localized if the disorder is strong enough. This agreed with Anderson's 1958 analysis. Third, all the wave functions of a two-dimensional disordered crystal are localized, albeit weakly so. This striking and surprising result is often called *weak localization* to distinguish it from Anderson's original prediction of *strong localization*. Finally, in contradiction to Mott, they deduced that there was no jump to a minimum metallic conductivity at the transition from localized to delocalized states in three dimensions. The conductivity simply rose smoothly from zero.

Experimental confirmation of the Gang of Four scaling theory came surprisingly quickly. As he had done with George Feher, Anderson now made a point of making a weekly visit to the laboratory of Douglas Osheroff. Osheroff had done spectacular work studying the exotic phases of liquid ^{3}He (see Chapter 10) but he was now studying the electrical resistance of ultrathin metal films at very low temperature.[39] At one point, Osheroff graphed his data and got some very oddly shaped curves.[40]

As he later reported,

Phil Anderson came into my lab and asked what I had been doing. I showed him the strange curves we had been obtaining and he said without any hesitation, "Why, that's a logarithm," and asked if he could borrow some of the data while he sat through a seminar. When he returned he had replotted our data in a form that indeed made it look like a logarithmic dependence. He then

---

[38] E. Abrahams, P.W. Anderson, D.C. Licciardello, and T.V. Ramakrishnan, "Scaling Theory of Localization," *Physical Review Letters* **42**, 673–5 (1979).

[39] Osheroff won a share of the 1996 Nobel Prize for Physics for his work on helium.

[40] Voltage (supplied by a battery) drives the current (flow of electrons) in a metal the way pressure drives the flow of water in a pipe.

pronounced that we had discovered weak localization in two dimensions, as predicted by his unpublished theory. [41]

Back-to-back theory and experiment papers discussing weak localization in two dimensions appeared in print only six weeks after the Gang of Four paper was published.[42]

Anderson followed up by co-authoring a paper that derived the Gang's scaling theory using methods and concepts introduced a decade earlier by his Harvard friend, Rolf Landauer.[43] This new paper delighted Phil, not least because it drew attention to Landauer's forgotten work and thereby helped launch the field of *mesoscopics*, a research topic focused on metals and semiconductors with sizes in the 100–1000 nanometer range.[44]

It is a testament to Anderson's taste in choosing problems that the Gang of Four paper motivated many physicists to think creatively about disorder-induced localization. Inside solid-state physics, a minor industry sprung up to study how the Coulomb repulsion between electrons (ignored by Anderson and the Gang of Four) influenced localization.[45] It also soon emerged that electron localization was essential for understanding a remarkable phenomenon called the *quantum Hall effect*. This manifestation of quantum mechanics on the macroscopic scale is rivaled only by superconductivity for its fundamental significance.[46]

---

[41] Douglas D. Osheroff, "The Nature of Discovery in Physics," *American Journal of Physics* **69**, 26–37 (2001).

[42] P.W. Anderson, E. Abrahams, and T.V. Ramakrishnan, "Possible Explanation of Nonlinear Conductivity in Thin-Film Metal Wires," *Physical Review Letters* **43**, 718–20 (1979); G.R. Dolan and D.D. Osheroff, "Non-Metallic Conduction in Thin Metal Films at Low Temperatures," ibid. **43**, 721–4 (1979).

[43] P.W. Anderson, D.J. Thouless, E. Abrahams, and D.S. Fisher, "New Method for a Scaling Theory of Localization," *Physical Review B* **32**, 3519–26 (1980). This paper generalizes Rolf Landauer, "Electrical Resistance of Disordered One-Dimensional Lattices," *Philosophical Magazine* **21**, 863–7 (1970).

[44] Supriyo Datta, *Electronic Transport in Mesoscopic Systems* (Cambridge University Press, Cambridge, UK, 1995).

[45] A.L. Efros and M. Pollack, *Electron-Electron Interactions in Disordered Systems* (Elsevier Science Publishers, Amsterdam, 1985).

[46] See, for example, Steven M. Girvin and Kun Yang, *Modern Condensed Matter Physics* (Cambridge University Press, Cambridge, UK, 2019), Chapter 12 and Chapter 16.

The Hall effect occurs when a current passes through a conductor immersed in a magnetic field.[47] The *quantum* Hall effect is special because the relevant conductor is simply a group of electrons trapped at the planar interface between two semiconductors that are capable of moving only in the two-dimensional space of that plane. Disorder-induced electron localization is unavoidable at such interfaces. In fact, *precisely because localization is present*, the quantum Hall effect provides a direct measurement of what is called the "quantum of resistance," $h/e^2 = 25812.807$ ohms. The variations in disorder from sample to sample turn out to be irrelevant and experimenters find that their measurements of h/e^2 are so precise (1 part in 10^{10}) that they are used to define the international standards for electrical units.[48]

Outside of solid-state physics, sustained efforts over many years have demonstrated that Anderson localization can arrest the propagation of light, sound, and other classical waves through disordered media.[49] In an entirely different experimental regime, atomic physicists have observed the localization of the quantum waves associated with entire atoms. These experiments use laser light to create a disordered medium which impedes and ultimately localizes the otherwise diffusive motion of extremely dilute atomic gases.[50]

[47] The Hall effect is the spontaneous appearance of an electric field in a current-carrying conductor which points in a direction that is perpendicular to both the current flow and an applied magnetic field.

[48] Klaus von Klitzing, "Quantum Hall Effect: Discovery and Application," *Annual Reviews of Condensed Matter Physics* **8**, 13–30 (2017).

[49] Ad Lagendijk, Bart van Tiggelen, and Diederick Wiersma, "Fifty Years of Anderson Localization," *Physics Today* **62**, 24–9 (2009).

[50] S.S. Kondov, W.R. McGehee, J.J. Zirbel, and B. DeMarco, "Three-Dimensional Anderson Localization of Ultracold Matter," *Science* **334** 66–8 (2011); F. Jendrzejewski, A. Bernard, K. Müller, P. Cheinet, V. Josse, M. Piraud, L. Pezzé, L. Sanchez-Palencia, A. Aspect, and P. Bouyer, "Three-Dimensional Localization of Ultracold Atoms in an Optical Disordered Potential," *Nature Physics* **8**, 398–403 (2012).

At this writing, there are no truly practical applications of Anderson localization. Some believe it may prove useful to reliably switch a material back and forth between a conductor and an insulator. Others imagine using it to localize spins to serve as quantum "bits" for quantum information processing and storage. What seems clear is that every new generation of physicists will find reasons to re-visit Anderson localization and its consequences.

9

The Love of His Life

The Bell Labs Theory Group tea room buzzed with excitement in February of 1957 when a preprint arrived from the University of Illinois. Its authors, John Bardeen, Leon Cooper, and J. Robert Schrieffer (BCS), claimed to have identified the microscopic origin of *superconductivity*, a mysterious solid-state phenomenon that had defied explanation for forty-five years.[1] If the BCS theory was correct, they would have achieved the physics equivalent of climbing Mount Everest, a similarly long-sought goal that had been achieved just four years earlier.[2]

Superconductivity was not of great interest to most Bell Labs scientists at the time. But Conyers Herring, Gregory Wannier, and Phil Anderson well remembered their former colleague Bardeen's intense interest in the subject. The 48-year-old Bardeen was at the height of his powers and his new theory looked very promising. It certainly seemed so to Anderson and he wound up seriously engaged with it for the next seven years.

What is Superconductivity?

Superconductivity is an extraordinary phenomenon where the electrical resistance of a crystalline solid drops abruptly to zero when its temperature falls below a material-dependent *critical*

---

[1] Jean Matricon and Georges Waysand, *The Cold Wars: A History of Superconductivity* (Rutgers University Press, Piscataway, NJ, 2003).

[2] It is conventional to write "BCS" when referring *either* to the three physicists Bardeen, Cooper, and Schrieffer *or* to the theory of superconductivity they produced.

temperature T_C.[3] The resistance returns to its normal value when the temperature of the crystal rises above T_C. Because there is no electrical resistance when $T < T_C$, an electric current established in a closed loop of superconducting wire circulates undiminished without a source of power to maintain it. The best estimate of the lifetime of such a persistent current is 10^{12} years.[4]

Superconductivity is a very low temperature phenomenon. When the BCS preprint arrived at Bell Labs, the highest critical temperature known was 18 K for the compound Nb_3Sn.[5] Not by accident, this record had been set only a few years earlier by Bernd Matthias, the ferroelectric-turned-superconductor expert at the Labs.[6] Evidently, only physicists with expensive refrigeration equipment could study superconductivity.

On the other hand, speculation is cheap and proposals for the microscopic origin of superconductivity flowed from the pens of many of the best theoretical physicists in the world. Albert Einstein, Niels Bohr, Werner Heisenberg, Lev Landau, Max Born, Felix Bloch, Edward Teller, and John Slater all made suggestions.[7] None of them was correct. Richard Feynman described his attempt at an International Congress of Theoretical Physics in the fall of 1956

[3] A superconductor also expels all magnetic fields from its interior. This phenomenon is important for understanding superconductivity in detail, but it is not essential here.

[4] The source of this theoretical estimate is A.M. Goldman, "Lifetimes of Persistent Currents in Superconducting Loops Interrupted by Josephson Junctions," *Journal of Low Temperature Physics* **3**, 55–63 (1970). The best direct measurement quotes a lower bound of 10^5 years. See J. File and R.G. Mills, "Observation of persistent current in a superconducting solenoid," *Physical Review Letters* **10**, 93–6 (1963).

[5] 18 K is −427.3 °F. Absolute zero, the lowest temperature possible, is 0 K or −459.67 °F.

[6] Matthias had a career-long goal to discover superconductors with the highest possible critical temperatures.

[7] J. Schmalian, "Failed Theories of Superconductivity," in *BCS: 50 Years*, edited by Leon N. Cooper and Dmitri Feldman (World Scientific, New Jersey, 2011), pp.41–55.

and suggested that "the only reason we cannot do this problem of superconductivity is that we haven't got enough imagination."[8]

Anderson got excited about superconductivity after hearing John Bardeen outline some details of his theory at Princeton University.[9] Some puzzling aspects needed clarification, but to Phil, the BCS approach explained too many experimental observations in a natural way to be seriously wrong. Almost immediately, he put aside his disorder-induced localization manuscript and set to work studying the BCS preprint in earnest.

Many-electron Physics

Anderson's new interest in superconductivity thrust him into an exploding revolution in theoretical physics focused on the "quantum many-electron problem." The language of this revolution, if not the details of its methods, are important for what follows, so a brief discussion is appropriate before examining Anderson's contributions.

In principle, it is possible to calculate all the physical properties of a many-electron system using a wave function Ψ which simultaneously describes all the system's electrons. Unfortunately, there is no method known to calculate Ψ which takes *exact* account of the Coulomb electric force of repulsion between every pair of electrons.[10] This force poses problems because its magnitude is quite large when the distance between two electrons is small and then decreases to zero quite slowly as the distance between the two electrons increases.[11]

---

[8] R.P. Feynman, "Superfluidity and Superconductivity," *Review of Modern Physics* **29**, 205–12 (1957). Anderson was present at Feynman's presentation.

[9] D. Pines, "Superconductivity: From Electron Interactions to Nuclear Superfluidity," in *BCS: 50 Years*, edited by Leon N. Cooper and Dmitri Feldman (World Scientific, New Jersey, 2011), pp. 85–105.

[10] This statement omits purely numerical methods because they rapidly become impractical as the number of electrons in the system increases.

[11] By "slow" decrease to zero, we mean that the magnitude of the Coulomb force (which varies as $1/d^2$ where d is the distance between two electrons) is

In the early days of quantum mechanics, the creativity and imagination of theoretical physicists like Hans Bethe, Lev Landau, and Rudolf Peierls helped create solid-state physics without grappling directly with the many-electron wave function. However, another group of physicists with different tastes spent the years 1930–1955 developing a sequence of methods to calculate Ψ approximately for atoms, molecules, and solids.[12] Each wave function in this sequence provided a more accurate description of the wave function than the one that preceded it.

The *Hartree–Fock* method was an early entry into the many-electron sweepstakes. This approach—common in quantum chemistry—takes account of two effects: (1) the average electric force each electron experiences due to the presence of the other electrons; and (2) the quantum mechanical phenomenon of *exchange* that Heisenberg had used to explain ferromagnetism (Chapter 6). For an N-electron system, the Hartree–Fock approximation writes the many-electron wave function Ψ_{HF} as a product of N different "orbitals" and assigns one electron to each orbital.

The aspect of exchange built into the Hartree–Fock approximation concerns the behavior of Ψ_{HF} when any two identical particles exchange their positions in space. For the class of particles called *fermions*, quantum mechanics requires that the exchange process change Ψ_{HF} to the negative of itself. For the class of particles called *bosons*, the exchange process leaves Ψ_{HF} unchanged. There are no other possibilities. Electrons are fermions and one consequence of that fact is the single occupancy rule discussed earlier which assigns electrons to the quantum states of a many-fermion

not nearly as close to zero (at the same value of d) as almost every other distance-dependent force found in Nature.

[12] See, for example, John C. Slater, *Solid-State and Molecular Theory: A Scientific Biography* (John Wiley, New York, 1975) and Henry F. Schaefer III, *Quantum Chemistry: the Development of Ab Initio Methods in Molecular Electronic Structure Theory* (Clarendon Press, Oxford, UK, 1984).

system. Helium atoms are bosons and any number of helium atoms can occupy any of its allowed quantum states.[13]

Up and down spins are present in equal numbers in non-magnetic many-electron systems.[14] However, because of exchange, Ψ_{HF} tends to keep electrons with *parallel spins* away from each other. This reduces the repulsive electric force between them. However, Ψ_{HF} does <u>not</u> keep electrons with *antiparallel spins* away from each other. The implications of this are particularly clear in a model system called the *electron gas* which is designed to capture the essential physics of a metal.

Figure 9.1 shows the energy band diagram of a typical metal. This differs from the corresponding diagram for a semiconductor (Figure 7.5) because the conduction band here is partially filled with electrons. Alan Wilson had used Figure 7.5 to explain why a pure semiconductor does *not* conduct electricity and he used Figure 9.1 to explain why a metal does. In a metal, an applied voltage readily accelerates electrons in the conduction band by

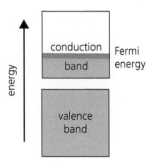

Figure 9.1 Energy band diagram for a metal. Yellow indicates states occupied by electrons. The valence band is fully occupied and the conduction band is partially occupied. The Fermi energy (green line) is the energy of the most energetic occupied electron state. Compare with the corresponding diagram for a semiconductor (Figure 7.5).

[13] This statement refers to ^{4}He, the isotope of helium where the nucleus consists of two protons and two neutrons.

[14] A magnet is precisely a system where the population of up and down spins is not equal.

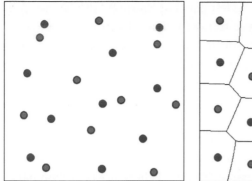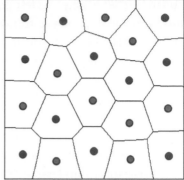

Figure 9.2 Snapshots of the spatial distribution of electrons in an electron gas where equal numbers of spin-up electrons (red) and spin-down electrons (blue) repel one another as they move through a uniform distribution of positive charge (white). Left panel: Hartree–Fock approximation. Right panel: Random-phase approximation. Note that each mobile electron carries with it an imaginary cell (solid lines) which encloses just enough positive charge to cancel the negative charge of the electron.

promoting them from occupied states just below the Fermi energy (the energy of the most energetic occupied electron state) to unoccupied states just above the Fermi energy.

The electron gas model replaces the positively charged ions of a real metal by a uniform distribution of positive charge.[15] Even with this simplification, the wave function Ψ of the electron gas cannot be calculated exactly. Happily, a sequence of approximate calculations can be done, beginning with the Hartree–Fock approximation (HFA). The left panel of Figure 9.2 is a snapshot of the spatial distribution of spin-up electrons (red) and spin-down electrons (blue) for an electron gas as calculated using the HFA.

Because of the exchange effect, no red electron ever gets very close to another red electron and no blue electron ever gets very close to another blue electron. The HFA deals very inexactly with the Coulomb repulsion between a spin-up electron and a spin-down

---

[15] Each atom of a metal becomes an ion because some of its localized electrons detach to occupy delocalized band states.

electron. That is why the left panel of Figure 8.2 displays several instances where a red electron and a blue electron wind up very close to each other. We do not expect close encounters of this kind in reality because, as noted above, the Coulomb repulsive force becomes very large when that happens.

A significant improvement to the HFA is the *random phase approximation* (RPA). David Pines and his PhD supervisor at Princeton, David Bohm, introduced this method in 1951–1953.[16] A few years later, Pines was one of the summer visitors who labored with Phil Anderson to understand George Feher's doped silicon data. Bohm was an early casualty of the Red Scare. Princeton did not renew his contract after 1951 and he had to leave the United States permanently to find a job as a physicist.[17]

Within the RPA, the Coulomb interaction produces two main effects in an electron gas. The first is a *collective excitation* where all the electrons slosh back and forth together at a characteristic frequency.[18] The second is the tendency of every electron to push away from itself all the electrons in its immediate vicinity. Because the positive charge of the electron gas is distributed uniformly, this standoffishness of every electron effectively partitions the system into a collection of charge-neutral cells, each of which is centered on one electron (right panel of Figure 9.2).

Because every electron-centered cell has zero net charge, the electric force between any two cells is extremely weak compared to the original Coulomb interaction between the electrons. In other words, electrons in different cells are barely aware of each other. At a stroke, this explained why the earliest (crude) theories

[16] See R.I.G. Hughes, "Theoretical Practice: The Bohm-Pines Quartet," *Perspectives on Science* **14**, 457–524 (2006).

[17] B.J. Hiley and F. David Peat, *Quantum Implications*, edited by B.J. Hiley and F. David Peat (Routledge, New York, 1987), p. 4.

[18] David Bohm, a dedicated Marxist his entire life, made an explicit connection between his political understanding of collectivism among people and his physical understanding of collectivism among charged particles. See Alexei Kojevnikov, "David Bohm and Collective Movement," *Historical Studies in the Physical and Biological Sciences* **33**(1), 161–92 (2002).

of metals accounted so well for their properties. These discussions simply *ignored* the Coulomb repulsion between every pair of electrons. The RPA showed why this was almost true.

In 1957, Lev Landau made a profound contribution to many-electron theory when he asked what happens when one slowly turns on the Coulomb interaction between a given electron and all of its neighbors. His answer was that the interactions alter or "renormalize" the properties of each "bare" electron and he coined the term *quasiparticle* to refer to every such "renormalized" electron.[19] He went on to discuss the effect of the weak residual interactions between quasiparticles and created what is called *Fermi liquid theory*. In the RPA, the charge-neutral cells in the right panel of Figure 9.2 play the role of Landau's quasiparticles.

Finally, physicists at the end of the 1950s realized that new theoretical methods developed after World War II to study quantum electrodynamics (the relativistic quantum theory of electrons and photons) could be repurposed to study the many-electron problem.[20] The resulting *quantum field theory* of many-electron systems was a complete game-changer. As one practitioner put it:

> 1957 was a magic year when an almost endless number of papers were published that led to a paradigm shift. Many classic problems that had been nagging theorists since the 1930s were solved completely...We learned a new language and...the way we think about many-body systems underwent a revolution.[21]

Much of the new language focused on the *many-particle Green function*, a mathematical quantity designed to monitor the reaction of a many-particle system when it is perturbed by the addition or

[19] L.D. Landau, "The Theory of a Fermi Liquid," *Soviet Physics JETP* **3**, 920–5 (1957).

[20] See, e.g., David Pines, *The Many-Body Problem* (W.A. Benjamin, Inc., New York, 1962).

[21] Allan Griffin, "Many-Body Physics in the 1960s: A Golden Age," Oral presentation at *Fifty Years of Condensed Matter Physics: A Symposium on the Occasion of Vinay Ambegaokar's Retirement*, June 16, 2007.

removal of one (or more) of its particles.[22] The Green function approach offered two distinct advantages. First, it allowed practitioners to calculate the most interesting properties of a many-electron system *without* first computing its wave function. Second, the method lent itself to systematic calculations using a diagrammatic form of perturbation theory invented originally by Richard Feynman for the quantum electrodynamics problem.

Feynman diagrams were a panacea. Every diagram was a graphical representation of a term in the desired perturbation series and the diagram itself provided the rules needed to write down a mathematical expression for that term. This fact alone made it possible to perform calculations in hours that might have taken days or weeks using previous methods. In short order, Feynman diagrams spread around the world and expanded from quantum electrodynamics into nuclear physics and many-body physics.[23] Russian physicists became particularly avid users and proselytizers of the new methodology.[24]

Equally important, it was possible to interpret each diagram as describing a specific sequence of processes involving the particles and their interactions. The most creative practitioners developed a feeling for the significance of individual diagrams and thereby transformed this tool for calculation into a new way to think about the physics of a problem. That is why the preface to an influential monograph written by three of Lev Landau's students asserts that the "basic advantage of the diagram technique lies in its intuitive character."[25]

Phil Anderson was never a great fan of quantum field theory or diagrammatic perturbation theory. Field theory was too formal

[22] Named for the early nineteenth-century mathematician George Green.
[23] David Kaiser, *The Dispersion of Feynman Diagrams in Postwar Physics* (University of Chicago Press, Chicago, 2005).
[24] See, for example, V.M. Galitskii and A.B. Migdal, "Application of Quantum Field Theory Methods to Many Body Problems," *Soviet Physics JETP* 7, 96–104 (1958).
[25] A.A. Abrikosov, L.P. Gor'kov, and I.E. Dzyaloshinski, *Methods of Quantum Field Theory in Statistical Physics* (Prentice-Hall, Englewood Cliffs, NJ, 1963). p. v.

for his taste and "understanding through diagrams" could hardly be said to describe his epistemology. For him, understanding came from simple models and simple physical pictures, not from a diagrammatic representation of the terms of a perturbation series. For most of his career, Anderson complained about the "tyranny of Feynman diagrams," which was his way of disparaging theorists who would rather follow the diagram rules and turn the computational crank than think hard about the physics.[26]

That being said, Anderson was not one to leave a labor-saving device unused. He spent days poring over the original articles and, over the years, he exploited diagrammatic methods whenever he thought they brought value. As a critic of rote calculations of any kind, he objected to field theoretic methods mostly because of their potential to seduce the unwary into thinking that the right answer always lay at the end of a diagrammatic rainbow.[27]

The BCS Model

A microscopic understanding of the origin of superconductivity begins with two facts about ordinary metals. The first, noted already in Figure 9.1, is that only the highest energy electrons near the Fermi energy contribute to an electric current and experience resistance. Second, a metal—indeed any crystal— exhibits vibrational excitations of its lattice of ions called *phonons*. Phonons are small-amplitude wave-like displacements of the ions made possible by the fact that every displaced ion always returns to its original lattice position as if it were attached to that position by a spring.

[26] E. Abrahams, "Some Reminiscences on Anderson Localization," in *PWA at Ninety: A Lifetime of Emergence*, edited by Premi Chandra, Piers Coleman, Gabi Kotliar, Nai Phuan Ong, Daniel L. Stein, and Clare Yu (World Scientific, New Jersey, 2016), pp. 33–7.
[27] Philip W. Anderson, "Brainwashed by Feynman?," *Physics Today* **53**(2), 11–12 (2000).

According to BCS, a resistive metal can become a superconductor at low temperature because:

1. A weak attractive force acts between some pairs of electrons near the Fermi energy. This force arises because negatively charged electrons attract and transiently displace positively charged ions. Because phonons are a natural language to describe ion motion, it is usual to identify the *electron–phonon interaction* as the origin of the force.
2. The weak attractive force always causes pairs of electrons near the Fermi energy to bind together to form stable quantum objects called Cooper pairs.
3. A many-electron wave function Ψ_{BCS} (guessed by BCS) formed by allowing a macroscopic number of electrons near the Fermi level to occupy identical Cooper pair states exhibits all the properties expected of a superconductor.
4. A superconductor reverts to a normal resistive metal when $T > T_C$ because the thermal energy at T_C breaks apart Cooper pairs.

Most physicists regarded the BCS approach as a major breakthrough.[28] An example was Nikolai Bogoliubov, the head of the Theoretical Department of the Steklov Mathematical Institute in Moscow. A few months after seeing the BCS preprint, he reproduced their results using a theoretical method that did not require a guess for the many-electron wave function, Ψ_{BCS}.[29] Bogoliubov reported his results at Lev Landau's theoretical seminar at the Institute for Physical Problems in Moscow.[30] Quite quickly, Landau's group began to generate new and interesting

[28] Leon N. Cooper, "Remembrance of Superconductivity Past," in *BCS: 50 Years*, edited by Leon N. Cooper and Dmitri Feldman (World Scientific, New Jersey, 2011), pp. 3–20.

[29] N.N. Bogoliubov, V.V. Tolmachov, and D.V. Shirkov, "A New Method in the Theory of Superconductivity," *Fortschritte der Physik* **6**, 605–82 (1958).

[30] Lev P. Gor'kov, "Developing BCS Ideas in the Former Soviet Union," in *BCS: 50 Years*, edited by Leon N. Cooper and Dmitri Feldman (World Scientific, New Jersey, 2011), pp. 107–26.

predictions using the BCS model and the quantum field theory methods that were their specialty.

Despite the enthusiasm of the Russians, BCS did not lack for critics. According to Anderson:

> You could divide the theorists into two groups. Those that had worked on the problem before and either had developed their own theory or had not been able to develop a theory. They were opposed to BCS. Then there were those who came to the subject fresh and looked at the achievements of BCS. They were almost universally positive.[31]

A somewhat different reaction came from Richard Feynman, who had chastised the theoretical physics community for their (and his) lack of imagination in solving the superconductivity problem. Decades after the fact, a colleague asked Feynman what he had thought about the BCS model when it appeared. He replied that he had been psychologically unable to look at the paper for six months, and did so only after his physicist sister Joan had shamed him into it.[32]

Figure 9.3 Leon Cooper (left), John Bardeen (center), and J. Robert Schrieffer (right). Source: Keystone Press/Alamy.

---

[31] Interview of PWA by Alexei Kojevnikov, March 30, 1999, Niels Bohr Library & Archives, American Institute of Physics, College Park, MD.

[32] John J. Hopfield, "Whatever Happened to Solid State Physics," *Annual Reviews of Condensed Matter Physics* **5**, 1–13 (2014).

Feynman's mention of psychology is revealing. Although he practiced theoretical physics at the highest conceivable level, his failure to understand the origin of superconductivity amounted to an existential threat. The associated emotional turmoil—driven by the fusion of identity with profession—is similar to that experienced by creative writers, musicians, and artists who suffer failure.[33]

Anderson and BCS

Anderson has described his relationship with BCS as "the scientific love of my life."[34] A testament to that ardor is the thirty papers he published on the subject between 1958 and 1966. Many of these focus on analysis, applications, or extensions of the original discussion given by BCS. The remainder aimed to improve on BCS for cases where their predictions failed to agree with experiment. We single out only a handful of papers from this rich trove.

Anderson was far from uninformed about superconductivity when the BCS preprint appeared. For one thing, he had continued chatting regularly with Bernd Matthias even after his colleague had switched his interests from ferroelectricity to superconductivity. Several of Matthias' pre-BCS superconductivity papers thank Phil for discussions. However, the stubborn experimentalist never tired of dismissing the value of theory as an aid to his research program.[35] Theory could not help him identify new superconductors and no theory (BCS included) could predict critical temperatures accurately. Matthias and Anderson

---

[33] The psychology of failure in theoretical physics is an important theme in Leonard Mlodinow's *Some Time with Feynman* (Penguin Book, London, 2003).

[34] P.W. Anderson, "BCS: the Scientific Love of My Life," in *BCS: 50 Years*, edited by Leon N. Cooper and Dmitri Feldman (World Scientific, New Jersey, 2011), pp. 127–42.

[35] B.T. Matthias, "Superconductivity II. The Facts," *Science* **144**, 378–81 (1964).

sparred about the relationship between theory and experiment in physics for years.

Otherwise, Anderson familiarized himself with the elegant work of the German émigré theorist Fritz London. London argued persuasively that superconductivity was a macroscopic yet still deeply quantum phenomenon.[36] Anderson and the one-generation-older London were kindred spirits: both emphasized constant engagement with experiment and both carefully distinguished theoretical physics from mathematical physics.

London's attitude is clear from a remark he made comparing a book he wrote about superconductivity with a book on the same subject written by his former Berlin colleague Max von Laue:[37]

> My book takes particular care in explaining why certain equations are chosen as the expression of the basic assumptions of the theory, while von Laue's book merely takes those equations unquestioned and develops their mathematical consequences.[38]

Anderson plunged into superconductivity research to address an issue raised by Gregor Wentzel, a 60-year-old, Cuban-cigar smoking, quantum field theorist from the University of Chicago.[39] It bothered Wentzel that BCS did not satisfy a sacrosanct property of physical theories called *gauge invariance*.[40] This refers to a certain arbitrariness in the theory which nevertheless leads to exactly the same formulas for measurable quantities. Phil was confident that the BCS picture was correct and he anticipated that the technical

---

[36] Fritz London, *Superfluids, Volume I: Macroscopic Theory of Superconductivity* (Wiley, New York, 1950).

[37] Forty years earlier, von Laue discovered that x-rays diffract from a crystal. See Chapter 5.

[38] Kostas Gavroglu, *Fritz London: a Scientific Biography* (Cambridge University Press, Cambridge, 1995), Chapter 5, Note 102, pp. 261–2. M. von Laue, *Theory of Superconductivity* (Academic Press, New York, 1952).

[39] Peter G.O. Freund, Charles J. Goebel, Yoichiru Nambu, and Reinhard Oehme, *Biographical Memoir of Gregor Wentzel, 1898–1978* (National Academy of Sciences Press, Washington, DC, 2009).

[40] G. Wentzel, "Problem of Gauge Invariance in the Theory of the Meissner Effect," *Physical Review Letters* **2**, 33–4 (1959).

corrections needed to restore gauge invariance could not have any significant effect on any of the theory's essential predictions. A few months of brute force calculations confirmed this belief and he communicated his results to John Bardeen.

The gauge invariance problem turned out to arise from a subtle violation of electric charge conservation. Bardeen agreed, and there is a footnote crediting Anderson for this insight in the 30-page follow-up paper BCS submitted to *Physical Review* in July 1957.[41] When Anderson later wrote up his admittedly inelegant solution of the gauge invariance problem for publication, he was happy to acknowledge "extensive help and advice from J. Bardeen, who independently arrived at a qualitative understanding of much of the above."[42]

Anderson and Bardeen both understood that the ability of BCS to account for many different experimental aspects of superconductivity was a sure sign that the gauge invariance problem was not really a problem. This reliance on experiment rather than technical considerations to evaluate the usefulness of a theoretical model was a common feature of their scientific styles. They both also favored the simple over the complicated when it came to constructing a model.[43]

Differences in style between Anderson and Bardeen show up in the latter's penchant to divide a problem into sub-problems and, if necessary, to tick through a checklist of theoretical methods until he found one he could use consistently to bully a problem into submission.[44] For his part, Anderson had already shown an

[41] J. Bardeen, L.N. Cooper, and J.R. Schrieffer "Theory of Superconductivity," *Physical Review* **108**, 1075–204 (1957).

[42] P.W. Anderson, "Coherent Excited States in the Theory of Superconductivity: Gauge Invariance and the Meissner Effect," *Physical Review* **110**, 827–35 (1958).

[43] David Pines, "An Extraordinary Man: Reflections on John Bardeen," *Physics Today* **45**(4), 64–70 (1992).

[44] Lillian Hoddeson and Vicki Daitch, *True Genius: The Life and Science of John Bardeen* (Joseph Henry Press, Washington, DC, 2002), Chapter 17.

un-Bardeen willingness to mix and match (and occasionally invent) theoretical methods when it suited his needs. He also did not let mathematical niceties like convergence bother him if his intuition about a problem was strong enough.

Anderson's gauge invariance paper was about to appear in print when he had a "eureka" moment. He realized there was a way to extend the ideas of BCS to avoid any violation of gauge invariance.[45] The idea came to him in a flash, as many of his best ideas did. Unlike some mathematicians and scientists, Phil never dreamed the solution of a physical or mathematical problem.[46]

The paper based on the eureka insight had two parts. In the first part, Anderson used an idea suggested by his Bell labs colleague Harry Suhl and replaced the Cooper pair variables in the BCS model by a set of spin variables.[47] These variables did not represent real, physical spins, but they behaved mathematically as if they did. Call them *pseudospins*. The pseudospin language allowed him to bring all the intuition and experience he had gained with the antiferromagnet problem to bear on superconductivity. In just a few lines, he reproduced all the essential results of BCS.

The second part of this paper solved a BCS-type model where Anderson allowed the pairing force to include both phonon-induced attraction and Coulomb repulsion. The theory maintained strict gauge invariance at every stage and extended the random phase approximation to include the effects of exchange. Anderson ignored the Coulomb repulsion at first and found that his model exhibited a sound-like collective oscillation of the delocalized electrons. Like all sound waves, the frequency of this

[45] P.W. Anderson, "Random-Phase Approximation in the Theory of Superconductivity," *Physical Review* 112, 1900–16 (1958).

[46] See Jacques Hadamard, *The Mathematical Mind. The Psychology of Invention in the Mathematical Field* (Princeton University Press, Princeton, NJ, 1945).

[47] Others introduced similar spin variables at about the same time. Most elegantly, Yoichiro Nambu in "Quasi-Particles and Gauge Invariance in the Theory of Superconductivity", *Physical Review B* 117, 648–63 (1960).

oscillation approached zero in the limit of very long wavelength. He then added the Coulomb repulsion between electrons and discovered that the sound wave had disappeared. In its place was the collective sloshing mode familiar from the electron gas. This wave had the property that its frequency was *not* zero in the limit of very long wavelength.

Anderson's RPA approach to the theory of superconductivity established his *bona fides* in the field. An exchange of preprints revealed that Nikolai Bogoliubov had come to the same conclusions independently using field theory methods.[48] On the other hand, neither Anderson nor Bogoliubov realized (yet) that the Coulomb-induced raising of the frequency of the long-wavelength collective oscillation from zero to non-zero was the first example in physics of a new phenomenon. Later, particle physicists would call it the Higgs mechanism whereby particles with zero mass acquire a non-zero mass (see Chapter 10).

Anderson completed his RPA paper while spending two summer months of 1958 as a paid consultant to Charlie Kittel's research group at the University of California, Berkeley. Kittel had arranged for the Anderson family to live *gratis* in a beautiful home in the hills high above the Berkeley campus.[49] The view was magnificent on the days when the fog lifted. Another bonus was the opportunity for Phil and Joyce to visit their old Harvard friend, the musical satirist Tom Lehrer, who was performing at the *hungry i* nightclub in San Francisco.

Strolling across the Berkeley campus one day, Anderson ran into Jim Phillips, a postdoc who had spent the previous two years at Bell Labs. Phillips asked him why experiments showed that superconductors were indifferent to the presence of non-magnetic

[48] N.N. Bogoliubov, V.V. Tolmachev, and D.V. Shirkov, "A New Method in the Theory of Superconductivity," *Fortschritte der Physik* **6**, 605–82 (1958). See also V.M. Galitskii, "Sound Excitations in Fermi Systems," (in Russian) *Zhurnal Éksperimental'noĭ i Teoreticheskoĭ Fiziki* **34**, 1011–13 (1958).
[49] Interview of PWA by Alexei Kojevnikov, March 30, 1999, Niels Bohr Library & Archives, American Institute of Physics, College Park, MD.

impurities.[50] Anderson did not have an immediate answer, but it motivated him to think hard about the effect of impurities on metals more generally.

Back in Murray Hill that fall, Anderson created a theory for what he called a "dirty superconductor." This is a disordered metal with a sufficient number of non-magnetic impurities present that the distance an electron travels before it collides with such an impurity is much less than the size of a Cooper pair. He discovered that the presence of these impurities did not impede the ability of a metal to form these pairs. Magnetic impurities were another story.[51]

Anderson answered Phillips' question by showing that none of the BCS predictions change for a dirty superconductor as long as one constructs Ψ_{BCS} using the Cooper pairs appropriate to the dirty medium. He emphasized that his results did not hold if, for any reason, the wave function of each Cooper pair turned out not to be spherical.[52] This conclusion came back to haunt him thirty years later when he proposed a theory for a newly discovered class of superconductors with extraordinarily high transition temperatures.

Anderson's dirty superconductor research is notable also because it exposed him for the first time to the phenomenon of multiple discovery. Physicists from Russia had independently and simultaneously reached the same conclusions he had. But the difference in style could not have been starker. Anderson's paper

[50] Author correspondence with James C. Phillips.

[51] Anderson's argument relied on a subtle symmetry of BCS theory that is preserved in the presence of non-magnetic impurities but is lost in the presence of magnetic impurities. In more modern language, magnetic impurities are pair-breaking while non-magnetic impurities are not.

[52] In quantum mechanics, it is possible for the wave function for two bound particles to be spherically symmetric. This happens with the hydrogen atom (an electron bound to a nucleus), with a diatomic molecule (an atom bound to an atom), and with the Cooper pairs used by BCS (an electron bound to an electron).

consisted almost entirely of words.[53] The Russian paper was couched in the language of many-particle Green functions and diagrammatic perturbation theory.[54]

A Visit to Russia

Halfway through the summer of 1958, Charlie Kittel asked Anderson to attend a conference in Moscow that winter with himself, Bernd Matthias, and three other scientists. The larger context for this offer was the slow but steady lessening of tension between Russia and the West that had followed the 1952 death of Joseph Stalin.[55] A significant exchange of technical information occurred in 1956 when fourteen Americans attended a conference on nuclear and particle physics in the USSR. The accelerator expert Luis Alvarez reported that he found the average Russian on the street to be very friendly despite the fact that "it would have been more probable a year ago to find a bunch of Martians in that spot than a group of Americans."[56]

Like most Americans, Anderson was stunned in November 1956 when the Soviet Union invaded Hungary and crushed a popular uprising against that country's Communist government. Nevertheless, he shared Alvarez's view when he sharply distinguished the behavior of the Soviet government from the behavior of individual scientists who happened to live and work behind the Iron Curtain. For that reason, he had no philosophical objections to joining Kittel's delegation.

[53] P.W. Anderson, "Theory of Dirty Superconductors," *Journal of the Physics and Chemistry of Solids* 11, 26–30 (1959).

[54] A.A. Abrikosov and L. P. Gor'kov "On the Theory of Superconducting Alloys I. The Electrodynamics of Alloys at Absolute Zero," (in Russian) *Zhurnal Eksperimental'noĭ i Teoreticheskoĭ Fiziki* 35, 1558–71 (1958).

[55] Robert F. Byrnes, *Soviet-American Academic Exchanges, 1958–1975* (Indiana University Press, Bloomington, IN, 1976); William Taubman, *Khrushchev: The Man and the Era* (W.W. Norton, New York, 2003).

[56] Luis W. Alvarez "Excerpts from a Russian Diary," *Physics Today* 10 (5), 24–32 (1957).

On the other hand, Anderson was not particularly excited to attend a conference devoted to the physics of insulators. In the end, Kittel's suggestion that he might be able to arrange a meeting between Phil and Nikolai Bogoliubov was a powerful argument in favor of going. If that meeting were to happen, Anderson needed to be careful. "Bogoliubov" was the name he had given to his fluffy orange house cat to memorialize the bright orange shoes he had seen the Russian wear two years earlier at an international conference of theoretical physics in Seattle.[57]

November arrived and Anderson flew off to Moscow to give his invited talk. His subject was a theory for the ferroelectric phase transition in barium titanate he had worked out, but not published, several years earlier. An extended question period kept him at the speaker's podium for an hour and forty minutes.[58] Afterward, the theorist Vitaly Ginzburg introduced himself and, in the course of two hours of discussion, Phil learned that Ginzburg had published (in Russian) an identical theory of ferroelectricity ten years earlier.[59]

Ironically, Ginzburg was also the co-author (with Lev Landau) of a remarkable macroscopic theory of superconductivity. This 1950 paper eventually earned Ginzburg a one-third share of the 2003 Nobel Prize for Physics.[60] In many ways, the speculative Ginzburg–Landau theory was more flexible and useful than the

---

[57] This was the successor to the conference Phil had attended in Japan in 1953. P.W. Anderson, "BCS and Me," in *More and Different: Notes from a Thoughtful Curmudgeon* (World Scientific, Hackensack, NJ, 2011), p. 43.

[58] "Report on the All Soviet Union Conference on the Physics of Dielectrics", November 20–28, 1958. Archives of the Woods Hole Oceanographic Institute, Robert Hugh Cole Papers, MC–04, Box 2, Folder 2.

[59] Vitaly L. Ginzburg, "Some Remarks on Ferroelectricity, Soft Modes, and Related Problems" in *About Science, Myself, and Others* (Institute of Physics, Bristol, 2005), pp. 127–50. Letter from PWA to Joyce and Susan Anderson, November 21, 1958.

[60] V.L. Ginzburg and L.D. Landau, "On the Theory of Superconductivity" (in Russian) *Zhurnal Éksperimental'noĭ i Teoreticheskoĭ Fiziki* **20**, 1064 (1950). English translation in Vitaly L. Ginzburg, *On Superconductivity and Superfluidity. A Scientific Autobiography* (Springer, Berlin, 2009), pp.113–37.

BCS approach. However, most Western physicists either could not read Russian or, like Anderson, they were simply unaware of its existence. Those who knew about the theory generally reserved judgment until post-BCS experiments confirmed some of its predictions.[61]

Soon after they arrived in Moscow, Kittel arranged a visit for Anderson and himself to the Kapitza Institute for Physical Problems. The Theory Division there was home to the "Landau school" of theoretical physics, a unique organization built and ruled by the formidable Lev Davidovich Landau. Except for a harrowing year in a Stalinist prison (for alleged subversive behavior), Landau had spent the years since 1932 solving problems over the entire range of theoretical physics and training students to be versatile generalists unafraid to attack any theoretical problem posed to them.[62] Unfortunately, most of his group's post-World War II achievements were unknown to American physicists due to the communication barrier erected by the Cold War. Anderson was only dimly aware of Landau and his group when he arrived in Moscow.

Kittel and Anderson met with two of Landau's former PhD students: Alexei Abrikosov and Lev Gor'kov. Earlier in the year, Gor'kov had devised an elegant field theory method to treat problems like superconductivity. Now, Gor'kov announced, he had used his methods to derive the macroscopic Ginzburg–Landau equations from the microscopic BCS equations. This ended any suspicion that the Ginzburg–Landau theory was in any way suspect or untrustworthy as a method to study superconductivity.

The next day, Anderson presented his pseudospin and RPA versions of BCS superconductivity theory at Landau's weekly

[61] Interview of A. Brian Pippard by Lillian Hoddeson and Gordon Baym, September 14, 1982, Niels Bohr Library & Archives, American Institute of Physics, College Park, MD.

[62] See the essays by Boris Ioffe and S.S. Gerstein in *Under the Spell of Landau, When Theoretical Physics Was Shaping Destinies*, edited by M. Shifman (World Scientific, Singapore, 2013).

theoretical seminar. This was a notorious forum where the listeners gave no quarter to the speaker.[63] Landau broke protocol by asking Anderson to continue talking and answer questions well into a second hour. To Phil's relief, his host announced at the end that he was satisfied with his guest's theory.[64] What is certain is that Anderson gained a deep appreciation of Landau and his methods. Years later, he characterized himself "as committed a member of the Landau cult as any of his pupils" and named Landau—along with Richard Feynman—as the theoretical physicist he most admired.[65]

Landau shared Anderson's zeal to reduce a complex problem to its essence, but their styles of doing theoretical physics were otherwise very different.[66] Landau had a profound understanding of every type of physics—from classical hydrodynamics to elementary particle physics to solid-state physics—and a complete mastery of all the mathematical methods used in any of them. This allowed him to free himself from the ideas of others and focus with great clarity, distinctiveness, and creative originality on any problem that interested him. From there, he displayed the rare ability to find the shortest and most expedient way to a solution. Tragically, Landau was debilitated by an automobile accident at the age of 54 and he did no physics in the six years before his death in 1968.

In 1966, Lev Gor'kov was permitted to travel to the United States to visit the country's major centers of physics. After his visit to Bell Labs, Anderson wrote a memo to his managers:

[63] Boris L. Ioffe, "Landau's' Theoretical Minimum, Landau's Seminar, ITEP in the Beginning of the 1950s." Available at arXiv:hep-ph/0204295v1.

[64] November 28, 1958 letter from PWA to Joyce Anderson.

[65] P.W. Anderson, "A Theoretical Physicist," *Science* **211**, 158 (1981).

[66] E.M. Lifshshitz, "Lev Davidovich Landau (1908–1968)," *Soviet Physics Uspekhi* **12**(1), 135–45 (1969). See also F. Janouch, *L.D. Landau: His Life and Work*, CERN 79–03, 28 March 1979 (unpublished).

Figure 9.4 The Russian theorists Lev Landau (left) and Nikolai Bogoliubov (right). Source: American Institute of Physics Emilio Segrè Visual Archive.

> I felt confirmed in my impression that the great strength of the "Landau group" is its systematic, formal mathematical and theoretical attack and its wide-ranging but very conscious choice of subject matter, combined with superb intellect; the great weakness is the lack of specific connection with—and even interest in—actual experimental work and experimentalists.[67]

Eventually, Kittel arranged a tour for Anderson and himself of the Joint Institute for Nuclear Research in Dubna, eighty miles north of Moscow. This was a new research facility where Nikolai Bogoliubov served as director of the theory group. The visitors were told that Bogoliubov was ill that day, but the tour proceeded anyway. Anderson was sure the "guide" was a KGB employee. At one point, Bogoliubov's young collaborator Dmitry Shirkov materialized and whisked Anderson into an empty room with a

---

[67] P.W. Anderson, "Informal Comments on Visit of L.P. Gor'kov—July 11, 1966." AT&T Archives, Warren, NJ.

blackboard. They spent 15 minutes talking about collective excitations in superconductors before the guide found them.

Taming the Bad Actors

The predictions of the BCS paper were like catnip to experimentalists. Measurements from all over the world soon revealed a group of "good actor" metals where experiment and the BCS model agreed well: aluminum, cadmium, indium, tin, vanadium, and zinc. There were also "bad actors" like lead and mercury where this was not so. The BCS model came under scrutiny and it soon became apparent that some of its approximations begged for improvement.

The *model* for superconductivity proposed by BCS had two components: a greatly simplified Hamiltonian (an approximate total energy expression) and a guess for the many-body wave function. BCS carefully discussed the electron-phonon and many-body physics they deemed essential to superconductivity, but little of this appears explicitly in their final model. Nevertheless, a straightforward analysis of it yielded a set of mathematical expressions which (after fixing the values of a few parameters) allowed practitioners to extract all the *qualitative* trends seen for most superconductors.[68]

To understand the bad actors, Anderson and others felt it was necessary to replace the BCS model by a BCS-based *theory* of superconductivity.[69] By theory, they meant a mathematical formalism sophisticated enough to incorporate all the physics identified and understood by BCS (including all the bits omitted by them in their bare bones model) yet flexible enough to describe the material properties of different superconductors well enough

[68] See, e.g., G.W. Webb, F. Marsiglio, and J.E. Hirsch, "Superconductivity in the elements, alloys and simple compounds," *Physica C: Superconductivity and Its Applications* **514**, 17–27 (2015).

[69] P.W. Anderson, "It's Not Over Till the Fat Lady Sings," in *More and Different: Notes from a Thoughtful Curmudgeon* (World Scientific, Hackensack, NJ, 2011), pp. 81–6.

to account for the full range of the observed phenomenology.[70] If completely successful, such a theory would make *quantitative* predictions without recourse to adjustable parameters.[71]

The physics Anderson wanted to include more realistically was the interaction between the electrons and the phonons. By good luck, he acquired help in 1959 when David Pines, by then an assistant professor at Princeton, bequeathed to him a graduate student named Pierre Morel. Pines was moving to the University of Illinois, but Morel could not join him because he worked at the French Embassy in New York City as a scientific attaché.[72] Phil agreed to take over Morel's supervision and the French scientist drove out to Bell Labs every week for discussions.

A June 1960 international conference on many-particle physics gave Anderson the opportunity to discuss his reservations about BCS with Bob Schrieffer, one of the model's inventors. Schrieffer reciprocated by telling Phil about recent work from the Soviet Union where Gerasim Eliashberg had used Gor'kov's methods to tackle the electron-phonon interaction without making the simplifications imposed by BCS.[73] Unfortunately, the Russian's theory produced very complicated equations whose exact solution was not possible using the pen-and-paper methods of mathematical physics.

[70] P.W. Anderson, "Science: A 'Dappled World' or a 'Seamless Web'," *Studies in the History and Philosophy of Modern Physics* **32**, 487–94 (2001).

[71] This goal remains unrealized. See, e.g., P. B. Allen, "Electron-Phonon Coupling Constants," in *Handbook of Superconductivity*, edited by Charles P. Poole Jr. (Academic Press, San Diego, 2000), Chapter 9G, pp. 478–89.

[72] Morel's job as scientific attaché was to promote French science in the United States and to collect information about important scientific developments in the US that might be valuable to the French government.

[73] G.M. Eliashberg, "Interactions between Electrons and Lattice Vibrations in a Superconductor," *Soviet Physics JETP* **11**, 696–702 (1960). A closely related approach was developed independently by Yoichiro Nambu in "Quasi-particles and Gauge Invariance in the Theory of Superconductivity," *Physical Review* **117**, 648–63 (1960).

A long-distance collaboration began between Anderson and Schrieffer. Anderson and Morel derived a version of Eliashberg's theory that was simple enough to solve by hand. To achieve that solvability, they omitted from the theory certain electron processes with energies greater than the energy of a typical lattice vibration while simultaneously reducing the magnitude of the repulsive Coulomb force. The Bogoliubov group had shown that "renormalizing" the Coulomb interaction in this way mimicked the contribution from the omitted high-energy processes.[74] The quantitative predictions made by Morel and Anderson for twenty-one different metals stimulated a great deal of subsequent work.[75]

For his part, Schrieffer exploited the fact that computers were just starting to become common in physics research. The tool at his disposal was the first "on-line" digital computer, a machine developed by the defense contractor Thompson-Ramo-Woolridge (TRW) to help guide ballistic missiles.[76] Unlike the best punch card controlled computer at the time—the IBM 7090—the TRW machine allowed the user to stop the computation at any time, display intermediate results, check for convergence, alter input data, and even change the code before restarting. Schrieffer's results complemented those reported by Morel and Anderson.[77]

[74] N.N. Bogoliubov, V.V. Tolmachev, and D.V. Shirkov, "A New Method in the Theory of Superconductivity," *Fortschritte der Physik* **6**, 605–82 (1958), Section 6.3.

[75] P. Morel and P.W. Anderson, "Calculation of the Superconducting State Parameters with Retarded Electron-Phonon Interaction," *Physical Review* **125**, 1263–71 (1962).

[76] Johannes Knolle and Christian Joas, "The Physics of Cold in the Cold War—'On-Line Computing' Between the ICBM Program and Superconductivity," in *History of Artificial Cold, Scientific, Technological and Cultural Issues*, edited by Kostas Gavroglu (Springer, Dordrecht, 2014), pp. 119–32.

[77] C.J. Culler, B.D. Fried, R.W. Huff, and J.R. Schrieffer, "Solution of the gap equation for a superconductor," *Physical Review Letters* **8**, 399–402 (1962). Similar work appears in J.C. Swihart, "Solutions of the BCS Integral Equation and Deviations from the Law of Corresponding States," *IBM Journal of Research and Development* **6**(1), 14–23 (1962).

Not long afterward, Anderson and Schrieffer published back-to-back papers in *Physical Review Letters*, a short-form physics journal used to rapidly report new and important results. Taken together, these papers demonstrated that the qualitative ideas implicit in the BCS picture could be generalized and solved to yield quantitative agreement between theory and experiment for a "bad actor" superconductor.

Anderson's paper was a collaboration with two experimental colleagues at Bell Labs. The experiment involved applying a voltage across a sandwich formed from three thin films—a normal metal, an insulator, and a superconductor—and measuring the current that flowed through the layers of the sandwich.[78] The insulator layer was a barrier to the flow of electrons, but quantum mechanical tunneling allowed a fraction of them to pass through and emerge into the superconductor. The transmitted signal carried detailed information about the superconductor.

Schrieffer worked with a postdoc and a graduate student to program the TRW computer to solve their version of the Eliashberg equations for this situation.[79] The agreement between their computations and the experimental tunneling data was very good. This meant that the description of the superconductor provided by Eliashberg's BCS-inspired theory was similarly good.

Anderson and Schrieffer were not alone in their efforts. For most of the 1960s, hundreds of physicists—experimenters and theorists—worked to create a quantitatively accurate theory of superconductivity based on the ideas of BCS. A 1969 collection of review articles written by thirty-two experts used 1400 pages to report the state of the art.[80] It is a measure of Anderson's status among his colleagues that the editor of this collection commissioned

---

[78] J.M. Rowell, P.W. Anderson, and D.E. Thomas, "Image of the Phonon Spectrum in the Tunneling Characteristic between Superconductors," *Physical Review Letters* **10**, 334–6 (1963).

[79] J.R. Schrieffer, D.J. Scalapino, and J.W. Wilkins, "Effective Tunneling Density of States in Superconductors," *Physical Review Letters* **10**, 336 (1963).

[80] R.D. Parks (editor), *Superconductivity* (Marcel Dekker, New York, 1969).

him to contribute a summary chapter. The essay he submitted, "Superconductivity in the Past and the Future," is notable because it introduced a witty and opinionated writing style that many readers came to expect in his later review and opinion articles.

Anderson's summary chapter characterized the 1950 Ginzburg–Landau theory of superconductivity as a "staggering achievement of understanding and intuition...which marks the beginning of the modern era." But, he goes on to ascribe the lack of appreciation of this paper in the United States to "outright censorship" because

> 1950 was in the midst of the McCarthy era, a time of which one of the silliest manifestations was the banning of Russian scientific publications in the United States. Some were even dumped in the harbors. The issue of [the Russian journal] containing the paper of Ginzburg and Landau...was one of these.

This story quickly passed into BCS legend and many authors quote it to this day.[81] Unfortunately, there is no truth in it. Anderson's zeal to criticize the excesses of the Red Scare era led him to incautiously report rumors of the day as fact.

The truth is that the US State Department briefly banned materials authored by "controversial figures, Communists, fellow travelers, etc." from its overseas diplomatic missions.[82] However, there is no evidence that Russian language physics journals entering the United States suffered from a ban or dumping from cargo ships. By the end of January 1951, the libraries at Princeton University, Yale University, and the Library of Congress all had received the 1950 journal issue where the original Ginzburg–Landau paper appears.

---

[81] See, e.g., Jean Matricon and Georges Waysand, *The Cold Wars: A History of Superconductivity* (Rutgers University Press, New Brunswick, NJ, 2003), p. 129.

[82] Joseph and Steward Alsop, "State Department Book-Burning Irks US Diplomats Abroad", *Boston Globe*, June 14, 1953, p. C13.

The Beauty of BCS

In later years, Anderson began to use BCS as an exemplar of not only a successful theory, but a beautiful one. For him, the characteristics of a successful theory are "conceptual depth, computational simplicity, and a transparent ability to deal with qualitative anomalies that arise in experiments."[83] It is possible to disagree with this definition, but it seems clear from the evidence presented earlier that not only does the BCS theory satisfy these criteria, but Anderson likely had them in mind (at least subconsciously) as he developed his theories for antiferromagnetism, superexchange, and localization.

Beauty is a trickier matter and Anderson differs from virtually all other writers on this subject when he recognizes beauty in a theory when it exhibits reality, craftsmanship, maximal cross-reference, and simplicity.[84] Reality refers here to the observable world and Anderson insists that a beautiful physical theory must be in constant and successful engagement with experimental observations. This, of course, is the characteristic that drew him to BCS in the first place.

According to Anderson, "craftsmanship is always an element of beauty...but it must be non-trivial...and done well." Interestingly, he does not rate the original BCS *model* particularly highly in this regard because it fails to describe the behavior of the "bad actor" superconductors. Instead, it is the subsequent quantitative and broadly applicable BCS theory developed by Eliashberg and others that earns his praise for craftsmanship.

---

[83] P.W. Anderson, "Remarks at the Panel Discussion on 'd-Wave Superconductivity'," *Journal of the Physics and Chemistry of Solids* **54**, 1457–9 (1993).

[84] P.W. Anderson, "Some Ideas on the Aesthetic of Science." Lecture given at the 50th Anniversary Seminar of the Faculty of Science and Technology, Keio University, Japan, May 1989. Reprinted in P.W. Anderson, *A Career in Theoretical Physics* (World Scientific, Singapore, 1994), pp. 569–83.

Chapter 15 will make clear that maximal cross-reference is an important part of Anderson's philosophy of science. The aesthetic notion here is that a truly beautiful theory influences (and perhaps alters) parts of the scientific enterprise that are far removed from the theory's original context. BCS satisfies this criterion in spades. The theory was born from a desire to understand a mysterious phenomenon in metals, but it has substantially contributed to scientific understanding in fields as diverse as atomic physics, nuclear physics, quantum liquids, and the physics of neutron stars.[85]

Anderson's idea of simplicity is primarily a requirement that a beautiful theory should derive a maximum of information from a minimum number of ideas. From the perspective given in this chapter, one might conclude that the Cooper pair is the single most important idea associated with BCS. Later, it will become clear that the most striking properties of a BCS system follow from the fact that its many-body wave function is a macroscopic (yet entirely quantum) broken-symmetry object.

Anderson published papers on superconductivity until about 1964. The most important of these involved the Josephson effect (Chapter 10) and a class of superconductors commonly called "Type II". The latter were used to fabricate high-field magnets which later became important to particle accelerators and magnetic resonance imaging machines.[86] Anderson's contribution to understanding the behavior of Type II superconductors is probably the closest any of his theoretical work ever came to informing a problem with practical applications.

---

[85] See Part IV: BCS beyond Superconductivity of *BCS: 50 Years*, edited by Leon N. Cooper and Dmitri Feldman (World Scientific, New Jersey, 2011).

[86] P.W. Anderson and Y.B. Kim, "Hard Superconductivity: Theory of the Motion of Abrikosov Flux Lines," *Reviews of Modern Physics* **36**(1), 39–43 (1964). A current-carrying wire wound helically around a cylinder produces a magnetic field that points along the long axis of the cylinder.

Fifteen years elapsed before the discovery of superconductivity in exotic metal alloys like $CeCu_2Si_2$ and UBe_{13} drew Anderson back to the subject. The allure was that these materials did not behave at all like BCS materials. He worked hard to understand this class of superconductors, but had little success ("my greatest failure"). Then, out of the blue, came the 1987 discovery of superconductivity in ceramic materials with critical temperatures greater than 100 K. We will see in Chapter 14 that these materials captured Anderson's imagination completely.

10

The Cantabrigian

The British invasion of the United States in the 1960s worked in reverse for Phil Anderson.[1] A vacation visit sparked a fondness for England which intensified during a sabbatical year spent at the University of Cambridge. A collaboration with the Czech-born publisher Robert Maxwell demonstrated by example that an immigrant could flourish in British society. An enduring relationship finally developed when Anderson served eight years (part-time) as Professor of Theoretical Physics at the University of Cambridge. Phil and Joyce built two homes in England and the couple thought seriously of moving there permanently. Ultimately, they returned to the United States full time, but not before Anderson enjoyed several scientific successes and endured a painful professional disappointment.

First Experiences

Until the end of the 1950s, Bell Labs rarely paid to send its researchers to scientific meetings outside the United States.[2] Management eventually succumbed to pressure from below and members of the technical staff were permitted to travel abroad to attend conferences at the company's expense. This facilitated

---

[1] The "British invasion" was a cultural phenomenon of the 1960s when many aspects of British music, fashion, and literature became very popular in the United States.

[2] The Fulbright Scholar Program paid Anderson's expenses for his 1952–1953 trip to Japan.

Anderson's trip to Russia, as well as a trip to a superconductivity conference at the University of Cambridge in the summer of 1959.

The Cambridge conference provided an opportunity for the family to spend two weeks touring the English countryside.[3] They rented a car and traced an 800-mile loop around the southwest of England. Eleven-year-old Susan was particularly taken with the Dartmoor region, the setting for Arthur Conan Doyle's *The Hound of the Baskervilles*. An intense affection for England and its history took root as they lodged in local inns, encountered huge herds of sheep, hiked along miles of hedgerows, and made a point of visiting medieval castles.

During the conference, the University of Cambridge experimentalist Brian Pippard asked Anderson to consider spending a year at his university.[4] The possibility of experiencing academic life after a decade at Bell Labs crystallized the following spring when Stanford University offered him a tenured faculty position. He was tempted, but Joyce showed no enthusiasm for living in California. Memories of a forest fire she experienced as a child made her fearful of the Golden State's famous wildfires.[5]

Anderson asked Pippard if he could teach a graduate course if he spent a year at Cambridge. Pippard was positive and Bell Labs management agreed to a sabbatical leave. Later, Anderson learned that Nevill Mott—the chair of the Cambridge physics department— was prepared to move Heaven and Earth to make it possible for him to come. Not only did Phil's theory of disorder-induced localization intrigue Mott personally, his presence would enhance Mott's efforts to promote the growth of solid-state physics in his department.[6]

---

[3] Joyce Anderson, Air Travels diary, June 24, 1959 to July 7, 1959.

[4] Interview of PWA by Alexei Kojevnikov on March 30, 1999, Niels Bohr Library & Archives, American Institute of Physics, College Park, MD.

[5] The fire Joyce witnessed occurred during a summer trip with her grandparents to a fishing cabin in the upper peninsula of Michigan.

[6] Mott added groups in solid-state theory and experimental surface physics to existing experimental groups in low-temperature physics and metal physics.

The Andersons arrived in Cambridge in September 1961. Joyce rented a semi-detached house built by one of Charles Darwin's sons and engaged an academic tutor and a piano teacher for Susan.[7] Phil discovered that Britain did not ban fireworks and he soon lit up the local sky with rockets, Catherine wheels, Roman candles, and other pyrotechnics.[8] The couple delighted in throwing parties for American visitors and expatriates fueled by Phil's powerful martinis.

Anderson settled into an office next to Mott in the Cavendish Laboratory—the formal name of the Physics Department of the University of Cambridge. Surprisingly, he and Mott did no research together. They would chat occasionally about their mutual interests, but their contacts were mostly social.[9]

The visiting physicist found the University of Cambridge to be both similar to, and different from Harvard, his *alma mater*. It was similar in the sense that its faculty, academic departments, museums, and libraries were excellent and held in very high regard. A major difference was Cambridge's many Colleges. Each was an independent self-governing unit with an endowment and property assets, the latter including the residence buildings and beautiful chapels that gave the University its architectural magnificence. Most of the faculty affiliated with a College in addition to their academic department.

Mott and Pippard arranged for Anderson to become a Fellow at Churchill College, a new institution dedicated to promoting science and engineering at Cambridge. The short walk from Phil's rented house to Churchill College made it easy for him to attend

Compare Table 12.1 and Table 13.1 in Malcolm Longair, *Maxwell's Enduring Legacy. A Scientific History of the Cavendish Laboratory* (University Press, Cambridge, 2016), Section 13.2.

[7] Author correspondence with Susan Anderson.

[8] Letter from Joyce Anderson to Jaynet and Alan Holden, November 11, 1961. Courtesy of Philip W. Anderson.

[9] P.W. Anderson, "A Happy Warrior," in *Nevill Mott: Reminiscences and Appreciations*, edited by E.A. Davis (Taylor & Francis, London, 1998), pp. 205–8.

its weekly dinners where camaraderie and port wine were in ready supply. On the other hand, the walk passed through a manure-strewn field and tradition barred Joyce from attending the College dinners.

Most scientists regard a sabbatical as an opportunity to develop a new research expertise. This was not Anderson's aim. He wanted to experience the academic life he had rejected coming out of graduate school. On the other hand, two of his sabbatical activities inspired Nobel Prize-winning work done by two *other* physicists: Brian Josephson and Peter Higgs.

Spontaneous Symmetry Breaking

Brian David Josephson earned a share of the 1973 Nobel Prize for theoretical work that revealed new and unexpected aspects of superconductivity. However, when Josephson began attending Anderson's 1961–1962 course on solid-state physics and many-electron theory, he was just a second-year graduate student at the University of Cambridge beginning PhD research with the experimentalist Brian Pippard.

Josephson was well-known locally because he had published a single-author theoretical paper while still an undergraduate.[10] Anderson got to know him better because the normally shy Welshman often approached him after class to correct small errors or slips he made during his lectures. It discomfited Josephson that his teacher "regularly omitted plus and minus signs because he couldn't be bothered to get those bits right."[11]

Soon after the course ended, Anderson's edited lecture notes appeared as the book, *Concepts in Solids*.[12] Unlike the existing books on the subject, *Concepts* mostly ignored the bread-and-butter

---

[10] B.D. Josephson, "Temperature-Dependent Shift of γ Rays Emitted by a Solid," *Physical Review Letters* **4**, 341–2 (1960).

[11] Interview of Brian D. Josephson by the author, July 10, 2015.

[12] P.W. Anderson, *Concepts in Solids: Lectures on the Theory of Solids* (W.A. Benjamin, Reading, MA, 1963). The book is based on notes taken by Lu Jeu Sham, who went from Cambridge to a postdoctoral position with Walter Kohn. Together,

topics of solid-state physics like crystal structure, binding energy, lattice vibrations, electrical conduction, thermal and optical properties, semiconductors, etc.[13] Instead, it focused on the ground and excited states of electrons and spins in various solid-state situations. Anderson's discussions strike the contemporary reader as quite modern, largely because the scientific community ultimately adopted his perspectives on the subject.

Josephson's Nobel Prize-winning research makes essential use of the phenomenon of symmetry breaking. Earlier chapters used this term in several ways, one of which was to describe situations where the ground state of a system lacked a symmetry exhibited by its total energy function. Anderson's lectures covered this idea using the antiferromagnet and the superconductor as examples. Chapter 6 explained symmetry breaking for an antiferromagnet already. Here, we review that discussion and then demonstrate that a superconductor behaves similarly.

The Heisenberg model of a crystalline magnet assumes that every pair of nearest-neighbor spins S_i and S_k contribute an amount $-JS_i \cdot S_k$ to the total energy of a magnetic crystal. This energy exhibits continuous spin rotation symmetry, i.e., its numerical value does not change if all the spins rotate together as a unit. However, the antiferromagnetic $(J < 0)$ spin configuration shown in Figure 6.5 breaks this symmetry because the same collective rotation of all the spins transforms this configuration into a physically distinct configuration (with the same energy) where all the spins point in a different direction.

they invented *density functional theory*, a very important method used for realistic calculations of specific many-electron systems.

[13] The graduate level treatments of solid-state physics available at the time were F. Seitz, *Modern Theory of Solids* (McGraw-Hill, New York, 1940), R. E. Peierls, *Quantum Theory of Solids* (Clarendon Press, Oxford, UK, 1955) and Gregory Wannier, *Elements of Solid-State Theory* (Cambridge University Press, London, 1959). The same year that *Concepts in Solids* appeared, Charles Kittel produced the book that became the standard graduate textbook of solid-state physics for the next fifteen years: *Quantum Theory of Solids* (Wiley, New York, 1963).

Similar symmetry breaking occurs for a ferromagnet ($J > 0$). However, the antiferromagnet alone exhibits the special phenomenon of *spontaneous symmetry breaking*. As Anderson used this term, this meant that a glacially slow collective rotation of the broken symmetry state begins so that, after one complete rotation, all the spins of the antiferromagnet will have pointed (albeit briefly) in every allowed direction.[14] In other words, the system evolves in time in such a way as to reclaim the continuous spin rotational symmetry of the Heisenberg model energy.

By an accident of intellectual history, Anderson was lecturing to his class about symmetry breaking in antiferromagnets just as the phrase "spontaneous symmetry breaking" made its debut in the physics literature.[15] Particle physicists coined the phrase in connection with analogous behavior found in some models for the behavior of subatomic particles. The British theoretical physicist Jeffrey Goldstone pursued the particle/solid-state analogy further and, not long after, solid-state physicists began to refer to the symmetry-restoring uniform rotation of spins described above as a *Goldstone mode*.[16]

Concepts in Solids does not discuss symmetry breaking in a superconductor. However, it is clear from multiple student testimonies that Anderson discussed the idea in the classroom using his pseudospin formulation of BCS theory.[17] In that language, a

[14] Over his career, Phil was not entirely consistent in his definition of the term spontaneous symmetry breaking. See the letter by Thomas A. Kaplan under the title "Reflections on Broken Symmetry" in *Physics Today* **44** (2), 15 (1991).

[15] Marshall Baker and Sheldon Glashow, "Spontaneous Breakdown of Elementary Particle Symmetries," *Physical Review* **128**, 2462–71 (1962). For a history of this subject, see Laurie M. Brown and Tian Yu Cao, "Spontaneous Breakdown of Symmetry: Its Rediscovery and Integration into Quantum Field Theory," *Historical Studies in the Physical and Biological Sciences* **21**, 211–35 (1991).

[16] Jeffrey Goldstone discussed symmetry breaking in relativistic field theory in "Field Theories with Superconductor Solutions," *Il Nuovo Cimento* **19**, 154–64 (1960).

[17] B.D. Josephson, "The Relativistic Shift in the Mossbauer Effect and Coupled Superconductors," Trinity College Fellowship Dissertation, 1962, p. 15; John Waldram, "50th Anniversary of Brian Josephson's Nobel Prize Discovery," *CavMag, News from the Cavendish Laboratory*, January 2013, Issue 9, p. 8.

Heisenberg-type exchange term accounts for the energy of the superconductor. Just like the passage from Figure 6.4(a) to Figure 6.4(b), spontaneous breaking of pseudospin rotational symmetry occurs because all the pseudospins align along one arbitrarily chosen direction when $T < T_C$.

The pseudospin formulation of superconductivity is elegant, but not well-suited to discuss Brian Josephson's work. A better choice for that purpose is the method of Ginzburg and Landau, which uses a macroscopic wave function to characterize the supercon-ducting state. Like every wave function, ψ_{GL} possesses a *magnitude* and a *phase*, both of which can vary as a function of position and time. The magnitude of ψ_{GL} encodes the density of Cooper pairs. The phase of ψ_{GL} helps determine the state of motion of the Cooper pairs.

Phase is an angle-type variable that can take any value between 0 and 360 degrees. For clarity, imagine a huge number of tiny stopwatches, each assigned to one point in the body of supercon-ductor. The position of the sweep hand at each point identifies the phase of ψ_{GL} at that point. Figure 10.1 illustrates the solution found by Ginzburg and Landau for the case of an isolated super-conductor: the phase takes exactly the same value at every point. Hence, all the sweep hands point in the same direction.

The similarity of Figure 10.1 to the ordered spin arrangements in Figure 6.4(b) and Figure 6.5 is apparent.[18] Also, like the spin direction chosen by both ferromagnets and antiferromagnets, the energy of a superconductor does not depend on the common value of the phase angle it chooses. However, the strict analogy is to the antiferromagnet because an isolated superconductor exhibits a Goldstone mode where all the locked sweep hands in Figure 10.1 slowly rotate together. In other words, Figure 10.1

---

[18] When $T > T_C$, one might expect the phase of ψ_{GL} to vary randomly from point to point so the sweep hands in a diagram like Figure 10.1 would point in random directions like the spin orientations in Figure 6.4(a). This does not hap-pen because in the non-superconducting phase, the magnitude of ψ_{GL} is zero (there are no Cooper pairs anywhere) and the phase of ψ_{GL} has no meaning.

Figure 10.1 Stopwatch model for the phase of the wave function of a superconductor. Each sweep hand points to the value of the phase at one point in space.

illustrates spontaneous symmetry breaking where it is the continuous rotational symmetry of the phase angle that is lost (or broken). An equivalent statement rooted in the mathematics of the situation is to say that a superconductor breaks *gauge symmetry*.

A very important influence on Anderson's thinking about these issues was Yoichiro Nambu, a theoretical particle physicist at the University of Chicago. Nambu learned about the BCS approach to superconductivity from his (then) faculty colleague Bob Schrieffer. In October of 1960, Nambu visited Bell Labs and he and Anderson had extensive discussions about gauge invariance, collective excitations, and symmetry breaking in superconductors.[19] It was Nambu who made the particle physics community aware of the similarities between their problems and the problem of a superconductor. Nambu's work in this area—particularly his

[19] Letter from PWA to Y. Nambu, August 1 1960; Memo of October 24, 1960 from PWA to F.B. Llewellyn, "Visit of a Foreign National, Y. Nambu", AT&T Archives, Warren, NJ. This letter and memo establish that Nambu visited Bell Labs for the first time in 1960, rather than in 1959 as stated in P.W. Anderson, "Y. Nambu and Broken Symmetry," in *More and Different: Notes from a Thoughtful Curmudgeon* (World Scientific, Hackensack, NJ, 2011), pp. 115–19.

insight that the physical vacuum is a kind of condensed phase that breaks a symmetry of space-time—earned him a share of the 2009 Nobel Prize for Physics.[20]

Josephson Effects

Brian Josephson was fascinated by the idea of broken symmetry and wondered if the phase of a superconductor was detectable in the laboratory.[21] By a bit of serendipity, his experimental research with Brian Pippard led him to study a paper by Ivar Giaever of the General Electric Research Laboratory. Giaever had discovered it was possible to drive a current through the sandwich structure shown in Figure 10.2 (today called a Josephson junction) where a very thin layer (10–20 Å) of insulating material separates two superconductors.[22]

The insulator presents a barrier to the flow of electrons. Giaever's experiment showed that current flows through the sandwich anyway because of quantum mechanical tunneling. Josephson was also aware of a theoretical result due to Lev

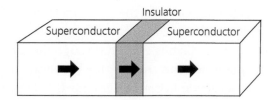

Figure 10.2 A Josephson junction (or sandwich) where a thin layer of insulator separates two superconductors. Arrows indicate the flow of current through the junction.

[20] Yoichiro Nambu, "Spontaneous Symmetry Breaking in Particle Physics; A Case of Cross-Fertilization," Nobel Lecture, 2008.

[21] B.D. Josephson, "The Discovery of Tunneling Supercurrents," *Science* **184**, 527–30 (1974).

[22] Ivar Giaever, "Electron Tunneling Between Two Superconductors," *Physical Review Letters* **5**, 464–6 (1960).

Gor'kov that the phase of superconductor was sensitive to the number of electrons it contained. Combining these two ideas suggested to him a novel and profound idea: the tunneling of a Cooper pair from one superconductor to another was possible if and only if the phase difference between the two superconductors was not zero. An observation of the resulting current amounted to using the phase of one superconductor as a reference to measure the phase of the other superconductor.

Fifty years after the events, John Waldram, another student in Anderson's class, recalled that:

> [Phil Anderson] gave a marvelous lecture course, explaining—among many other things—the nature of the phase [of a superconductor]…[Brian] talked incessantly to Phil, rapidly worked up a good working interaction, and then went away to cogitate—he realized that the…phase might be made observable in a tunneling experiment.[23]

Josephson made two predictions that are known today as the DC and AC Josephson effects. Both were so extraordinary that John Bardeen did not believe them at first.[24] The DC effect predicted a current (in the *absence* of an applied voltage) which was a function of the phase difference between the two superconductors. The AC effect predicted that a constant voltage applied across the junction induces an oscillating current through the junction.

Josephson submitted his results for publication just as Anderson's sabbatical year in Cambridge ended. Phil was enthusiastic about Josephson's predictions and eager to confirm them.[25] Back in Murray Hill, he had little trouble convincing the

[23] John Waldram, "50th Anniversary of Brian Josephson's Nobel Prize Discovery," *CavMag, News from the Cavendish Laboratory*, January 2013, Issue 9, p. 8.

[24] Donald G. McDonald, "The Nobel Laureate versus the Graduate Student," *Physics Today* **54** (7) 46–51 (2001).

[25] Philip W. Anderson, "How Josephson Discovered his Effect," *Physics Today* **23** (11) 23–9 (1970).

experimentalist John Rowell to try.[26] Anderson hung around the laboratory (of course) and according to Rowell, he not only provided theoretical guidance, he made several technical suggestions that were essential to their eventual observation of the DC Josephson effect.[27]

One year later, Anderson and Bell Labs experimenter Aly Dayem obtained a patent for a device based on the AC Josephson effect.[28] Today, technology based on Josephson's insights, e.g., superconducting quantum interference devices (SQUIDs), facilitate fundamental physics measurements as well as a wide range of sensor, amplifier, signal processing, and medical technology applications.[29]

Anderson devoted the winter of 1962–1963 to pondering the meaning of Josephson's results. These ruminations produced a set of summer school lectures that are deep, elegant, and focused on the essentials.[30] They contain none of the complicated mathematics that made his papers on disorder-induced localization and gauge invariance in superconductivity so difficult for others

[26] Rowell was already collaborating with Anderson on tunneling measurements of the "bad actor" superconductors (see Chapter 8).

[27] J.M. Rowell, "Superconducting Tunneling Spectroscopy and the Observation of the Josephson Effects," IEEE *Transactions on Magnetics* **MAG–23**, 380–9 (1987); J.M. Rowell, "Tunneling and the Josephson Effect," in *100 Years of Superconductivity*, edited by Horst Rogalla and Peter H. Kes (CRC Press, Boca Raton, FL, 2012), Section 3.11. The discovery paper is P.W. Anderson and J.M. Rowell, "Probable Observation of the Josephson Superconducting Tunneling Effect," *Physical Review Letters* **10**, 230–2 (1963).

[28] Philip W. Anderson and Aly H. Dayem, "Superconducting Device of Varying Dimension Having a Minimum Dimension Intermediate Its Electrodes," US Patent 3,335,363. For technical reasons, the device they proposed—known today as the Dayem bridge—does not have many practical applications. See K.K. Likharev, "Superconducting Weak Links," *Reviews of Modern Physics* **51**, 101–59 (1979).

[29] Horst Rogalla and Peter H. Kes (editors), *100 Years of Superconductivity* (CRC Press, Boca Raton, FL, 2012), Chapter 5.

[30] P.W. Anderson, "Special Effects in Superconductors," in *Lectures on The Many-Body Problem*, volume 2, edited by E.R. Caianiello (Academic Press, New York, 1964). pp. 113–35.

to read and understand. In their place, one finds simple analogies, like one which relates the behavior of a Josephson junction to a swinging pendulum. These lectures are significant also because, at the very end, Anderson discusses for the first time how symmetry breaking has similar consequences for crystalline solids, ferroelectrics, ferromagnets, antiferromagnets, and superconductors. Chapters 12 and 13 expand on this idea.

Just at this time, Anderson received a copy of Josephson's Fellowship dissertation and discovered that the two of them had come to most of the same conclusions independently. Because he "feared his own renown would rob Josephson of the credit he deserved," Anderson delayed publishing his summer school lectures in an easily accessible place.[31] He was similarly gracious when he accepted an award on Josephson's behalf at a 1970 conference:

> Apart from [some] ideas about broken symmetry and some minor points acknowledged explicitly in [Josephson's] paper, I want to emphasize that the whole achievement from the conception to the explicit calculation to the publication was completely Josephson's.[32]

Phil's private opinion differed somewhat. Quite soon after Josephson published his original article, he mused in his Bell Labs notebook that "I still don't like to saddle BJ with credit for the whole thing." Ten years later, he and Joyce were enduring a cold and wet stay in Cornwall in the far southwest of England when the Swedish Academy of Sciences awarded one-half of its 1973 Prize for Physics to Josephson. Ivar Giaever and the experimentalist Leo Esaki shared the other half. Anderson was bitterly

[31] Letter from PWA to Gloria Lubkin (Senior Editor, Physics Today), November 15, 1973, Niels Bohr Library and Archives, American Institute of Physics, One Physics Ellipse, College Park, MD, 20740.

[32] P.W. Anderson, "London Award Lecture: Brian Josephson and Macroscopic Quantum Interference," in *Proceedings of the 12th International Conference on Low Temperature Physics*, edited by Eizo Kanda (Keigaku Publishing Co., Ltd. Tokyo, 1971), pp. 1–17.

disappointed and Joyce wrote to close friends that "Phil has suffered over the Nobel Prize, which he feels he has as much right to
as the three who got it—all very well to say it doesn't matter, but
during the announcements it weighs on him."[33]

Phil's feelings occasionally spread beyond his circle of family
and close friends. Soon after the Nobel Prize announcement, he
informed Bell Labs management that Josephson discovered how
to use current to measure the phase of a superconductor, but
that he (Anderson) had discovered broken symmetry and the fact

Figure 10.3 Brian Josephson, circa 1964–1965. Source: American Institute
of Physics Emilio Segrè Visual Archive.

---

[33] Letter from Joyce Anderson to Jaynet and Alan Holden, October 24, 1973.
Courtesy of Philip W. Anderson.

that phase is an observable physical quantity.[34] In a talk to non-physicists a few months later, Phil identified himself as "one of the co-discoverers of the Josephson effect." He repeated this claim in print when he later sought election as Councilor-at-Large to the American Physical Society.[35]

A portrait of Josephson published by Anderson in 2011 is quite brutal:

> His hold on reality was always somewhat tenuous. He loped with a one-sided gait along the streets of Cambridge, sometimes talking to himself.... After the great discovery, he became more and more nervous, bothered by hallucinations and disturbed enough to spend some time in a nursing home.... Then, all of a sudden, the tension snapped—he stopped working on physics... and took up wholly with Transcendental Meditation and the sponsoring of mediums and poltergeists.... Broken symmetry and its consequences solved so much—is it any wonder he reached for more and melted his wings?[36]

It is true that Josephson abandoned solid-state physics in the early 1970s and switched his research to studying human consciousness and paranormal behavior.[37] A public exchange of letters between Josephson and Anderson on the subject of psychokinesis (moving objects with the mind) exposed an irreconcilable breach in their points of view.[38]

---

[34] Letter from PWA to Gloria Lubkin (Senior Editor, Physics Today), November 15, 1973, Niels Bohr Library and Archives, American Institute of Physics, College Park, MD.

[35] Transcript of a 1974 public lecture, AT&T Archives, Warren, NJ; Bulletin of the American Physical Society, Series II **24**, 787 (1979).

[36] P.W. Anderson, "A Mile of Dirty Lead Wire. A Fable for the Scientifically Literate," in *More and Different: Notes from a Thoughtful Curmudgeon* (World Scientific, Hackensack, NJ, 2011), pp. 50–61. This essay is a chapter from an unfinished history of superconductivity that Anderson worked on in the late 1970s and early 1980s.

[37] See, e.g., B.D. Josephson and V.S. Ramachandran (eds.) *Consciousness and the Physical World* (Pergamon Press, Oxford, UK, 1980).

[38] Brian D. Josephson and Philip W. Anderson, "Has Psychokinesis Met Science's Measure?," *Physics Today* **45**(7), 15 (1992).

These and other comments by Anderson reveal a combination of sadness, disappointment, and perhaps even anger. Josephson was never Anderson's research student and both insist that their relationship never rose to the level of mentor and mentee.[39] Nevertheless, the tone of Anderson's remarks and the fact that elsewhere in the quoted essay he attests to Josephson's brilliance suggests that Brian meant something more to him than he was ever willing to admit.

A plausible guess is that Anderson was disillusioned by Josephson. This obviously brilliant fellow had turned his back on his theoretical physics gift, he had failed to fulfill the promise of Phil's personal interest in him, and he had embarrassed his profession by championing ideas regarded as beyond-the-pale by his physics colleagues.

The Higgs Mechanism

In July 2012, worldwide publicity accompanied the discovery of the subatomic *Higgs particle*. This discovery concluded a saga that began in 1963 when the British physicist Peter Higgs acted on a suggestion made by Anderson in a paper aimed at particle physicists.[40] Anderson had learned from casual conversations with Cambridge colleagues that there were serious problems with many of the theories designed to provide a consistent description of the strong forces present inside the atomic nucleus.[41]

The theories in question had two things in common: they all possessed one or another mathematical symmetry and they all employed a zero-mass particle to mediate a force by shuttling

---

[39] Letter from PWA to John Horgan, January 11, 1995. Princeton University Archives, Department of Rare Books and Special Collections, Princeton University Library. Private communication with Brian Josephson, September 12, 2019.

[40] P.W. Anderson, "Plasmons, Gauge Invariance, and Mass," *Physical Review* **130**, 439–42 (1963).

[41] Frank Close, *The Infinity Puzzle* (Basic Books, New York, 2011).

back and forth between other particles. The poster child for this behavior was quantum electrodynamics. This theory possessed gauge symmetry and used the zero-mass photon to mediate the Coulomb force between charged particles. The problem was that the particle physics community had good reason to believe that the particles responsible for conveying some nuclear-related forces were *not* massless. How could one generate mass for these shuttle particles without destroying the desirable symmetry of the theory?

Zero-mass particles of another sort engaged a different group of particle physicists—those pursuing the idea of spontaneous symmetry breaking in subnuclear physics. These theorists inevitably found Goldstone modes, just as Anderson had done for the antiferromagnet and the superconductor. However, in their particle physics manifestation, Goldstone modes are exactly zero-mass particles!

With important input from the British theorists John Ward and John G. Taylor, Anderson realized that his RPA work on the behavior of electrons in a superconductor allowed him to propose a mechanism to give mass to a zero-mass particle. The common element between the two problems was the Goldstone mode associated with spontaneous symmetry breaking. For him, this mode was the collective motion of electrons in the limit of a very long wavelength oscillation. As discussed in Chapter 9, he found the frequency of this mode to be *zero* if he ignored the Coulomb interaction between the electrons and *non-zero* if he included that interaction. By analogy, Anderson suggested that the mass of the Goldstone mode particle would increase from zero to non-zero if it interacted with the field associated with some other zero-mass particle, e.g., the photon.

Peter Higgs, then a Lecturer at the University of Edinburgh, read Anderson's paper and realized that his proposed mechanism to synthesize a massive particle from two massless particles was probably correct. He incorporated the basic idea into a field theory with spontaneous symmetry breaking and showed that the mechanism still worked when he modified the theory so it was consistent with Einstein's theory of relativity.[42] The latter is necessary for any theory designed to explain particle physics phenomena. Besides confirming Anderson's assertion about mass generation, the completeness of Higgs' model permitted him to predict the existence of another massive particle. This is the "Higgs particle" which was discovered in 2012.

Like Nambu's work on spontaneous symmetry breaking, the Anderson–Higgs mechanism for the generation of mass exports an idea from solid-state physics to particle physics. Both are beautiful displays of the intellectual unity of modern physics. The same mathematics applies to the Higgs particle and to a Cooper pair even though the energy of the former is one

Figure 10.4 Peter Higgs in 1954. Source: University of Edinburgh.

[42] Peter W. Higgs, "Broken Symmetries and the Masses of Gauge Bosons," *Physical Review Letters* **13**, 508–9 (1964).

hundred trillion times greater than the energy of the latter. As one commentator put it, "By thinking hard about a piece of metal, Anderson had divined the solution to a puzzle about fundamental particles."[43]

Physics Physique Fizika

In the mid-1960s, Phil Anderson and Bernd Matthias co-edited a physics journal with the peculiar name *Physics Physique Fizika, an International Journal for Selected Articles Which Deserve the Special Attention of Physicists in all Fields.* The two Bell Labs scientists did all the reviewing of submitted manuscripts and they made all the acceptance/rejection decisions themselves. In other words, they curated *Physics* rather than edited it. Authors were paid ten cents a word if their paper appeared in print.

Physics got started because Anderson complained to Matthias one day about the problems with physics journals.[44] The referees often did a bad job, the editors took too long to make decisions, and too much wrong physics got published. Matthias asked Anderson how he might run his own journal and then contacted a person he had met socially, Robert Maxwell, the founder of Pergamon Press.

Maxwell was a remarkable figure.[45] Born and raised in Czechoslovakia, he has been variously described as brilliant, capricious, transformative, ingenious, dishonest, visionary, and predatory. During World War II, he earned British citizenship by fighting for the British army and serving as an intelligence officer. After the war, he entered the publishing business and soon founded Pergamon Press.

---

[43] Shivaji Sondhi, "The Tao of Modern Physics," *The Indian Express*, November 15, 2013.

[44] Interview of PWA by Alexei Kojevnikov, June 29, 2000, Niels Bohr Library & Archives, American Institute of Physics, College Park, MD.

[45] Brian Cox, "The Pergamon Phenomenon 1951–1991: Robert Maxwell and Scientific Publishing," *Learned Publishing* 15(4) 273–8 (2002).

Maxwell understood that the peculiar economics of publishing scientific journals made it possible for a nimble and aggressive individual to make a great deal of money.[46] He did this by appealing to the egos of scientists and signing contracts with dozens of them to start and edit new journals. The timing was right because the tremendous expansion of science in the postwar era provided a built-in audience for his products.

Anderson and Matthias travelled to England and negotiated a deal to create their physics journal over a lavish dinner at Maxwell's 51-room mansion. At the end of the evening, Maxwell startled his guests by flipping them a gold coin. Neither guest was aware that the imprint of the press was a reproduction of a Greek coin from the Asia Minor city of Pergamon.

The neophyte editors sent out more than 500 solicitation letters to their professional colleagues and had no trouble attracting high-quality manuscripts. The very first issue of *Physics* featured an article by the soon-to-be Nobel Prize-winning particle physicist Murray Gell-Mann. The third issue contained an article about the completeness of quantum mechanics by John S. Bell, a physicist not known to either editor. Anderson accepted Bell's paper because the subject interested him personally. By the end of 2019, it had earned over 14,000 citations.[47]

Anderson and Matthias discontinued *Physics* after four years (1964–1967) and nineteen issues. They did this, not because it lacked submissions, but because they had promised contributors rapid publication and Maxwell kept increasing the time between the appearances of successive issues. Anderson was philosophical

---

[46] Robert Buryani, "Is the Staggeringly Profitable Business of Scientific Publishing Bad for Science?," *The Guardian*, June 27, 2017. Available at https://www.theguardian.com/science/2017/jun/27/profitable-business-scientific-publishing-bad-for-science. Accessed August 11, 2019.

[47] J.S. Bell, "On the Einstein-Podolsky-Rosen Paradox," *Physics* 1(3), 195–200 (1964). Google Search, January 1, 2019.

about his adventure in publishing. He had enjoyed running the journal and Maxwell was a truly larger-than-life character.[48]

The Half-Time Professor

Half-way through Phil's Cambridge sabbatical, Nevill Mott was knighted and a celebratory feast was held in his honor. Mott addressed the revelers and, among other things, he told his listeners that it would be nice if Anderson could be convinced to stay at Cambridge permanently.[49] Mott reiterated this sentiment every time he and Anderson met until his quarry succumbed, at least in part. Phil agreed to accept a half-time appointment as Professor of Theoretical Physics at the University of Cambridge. Bell Labs was amenable and, immediately after his election to the National Academy of Sciences, Anderson began a new phase of his life and career.[50]

From 1967 to 1975, Phil and Joyce Anderson lived in Cambridge from October to March (the Fall and Winter academic terms), vacationed in April, and lived in New Jersey from May to September.[51] A few years earlier, Joyce had functioned as principal

[48] The success of Pergamon Press led Maxwell to make a series of attention-grabbing but questionable acquisitions, including soccer teams, television stations, and the Daily Mirror newspaper group. He sold Pergamon Press in 1984, but soon found himself mired in a series of financial scandals. Maxwell drowned in 1991 while vacationing on his yacht near the Canary Islands. In the aftermath, British banking authorities alleged that he had looted pension funds under his control to acquire the cash he needed to keep his businesses afloat. Richard O'Mara, "Maxwell Leaves a Legacy of Scandal," *Baltimore Sun*, December 22, 1991.
[49] Interview of PWA by Alexei Kojevnikov, November 23, 1999, Niels Bohr Library & Archives, American Institute of Physics, College Park, MD.
[50] Bell Labs had previous experience with a half-time staff scientist in the person of mathematician John Tukey. His other half-time appointment was at Princeton University. David R. Brillinger, "John Wilder Tukey (1915–2000)," *Notices of the AMS*, **49**(2) 193–201 (2002).
[51] The academic year at the University of Cambridge consists of three eight-week terms: Michaelmas (October and November), Lent (mid-January to mid-March), and Easter (mid-April to mid-June).

designer and general contractor when she and Phil built a new home in New Vernon, New Jersey. She now put her stamp on the final aspects of the design and materials for a three-story row house near Newnham College which they soon occupied.

Susan Anderson did not accompany her parents to Cambridge. She was beginning her sophomore year at the University of Rochester and still navigating her freedom from her parents' control.[52] Joyce had demanded academic excellence from her daughter and was otherwise the dominant force in her life. She organized much of Susan's routine, shared with her a powerful love of Nature, and instilled an appreciation of culture and civic responsibility. Joyce was also concerned about appearances, a vestige of her embarrassment about her own family circumstances.

Phil was a more distant presence. He was affectionate, taught Susan chess and Go, and shared his interest in subjects like geology. Family dinners were verbal, with much talk about politics, books, and news about friends around the world. Walks were an important family ritual, as was tending the vegetable garden. Nevertheless, Phil travelled frequently and he left much of the child-rearing (and all of the discipline) to his wife.

The zeitgeist of the late 1960s was in full swing and Susan got caught up in sex and drugs and rock-and-roll. She dropped out of college, joined a commune in California, and fell under the influence of the charismatic spiritual leader Swami Satchidananda. Five-and-a-half years later, she was working as a typist and taking art classes at night when Phil won the Nobel Prize. This inspired her to have confidence in herself and she went on to earn a degree from Boston's School of the Museum of Fine Arts. Painting and drawing became the center of her life from that moment on.

Her father's diffidence and her mother's need for control led to Susan's long-term relationship with her parents to be less than

[52] Interview of Susan Anderson by the author, March 3, 2016.

ideal. Nevertheless, she made a point of visiting them several times a year. It took a debilitating stroke suffered by Joyce in the summer of 2009 to bring father and daughter closer together than at any other time in their lives.

Back in England, the Andersons made it their business to explore the country and immerse themselves in its culture. Many weekends were spent at rustic inns near and far. One favorite destination was the artist's colony in the village of St. Ives in the far southwest of the country. At Cambridge, they attended lectures on archaeology, art, architecture, ethology, and economics. The latter interested Phil particularly and the couple became friendly with the economists (Baron) Richard Kahn, James Meade, and John Kenneth Galbraith, who was visiting from Harvard. They met the celebrated cellist Jacqueline du Pré and attended intimate concerts where she and the oboist Léon Goossens performed. They also became habitués of the Cambridge Arts Theatre where first-rate actors came from London to perform the plays of William Shakespeare, Harold Pinter, and Tom Stoppard.

Mott made Anderson the Head of the Solid-State Theory group. The previous group Head, the amiable and self-effacing Volker Heine, continued to handle the administrative matters. The two got on well and met for lunch once a week to discuss Group matters.[53] Anderson devoted most of his time to research and, naturally enough, he assigned problems to PhD students and postdocs drawn from magnetism, superconductivity, many-electron theory, and localization.

Anderson supervised a total of thirteen PhD students during the eight years he worked half-time at Cambridge. Readers familiar with his oeuvre may be surprised by the problem he suggested to his first PhD student, Hugh Fricker. The task was to design and implement a numerical scheme to solve the Schrödinger equation approximately for small molecules.[54] Fricker did not pursue

---

[53] Interview of Volker Heine by the author, July 11, 2015.
[54] Hugh S. Fricker, "The Quantum Defect Method and the Calculation of Molecular Wave Functions," PhD Thesis, 1970, University of Cambridge.

research as a vocation, choosing instead to devote his post-PhD career to secondary education. Fifty years after the events, his memories of Anderson were fond:

> Anderson was always very kind, approachable, and helpful. At the time I think I took that for granted, but I have come to appreciate it with the clarity of hindsight. Around the Cavendish, he came across as a wise, civilized man, with a gentle, wry sense of humor, wide interests, and not remotely obsessive about his work. He seemed to have a strong domestic hinterland, and many a conversation with him would end, as 5 p.m. approached, with Phil glancing at his watch and "Gee, I must go and meet my wife."[55]

Anderson's last PhD student at Cambridge was Duncan Haldane, a half-Scot, half-Slovenian native of London who combined an interest in physics with enthusiasm for the Socialist Workers Party. In his final year as an undergraduate at Cambridge, Haldane took a class from Anderson called Advanced Quantum Mechanics where the instructor "talked about the problem of localization by disorder and other inspiring... and deeply conceptual quantum mechanical problems different from the [bland] diet... the more conventional classes had been feeding us."[56]

Haldane resolved to stay at Cambridge and earn his PhD working with Anderson. His thesis work used a model devised by his advisor to study the peculiarities of compounds synthesized from elements whose chemical valence appeared to fluctuate.[57] Later, Haldane joined Anderson on the physics faculty at Princeton and earned a share of the 2016 Nobel Prize for Physics for his work on many-body magnetism.

Anderson involved himself in several important changes at the Cavendish. Most important, perhaps, he helped convince the

---

[55] Interview of Hugh S. Fricker by the author, July 3, 2015.
[56] F. Duncan M. Haldane, Nobel Biography, 2016.
[57] F. Duncan M. Haldane, "Extension of the Anderson Model and a Model for Mixed Valence Rare Earth Materials," PhD Thesis, 1987, Cambridge University.

solid-state theorist Samuel Edwards to accept a professorship there. Edwards, a laconic Welshman who had been a PhD student of Julian Schwinger at Harvard, was one of the first theorists to specialize in *soft matter physics*, a research field focused on "squishy" systems like polymers, gels, liquid crystals, and biological materials. Two years later, Phil and Sam collaborated on a groundbreaking theory of a class of magnetic alloys called *spin glasses* (see Chapter 12).

Anderson and Volker Heine reflected on the state of their field and decided to change the name of their research group to something that better described its activities from magnetism to quantum liquids to colloids. The title of a little-read, ten-year-old Swiss journal, *Physics of Condensed Matter*, seemed to fit the bill. Moreover, a 1000-page book called *The Theory of Condensed Matter* documented a two-month course of lectures on the subject presented in 1967 at the International Centre for Theoretical Physics in Trieste, Italy.[58] Hence, beginning in 1972, the Solid-State Theory group became the Theory of Condensed Matter (TCM) group.

The new name had the great virtue of providing a common identity for individuals who had similar training and similar outlooks, e.g., traditional solid-state physicists, many-body theorists, low-temperature physicists, and physicists who used the methods of statistical mechanics to study matter. Quite rapidly, the name gained broad acceptance.[59] Today, it is rare to read or hear about solid-state physics at all.

Anderson was always making small changes to the graduate course that had inspired Brian Josephson's work. However, during the 1973–1974 academic year, he took a sabbatical leave from teaching, retreated to a vacation home he and Joyce had built (see "Port Isaac" later in this chapter) and completely revamped his

[58] *The Theory of Condensed Matter*, edited by F. Bassani, G. Caglioti, and J. Ziman (International Atomic Energy Agency, Vienna, 1968).

[59] Joseph D. Martin, "What's in a Name Change?," *Physics in Perspective* **17**, 3–32 (2015).

lecture notes to reflect the evolution of his thinking about many-body physics over the previous decade. What emerged was the first draft of what would later become the book, *Basic Notions of Condensed Matter Physics*. His goal was to identify and illustrate a few general principles which he felt brought unity and depth to the subject.

Unfortunately, and notwithstanding the inspiring effect of his teaching on Brian Josephson and Duncan Haldane, most students found that Anderson's lectures lacked organization and clarity. Two accounts among many are:

> Phil was like an oracle. He would stand up at the blackboard thinking for a great while and then write. It was very difficult for students to follow his train of thought. Mott had great intuition, which he dressed up with hand waving. Phil had truly brilliant intuition, which he dressed up with mathematics.[60]
>
> I went to his lectures every year for three years. . . . Anderson is a guy who thinks in a different way and, for that reason, it is very difficult to follow his lectures. For that reason also, Phil is a very bad teacher in the traditional sense. But he is a very good teacher because he makes you think.[61]

Anderson did not speak out often on undergraduate matters. One occasion when he did was to respond to a new teaching philosophy proposed by Brian Pippard.[62] This happened in the spring of 1972, right after Pippard succeeded Nevill Mott as Cavendish Professor and Head of the Department of Physics.[63] According to Pippard, the teaching of physics had become increasingly wedded to the exact mathematical analysis of simple unrealistic problems. In his view, students needed more instruction in tackling

---

[60] Interview of Roger Haydock by the author, March 3, 2015.

[61] Interview of Pedro Echenique by the author, July 16, 2015.

[62] A. B. Pippard, *Reconciling Physics with Reality* (Cambridge University Press, Cambridge, UK, 1971).

[63] The Cavendish Professor served as Head of the Department until Brian Pippard separated the two positions in 1979.

practical problems of the sort they would encounter in the outside world.[64] The curricular revisions he proposed de-emphasized
fundamental concepts in favor of methods of approximation
designed to produce practical solutions.

Anderson sent a three-page letter to his Cambridge colleagues.[65]
He began by taking exception "with almost the whole of" Pippard's
proposed teaching philosophy and argued that "it is especially
important to teach physics at as fundamental a level as possible to
those students for whom undergraduate work will be a terminal
education." He noted that "students are eager to learn the unifying principles behind the qualitative phenomenon they have been
taught" and averred that "theory should be used to unify and cor-
relate the experimental facts, rather than each experimental subject given its own theoretical discussion." He conceded that
"formal theoretical courses need more physical examples" but
concluded that "the most important thing to do for [students] is to
leave them with as much understanding of the depth and univer-
sality of the logical order of the subject as we possibly can."

Anderson articulated his philosophy of teaching physics just as
his research group was changing its name from Solid-State Theory
to Theory of Condensed Matter and just before he radically restruc-
tured his graduate course to de-emphasize calculation in favor of
identifying basic principles. It is apparent that all of these actions
sprang from the same emerging impulse to focus his attention on
fundamental concepts with the maximum breadth of application.

Superfluid Helium

Before and during his tenure at Cambridge, Anderson carried on
a fruitful relationship with the second element of the periodic

---

[64] J.G. Crowther, *The Cavendish Laboratory 1874–1974* (Science History
Publications, New York, 1974), p. 440.

[65] P.W. Anderson, *Subject: "A Tentative Proposal for Part II Physical Sciences"*, Spring
1972, AT&T Archives, Warren, New Jersey.

table, helium. Despite its dull-sounding status as an "inert gas," helium has fascinated physicists for years. Unlike any other substance, it resists freezing and remains a liquid down to the lowest temperatures known. Virtually all naturally occurring helium is ^{4}He, an atom where two electrons orbit a nucleus composed of two protons and two neutrons. Its lighter and much rarer cousin, ^{3}He, differs only because its nucleus contains only one neutron.

Physicists have known since 1937 that liquid ^{4}He undergoes a phase transformation upon cooling. Below 2.19 K, the liquid becomes a *superfluid* and exhibits remarkable, viscosity-free behavior.[66] For example, an empty test tube partially submerged upright in a beaker of superfluid slowly fills because the superfluid spontaneously creeps up the outer surface of the tube and then down the inner surface of the tube. This "siphon effect" stops only when the fluid level in the tube coincides with the fluid level in the beaker.

Superfluidity is a consequence of a phenomenon called *Bose-Einstein condensation*, where a macroscopic number of ^{4}He atoms occupy the lowest quantum energy state of the liquid. This is possible because ^{4}He atoms are *bosons*, which Chapter 9 defined as particles where the many-body wave function does not change sign when two particles exchange their positions.

The absence of viscosity in a superfluid is analogous to the absence of electrical resistance in a superconductor. Indeed, there is a deep similarity between superfluidity and superconductivity.[67] In 1966, this knowledge led Phil Anderson to exploit his understanding of superconductivity to predict various flow effects in superfluid ^{4}He.[68] Figure 10.5 shows his whimsical use of a steam

---

[66] The names Helium I and Helium II are often used, respectively, for the high-temperature non-superfluid phase and the low-temperature superfluid phase of ^{4}He.

[67] See, e.g., A.J. Leggett, *Quantum Liquids. Bose Condensation and Cooper Pairing in Condensed Matter Systems* (Oxford University Press, Oxford, UK, 2006).

[68] P.W. Anderson, "Considerations on the Flow of Superfluid Helium," *Reviews of Modern Physics* **38**, 298–310 (1966).

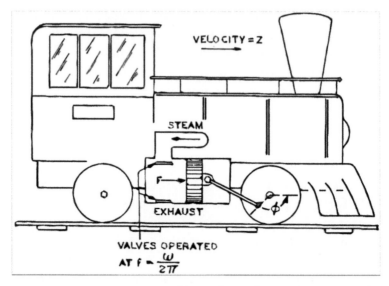

VELOCITY = Z

STEAM

EXHAUST

VALVES OPERATED
AT $f = \frac{\omega}{2\pi}$

Figure 10.5 Steam engine used by Anderson ["Considerations on the Flow of Superfluid Helium," *Reviews of Modern Physics* 38, 298–310 (1966)] to illustrate an analog of the AC Josephson effect for superfluid ⁴He. Source: American Physical Society.

locomotive engine to discuss a superfluid analog of the AC Josephson effect (Figure 10.5). A 2015 review of the experimental situation for ⁴He illustrates the correctness of his ideas and speaks to their lasting value.[69]

Beginning in the 1950s, macroscopic quantities of ³He became available for scientific study as a by-product of the decision of the United States and the Soviet Union to manufacture and test hydrogen bombs.[70] For physicists, a key point was that ³He atoms are fermions (like protons and neutrons) and not bosons. This led nuclear physicists to view a sample of liquid ³He as an enormous atomic nucleus—a dense system of fermions dominated by pair-

[69] Eric Varoquaux, "Anderson's Considerations on the Flow of Superfluid Helium: Some Offshoots," *Reviews of Modern Physics* **87**, 803–54 (2015).
[70] David M. Lee, "The Extraordinary Phases of ³He," *Reviews of Modern Physics* **69**, 645–65 (1997).

wise interatomic forces. Condensed matter physicists tended to see liquid ^{3}He as the prototype of a quantum system described by Lev Landau's Fermi liquid theory.

The fermion character of ^{3}He precluded the possibility of Bose-Einstein condensation into a superfluid phase. However, as the 1950s ended, it occurred to a few individuals that ^{3}He might become a superfluid by another mechanism. Anderson was one of these people and he asked his Princeton graduate student Pierre Morel (the same person who worked on the "bad actor" super-conductors discussed in Chapter 9) to generalize the BCS theory of superconductivity to the case when the two particles of a Cooper pair orbited one another and therefore possessed a non-zero *orbital angular momentum, L*.[71] If the two Cooper pair particles were neutral ^{3}He atoms rather than electrically charged elec-trons, the final BCS many-atom wave function would exhibit superfluidity rather than superconductivity.

Morel was just starting to get results when the nuclear theor-ist Keith Brueckner happened to visit Bell Labs. To his surprise, Anderson learned that Brueckner was doing very similar work with one of his graduate students. A brief negotiation led to a 1960 paper co-authored by all four of them.[72] By their estimate, liquid ^{3}He would enter a BCS-like cooperative state below a critical tem-perature of about $T_C = 0.1$K. Simultaneously and independently, a student of Lev Landau and two nuclear/particle theorists at Berkeley made similar predictions.[73] A detailed follow-up paper

[71] The magnitude of the classical angular momentum vector *L* for two par-ticles with mass *m* orbiting one another at a distance *r* with speed *v* is $L = mvr$. The direction of *L* is perpendicular to the plane of the orbit. BCS assumed that the two electrons of a Cooper pair do not orbit one another. The resulting pair wave function is spherically symmetric with $L = 0$.

[72] K.A. Brueckner, Toshio Soda, Philip W. Anderson, and Pierre Morel, "Level Structure of Nuclear Matter and Liquid ^{3}He," *Physical Review* **118**, 1442–6 (1960).

[73] L.P. Pitaevskii, "On the Superfluidity of ^{3}He," *Soviet Physics JETP* **37**, 1267–75 (1960).

V.J. Emery and A.M. Sessler, "Possible Phase Transition in Liquid ^{3}He," *Physical Review* **119**, 43–9 (1960).

by Anderson and Morel discussed hypothetical superfluids with different values for the Cooper pair angular momentum. They also lowered their estimate of the superfluid transition temperature for ^{3}He to $T_C = 0.02$K.[74]

The 1960s came and went with no evidence for a superfluid phase of ^{3}He. Research focused mostly on its Fermi liquid character.[75] Anderson played almost no role in this activity.[76] He also did not join the many solid-state theorists from this period who devoted their efforts to detailed studies of the properties of metals.[77] His main focus was superconductivity and a magnetic impurity problem discussed in Chapter 11 that the Swedish Academy of Science later singled out (along with disorder-induced localization) when they chose Anderson to share in the 1977 Nobel Prize.

Then, in July 1972, Anderson was struggling to endure a deadly dull conference (pinball was the only recreation at the conference venue) when big news began to spread through the assembled physicists. Experimenters at Cornell University had discovered not one, but two distinct superfluid phases of ^{3}He (called the A phase and the B phase) at temperatures below 0.003 K.[78]

Anderson and his Bell Labs colleague Chandra Varma quickly offered some speculations about the two superfluid phases.[79]

[74] P.W. Anderson and P. Morel, "Generalized BCS States and the Proposed Low-Temperature Phase of Liquid ^{3}He," *Physical Review* **123**, 1911–1934 (1961).

[75] J.C. Wheatley, "Experimental Properties of Pure ^{3}He and Dilute Solutions of ^{3}He in superfluid ^{4}He at Very Low Temperatures. Application to Dilution Refrigeration," in *Progress in Low Temperature Physics*, volume VI, edited by C.J. Gorter (North-Holland, Amsterdam, 1970), pp. 77–161.

[76] The lone exception was a paper Anderson published in his own journal: P.W. Anderson, "Does Fermi Liquid Theory Apply to ^{3}He?," *Physica* **2**, 1–3 (1965).

[77] See, e.g., A.A. Abrikosov, *Fundamentals of the Theory of Metals* (Academic Press, New York, 1972).

[78] D.D. Osheroff, W.J. Gully, R.C. Richardson, and D.M. Lee, "New Magnetic Phenomena in Liquid ^{3}He below 3 mK," *Physical Review Letters* **29**, 920–3 (1972).

[79] P.W. Anderson and C.M. Varma, "Properties of a Possible Superfluid State of ^{3}He," *Nature* **241**, 187–9 (1973).

Simultaneously, British theorist Anthony Leggett published the first of several papers on the subject. Phil and others soon recognized that Leggett had analyzed the situation correctly and imaginatively. Among other things, he took a page from Anderson's playbook when he suggested that many of the unusual properties of ^{3}He were a consequence of what he called *spontaneously broken spin-orbit symmetry*.[80] This meant that the angle between the orbital angular momentum vector L and the spin vector S was exactly the same for every Cooper pair in the superfluid.

Comparison with experiment convinced Leggett that the Cooper pairs in the A phase were one of the varieties of pairs that Anderson and Morel (AM) had studied a decade earlier. The Cooper pairs in the B phase were different and Leggett soon appreciated that they had a form AM had missed, but which Roger Balian and Richard Werthamer (BW) had described a few years later.[81] The entire story was quite lovely except that the BW state always had the lowest energy. Why did the AM state show up at all?

Anderson discussed this puzzle with Bill Brinkman, a Bell Labs theorist then serving as the Head of the Infrared Physics and Electronics Department. Brinkman recalled a paper he had read which exploited a positive feedback mechanism for a related problem. Phil did a quick-and-dirty calculation and concluded that a spin-feedback mechanism could lower the energy of the AM state relative to the BW state. He wrote up a draft manuscript, dropped it on Brinkman's desk with the scrawled remark "let's publish!", and promptly left for Cambridge.[82]

[80]   A.J. Leggett, "Interpretation of Recent Results on ^{3}He below 3 mK: A New Liquid Phase?," *Physical Review Letters* **29**, 1227–30 (1973).

[81]   R. Balian and N.R. Werthamer, "Superconductivity with Pairs in a Relative p Wave," *Physical Review* **131**, 1553–64 (1963). Richard Werthamer was one of Phil's Bell Labs colleagues at the time. Very similar work published in Russian by Yu. A. Vvodin was little noticed in the West.

[82]   Interview of William Brinkman by the author, March 19, 2016.

Brinkman was dubious about co-authoring a paper with Phil until he had checked his senior colleague's assertions using his own methods. Brinkman's first detailed calculations contradicted Phil's conclusions and he made an anxious telephone call to Cambridge to break the bad news. Phil listened, but didn't ask any questions. He then made a suggestion for a revised calculation. Brinkman did so and his final results showed that Anderson's mechanism did indeed provide a natural explanation for the appearance of the A phase in the Cornell experiment. Today, their joint work has the status of a classic in the field.[83] Brinkman always wondered how Phil was able to intuit the flaw with his original, rather complicated calculation and suggest the path to success. Other Anderson collaborators would have the same experience in the future.

In 1996, the members of the Swedish Academy of Sciences recognized the discovery of superfluidity in ^{3}He when they awarded a Nobel Prize to the Cornell experimental team. They followed up in 2003 by awarding a share of that year's Nobel Prize to Tony Leggett for his elegant theoretical contributions to the subject. Leggett's Nobel lecture characteristically emphasized the contributions of others. Specifically, he noted, "it is impossible not to mention one name in particular, that of Phil Anderson, who with various collaborators contributed so many vital insights during these years and later."[84]

Years later, Anderson recalled with fondness the ^{3}He phase of his life. In his memory, it was "very competitive in experiment and in theory, but the competition remained friendly and

[83] P.W. Anderson and W.F. Brinkman, "Anisotropic Superfluidity in ^{3}He: A Possible Interpretation of Its Stability as a Spin-Fluctuation Effect," *Physical Review Letters* 30, 1108–11(1973); P.W. Anderson and W.F. Brinkman, "Theory of the Anisotropic Superfluidity in ^{3}He," in *The Helium Fluids*, edited by Jonathan G.M. Armitage and Ian E. Farquhar (Academic Press, London, 1975), Chapter 8. Because of Brinkman's contributions, what we have called the AM phase is usually called the ABM phase in the helium literature.

[84] Anthony J. Leggett, "Superfluid ^{3}He: The Early Days as Seen by a Theorist," Nobel Lecture, December 8, 2003.

constructive."[85] A similarly very competitive atmosphere arose in the field of high-temperature superconductivity in the late 1980s and 1990s. Unfortunately, we will see that Anderson's contribution to that competition was not always friendly and constructive (see Chapter 14).

Port Isaac

In 1975, Phil and Joyce began to tire of the rootlessness associated with their bi-continental life. They might have remained permanently in England, but Phil's continued unhappiness with Brian Pippard's curriculum reforms did not bode well for the future. An opportunity appeared for a full-time professorship at Harvard, but it foundered on the details. Therefore, he retained his half-time appointment at Bell Labs and negotiated a half-time appointment at Princeton University, which lay only one hour south of the Labs.

The Andersons did not walk away empty-handed from the country they had grown to love. Inevitably, the experience strengthened their prior tendency to view world affairs from a broader perspective than most Americans. Their table manners improved and they relished the company of anyone who would listen to them talk about church-crawling, brass-rubbing, or moor-walking.

Most important, for thirty years beginning in 1973, they owned a vacation home in the small English village of Port Isaac (Figure 10.6). The village (pop. 1000) sits in a steep, green valley on the north shore of Cornwall, a stunningly beautiful peninsula in the far southwest of Great Britain. D. H. Lawrence called it "very

[85] Interview of PWA by P. Chandra, P. Coleman, and S. Sondhi on October 15, October 29, and November 5, 1999, Niels Bohr Library & Archives, American Institute of Physics, College Park, MD.

Figure 10.6 The village of Port Isaac, Cornwall, UK where Phil and Joyce Anderson owned a vacation home from 1973 to 2003. Source: Susan Anderson.

primeval: great, black, jutting cliffs and rocks, like the original darkness, and a pale sea breaking in, like dawn...”[86]

Port Isaac enchanted the Andersons with its narrow winding streets, whitewashed houses, and spectacular views. Every day, they could watch fishermen unload their catches near the town center, continuing a trade that began there in the Middle Ages. Some readers will recognize the village as the setting for the long-running British television series *Doc Martin*.

The Anderson cottage perched at the edge of a steep cliff over-looking the Atlantic Ocean. It was less than 1000 square feet in area, but there was a roomy kitchen, a living room with a fire-place, and three small bedrooms on the second level. When the weather was good and the tide was low, they followed paths down to the ocean and indulged their passion for geology by col-lecting rocks from beaches up and down the coast.[87] Whenever possible, they hiked along the Southwest Coast Path and paddled a canoe in the nearby Fowey and Camel rivers. Daughter Susan joined in on the occasions of her visits.

---

[86] Quoted in Elizabeth Neuffer, “Along the Rugged Rim of Cornwall,” *New York Times*, May 26, 1985, p. 16.

[87] Author correspondence with Christopher Key, December 17, 2015.

When rain and frequent gale force winds from the ocean made hiking dangerous, the Andersons caught up with their reading, wrote letters, listened to recordings of classical music, and played board games. Visitors came often and, as had happened in New Jersey and Cambridge, the couple gained a reputation as graceful hosts with a penchant for martinis. When they finally sold the cottage in 2003, they returned their entire collection of Atlantic rocks to the ocean.

11

Hidden Moments

Swedish theoretical physicist Per-Olov Löwdin introduced the physics laureates at the 1977 Nobel Prize award ceremony.[1] His task was to summarize the work of Philip Anderson, Nevill Mott, and John Van Vleck in terms a layperson could understand. This called for a metaphor and Löwdin chose to compare the electrons in a solid to the dancers in a ballet.

Anderson's research presented a challenge because he had been recognized for two quite different achievements: the discovery of disorder-induced localization and a theory of magnetic impurities in metals. For the first of these, Löwdin imagined the principal dancers brought to a halt by disorder in the *corp de ballet*.[2] For the second, he spoke of the exchange interaction between electrons as a *pas de deux* and identified the hero of the ballet as "a metal atom with strong personal magnetism whose special properties may vary strongly with the environment."

Löwdin's description aside, the consensus among solid-state physicists is that Anderson's magnetic impurity research is first rate, but not so brilliantly novel or as deeply transformative as his discovery of disorder-induced localization. Accordingly, this chapter treats his magnetic impurity work as merely the first step of a journey. The journey begins with his attempt to understand some peculiar experimental results. It concludes with his anticipation of an entirely new technique of theoretical physics called the *renormalization group*. At the end, we speculate about why the

---

[1] Per-Olov Löwdin, Award Ceremony Speech, Nobel Prize for Physics, 1977.
[2] The *corps de ballet* are members of a ballet troupe who dance as a group.

Nobel committee chose to honor Anderson for two pieces of research that were not only unrelated as physics problems, but unequal in their exceptionality.

Magnetic Moments in Metals

Anderson's Nobel-honored magnetism research posed a question that is fundamental to the existence of all magnetic matter. Does a magnetic atom retain its magnetism when it becomes part of a solid? Nature's answer to this question is yes when the solid is an electrical insulator. The answer is much less clear when the solid is an electrical conductor.

Physicists say that an atom with a net spin S possesses a non-zero *magnetic moment*. Immersing this "magnetic atom" in a non-magnetic metal raises several questions. Will the unpaired spins hop off the atom and disappear into the conduction band of the metal? Will spins from the metal hop onto the atom and pair with antiparallel spins to yield zero net spin? In either case, the immersed atom loses its magnetic moment and thus its ability to create a magnetic field.

A few months before he began his sabbatical in Cambridge, Anderson invented a mathematical model to study the fate of a magnetic atom immersed in a non-magnetic metal. An exact solution of the model was beyond his capabilities, so he used the Hartree–Fock approximation to calculate measurable quantities. The satisfactory agreement he found with existing experimental data led him to construct a consistent and physically reasonable account of magnetic impurities in metals.

Anderson was aware that an increasing number of his theoretical colleagues were solving many-electron problems using the new methods of quantum field theory. This meant that a competitor might soon use an approximation superior to Hartree–Fock to study the magnetic moment problem. He wondered in his published paper whether "a real many-body theory would give answers radically different" from his own.[3] He guessed that it would not and that only numerical modifications were likely.

Subsequent events showed that Anderson's guess was *wrong*. His model did a good job of describing the basic physical situation. It also made a prediction that subsequent analyses never contradicted: an immersed magnetic moment tends to align the spins of nearby conduction electrons antiparallel to its own spin. Be that as it may, the Hartree–Fock approximation was simply not adequate to capture all the physics implied by the model.

The interaction of a magnetic atom with a sea of delocalized electrons turned out to be a problem of very great subtlety. Accordingly, his model became a touchstone and testing ground for virtually every new development in the exploding enterprise of many-body theory. As one textbook writer later put it, "some of the great conceptual advances in theoretical physics had their inception in [this] seemingly modest subject.... Anderson opened an entirely new field of investigation."[4]

The Anderson Impurity Model

Like 60 percent of his roughly 500 publications, Anderson's paper on magnetic moment formation in metals has no co-authors.[5] However, that does not mean that no other physicists played a role in its genesis. At least three independent influencers led Anderson to create the "impurity model" which today bears his name.

Initially, Bernd Matthias challenged him to explain the results of his experiments on magnetic impurities in superconductors. Then, the generality of Nevill Mott's ideas about Coulomb-induced electron localization crystallized in his mind as he

[3] P.W. Anderson, "Localized Magnetic States in Metals," *Physical Review* **124**, 41–53 (1961).

[4] Daniel C. Mattis, *The Theory of Magnetism I* (Springer-Verlag, Berlin, 1981), p. 35.

[5] P.W. Anderson, "Local Moments and Localized States," *Reviews of Modern Physics* **50**, 191–201 (1978).

struggled for a deeper understanding of his earlier work on super-exchange. Finally, he discovered the highly relevant work of the French theoretical physicist, Jacques Friedel.

Matthias' Influence

Phil Anderson and Bernd Matthias had become good friends at Bell Labs. Anderson found himself drawn to Matthias' powerful personality and unconventional nature. Matthias (Figure 11.1) was brash, impertinent, often very critical, and highly skilled at getting people to give him the things he wanted.[6] But he also had the ability to connect with people at a deep and personal level. As Phil put it:

> With his dark shock of hair, brilliant eyes, pale face, and slim, wiry, build there was always a kind of electricity in the complete

Figure 11.1 Bernd Matthias circa 1960. Source: American Institute of Physics Emilio Segrè Visual Archive.

---

[6] Paul R. Stein, "Bernd Matthias. A Personal Memoir," *Los Alamos Science* **2**, 87–95 (1982).

focus of his attention on the individual to whom he was speaking.[7]

All observers agree that Matthias' joy at discovering something new was particularly unabashed when his discovery owed nothing to theoretical input. Perhaps that is why Anderson once characterized his relationship with Matthias as "(mostly) love (some) hate."[8]

Matthias had searched for new ferroelectric materials when he first came to Bell Labs. After switching his research focus to superconductivity, he began a program to discover new superconductors as a way to study their systematics. As part of this effort, Matthias studied dilute alloys made by introducing small concentrations of magnetic impurity atoms into known elemental superconductors. He measured the magnetic moments of the immersed atoms and discovered that the impurity atom retained its magnetic moment in some cases and lost its magnetic moment in other cases. It was quite mysterious why any particular impurity/host system behaved the way it did. For the physics of magnetism, Matthias's data presented a puzzle every bit as striking as George Feher's data had presented a puzzle for the physics of doped semiconductors.

Mott's Influence

Not every moment of Anderson's 1958 summer visit to Charlie Kittel's research group at Berkeley (see Chapter 9) was devoted to superconductivity. For relaxation one day, Anderson read an experimental paper that related the electrical properties of transition metal oxides to their magnetic properties. These oxides were the materials to which he had applied the concept of superexchange almost ten years earlier. The new experiments reported

---

[7] P.W. Anderson, "BCS and Me," in *More and Different: Notes from a Thoughtful Curmudgeon* (World Scientific, Hackensack, NJ, 2011), p. 10.

[8] P.W. Anderson, "Commentary on 'Superconductivity (Two Opinions)'," in *A Career in Theoretical Physics* (World Scientific, Singapore, 1994), pp. 131–2.

that the conducting oxides were always ferromagnetic while the non-conducting (insulating) oxides were always antiferro-magnetic.[9]

Anderson knew this could not be an accident. Instead, it was a challenge (like localization) to once again *follow the data* and hopefully discover a big unifying idea about the oxides that everyone, including he, had missed. Antiferromagnetism was his specialty, so he thought first about insulators. An insulator is a semiconductor with a very large band gap (see Figure 7.5), but this information did not spark a new idea. On the other hand, he knew well another reason why a crystal might not conduct electricity.

Nevill Mott was the prophet of a new religion based on a single credo: electrons do not hop from atom to atom through a crystal if the electrostatic energy cost to doubly occupy an atomic orbital is too great. Anderson wondered if he could relate the physics of the Mott insulator to the superexchange mechanism for antiferromagnetism.

A clue came from Leslie Orgel, a theoretical chemist from the University of Cambridge who happened to be in Berkeley to give a seminar. Orgel spoke about molecules where non-magnetic ions surround a transition metal ion.[10] He emphasized that one should treat the entire molecule as a single, nearly covalently bonded entity. This was contrary to the thinking of most physicists at the time who treated the ions as point-like electric charges and ignored the possibility of chemical bonding.[11]

In a eureka moment, Anderson realized he could exploit Lev Landau's distinction between a particle and the "quasiparticle" to which it evolves when one takes account of its interactions with all the other particles in its environment. For a transition metal oxide, the relevant particles were the spins of the transition metal

    [9] R.R. Heikes, "Relation of Magnetic Structure to Electrical Conductivity in NiO and Related Compounds," *Physical Review* **99**, 1232–4 (1955).

    [10] J.S. Griffith and L.E. Orgel, "Ligand Field Theory," *Quarterly Reviews of the Chemical Society* **11**, 381–93 (1957).

    [11] Author correspondence with John B. Goodenough, November 13, 2015.

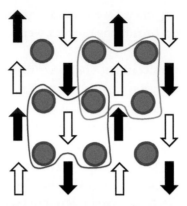

Figure 11.2 Cartoon of a transition metal oxide repeated from Figure 6.6. An oxygen ion (blue circle) sits midway between every pair of next-nearest neighbor transition metal ions. The spins of the latter have the same color (black or white), lie along diagonals in the diagram, and are oriented antiparallel to each other. The red and orange curves enclose neighboring spin quasiparticle complexes where O ions surround the spins on the transition metal ions.

ions (arrows in Figure 11.2). The corresponding "spin quasiparticles" were complexes (closed curves in Figure 11.2) composed of these spins interacting fully with the electrons and oxygen ions in their immediate vicinity.

With this definition, Anderson was able to show that the cost for a spin quasiparticle to hop between adjacent complexes was exactly the Coulomb energy cost U singled out by Mott. Moreover, treating this hop in perturbation theory (as he had done for the localization problem) *always* generated an antiferromagnetic superexchange interaction between the spins at the center of adjacent spin quasiparticle complexes. As he later put it, these insulators are antiferromagnets as "a consequence of their frustrated attempt to become metals [by electron hopping]."[12]

The summer ended with Anderson sensitized to the generality of Mott's mechanism. It was relevant for donor localization in

---

[12] P.W. Anderson, "Some Memories of Developments in the Theory of Magnetism," *Journal of Applied Physics* **50**, 7281–4 (1979).

doped semiconductors; it was relevant for superexchange in transition metal compounds; perhaps it was relevant for other problems.

Friedel's Influence

In the fall of 1959, Anderson attended a small magnetism meeting at Oxford University. He travelled to England via Paris, where he met Jacques Friedel, a seventh-generation scientist who had earned his PhD at the University of Bristol with Nevill Mott.[13] Anderson never took very long to decide if a physicist he had just met was worth his time and respect. Friedel easily met his standard and he paid close attention to his host's summary of his research successes. He was particularly interested to learn about Friedel's simple and elegant method to calculate the properties of metallic alloys.[14]

The French physicist had studied the case when the energy of a localized state of an immersed atom happened to coincide with the energy of a delocalized conduction band state of the host metal. Quantum mechanics required that the wave functions of these two states mix. As a result, Friedel found that the narrow energy level of the atomic state acquired an energy width Γ, much like the narrow spectral lines in Figure 4.3 acquired widths due to collisions.

Friedel's PhD student André Blandin spoke at the Oxford meeting about a model he and his advisor had developed to describe the formation of magnetic moments in dilute alloys.[15]

[13] Adrian P. Sutton and Olivier Hardouin Duparc, "Jacques Friedel: 11 February 1921—27 August 2014," *Biographical Memoirs of Fellows of the Royal Society* **61**, 3–21 (2015).

[14] J. Friedel, "On Some Electrical and Magnetic Properties of Metallic Solid Solutions,", *Canadian Journal of Physics* **34**, 1190–211 (1956).

[15] Author correspondence with Anthony Arrott, April 19, 2018. A. Blandin and J. Friedel, "Magnetic Properties of Dilute Alloys. Magnetic and Antiferromagnetic Interactions in Alloys of Noble Metals and Transition Metals" (in French), *Le Journal de Physique et le Radium* **20**, 160–8 (1959).

They identified the width Γ and the exchange energy J as the two most important parameters in the problem. The latter is the difference in energy between two parallel atomic spins and two antiparallel atomic spins. Their main prediction was that a magnetic atom immersed in a metal retains its magnetic moment—the spin remaining near the impurity in the broadened atomic state—if the ratio J/Γ was large enough. That is, if the energy cost to reverse a spin in an atom was large compared to the energy broadening of the atomic state.

The Local Moment Paper

Anderson spent the first half of 1961 constructing, solving, and writing up his own theory of magnetic moment formation in metals.[16] As was his habit by then, he eschewed calculations for any particular material and instead invented a model Hamiltonian. The first term of this energy expression described the delocalized states of the host metal. The second term used a single atomic orbital to represent the immersed atom. The third term imposed a Coulomb energy cost U *à la* Mott if two electrons of opposite spin occupied the same atomic orbital.[17] The final term used an energy V to quantify the hopping of an electron between the localized orbital and a delocalized state of the same energy.

Anderson's parameter V determined the width Γ of the broadened atomic level and his Hartree–Fock analysis focused mostly on the ratio U/Γ. For an alloy where this quantity happened to be large, the electrostatic energy cost to doubly occupy the atomic state was large compared to the kinetic energy savings associated with the electron spreading out into the broadened version of that state. This guaranteed single spin occupancy for the atomic state and thus a non-zero localized magnetic moment.

---

[16] P.W. Anderson, "Localized Magnetic States in Metals," *Physical Review* **124**, 41–53 (1961).

[17] Quantum theory forbids two electrons with parallel spin to occupy the same atomic state.

Conversely, an alloy where the ratio U/Γ was small would exhibit no magnetic moment because two antiparallel spins happily co-occupied the atomic state. Anderson estimated U/Γ for different magnetic atoms in different host metals and found he could rationalize most of Matthias' experimental results.

The magnetic moment paper is one of the best written of all of Anderson's scientific papers. It introduces the problem using Matthias' experimental results, discusses previous theory on the subject, writes down the model Hamiltonian, gives a qualitative discussion of two special cases, performs a Hartree–Fock analysis, extracts the important conclusions, and points out the limitations of his approximation.

Anderson gives full credit to Friedel for the idea of the broadened atomic state. However, he makes clear that Friedel was incorrect to identify the intra-atomic exchange energy J as the driving force for localization of the magnetic moment. Instead, Phil supported his (by now) well-developed physical intuition with numerical estimates to argue that it was Mott's intra-atomic Coulomb energy U that was relevant.[18] U was generally much larger than J, a fact that was crucial if the model was to account for Matthias' experimental data.

A small but important point is that Anderson wrote his model Hamiltonian for the magnetic impurity problem in the language of *second quantization*. This is a formulation of many-particle quantum mechanics which focuses on the number of particles that occupy the available quantum states. Second quantization is the natural language of quantum field theory and, for that reason, practitioners of that art were drawn to Anderson's model as a challenge to their methods.[19] In fact, because he employed a simple but non-trivial second-quantized Hamiltonian, Anderson's paper attracted

[18] Interview of PWA by Lillian Hoddeson on July 13, 1987, Niels Bohr Library & Archives, American Institute of Physics, College Park, MD.

[19] Silvan S. Schweber, *QED and the Men Who Made It: Dyson, Feynman, Schwinger, and Tomonaga* (University Press, Princeton, 1994), pp. 23–38.

a much greater audience than all of Friedel's quantum scattering theory papers on the same subject combined.

The Kondo Effect

Anderson's magnetic impurity model might have suffered the fate of most good but not great scientific papers (read profitably by a few and then forgotten) if not for a phenomenon known as the *Kondo Effect*. Chapter 7 pointed out that an important source of electrical resistance in metals comes from the collisions of electrons with ions displaced from their normal lattice position by small-amplitude vibrational motion. This source of resistance decreases as the temperature decreases until it reaches a minimum value at $T = 0$.

However, it had been known since the 1930s that the resistance of some dilute metal alloys exhibits a shallow minimum as the temperature decreases before *rising* to its final $T = 0$ value. In his widely used book, *Electrons and Phonons: the Theory of Transport Phenomena in Solids* (1960), John Ziman devoted several pages to this minimum and concluded that "its explanation is still one of the unsolved problems in the theory of metals."

This changed in 1964, when the Japanese physicist Jun Kondo surveyed the experimental data and concluded that the resistance minimum appeared only if the impurities in the alloy retained their magnetic moments. This led him to study the effect on the resistance of the perturbation $H_1 = -K\mathbf{S}\cdot\mathbf{s}$ where K is a constant, $\mathbf{S}$ is the spin of the impurity atom, and $\mathbf{s}$ is the net spin contributed by all the conduction electrons at the position of the impurity atom.

Kondo discovered that spin flip processes (where $\mathbf{S}$ and individual electron spins that contribute to $\mathbf{s}$ reverse directions simultaneously) produce a contribution to the electrical resistance which *increases* as the temperature decreases.[20] Adding this to the

[20] J. Kondo, "Resistance Minimum in Dilute Magnetic Alloys," *Progress of Theoretical Physics* **32**, 37–49 (1964).

usual disorder-induced resistance produced a total resistance with the desired shallow minimum. Importantly, this happens only if $K < 0$ in the Kondo energy H_1 defined just above. That is, the interaction between the local moment and the conduction electrons must be a*ntiferromagnetic*.

Kondo's work was surprising and exciting. This was *not* because he had explained the resistance minimum—a fact which interested only a tiny group of specialists. The excitement came from the fact that Kondo's correction term became *infinite* at $T = 0$. This got the attention of solid-state physicists everywhere because zero temperature corresponds to the ground (lowest energy) state of a physical system and a true infinity cannot occur for a measurable quantity like electrical resistance. Kondo's spin-flip resistance was a red flag. Something very peculiar was going on with the ground state of a localized magnetic moment immersed in a metal.

In passing, it is worth noting that that this calculation was characteristic of Kondo's style of doing theoretical physics. Like Anderson, he began by engaging seriously with experimental data and identifying trends or anomalies that had not been appreciated by others. With a problem identified, Anderson's first impulse was always to invent a simple mathematical model and analyze it using just enough mathematics to indicate what he regarded as the essential physics. Kondo's papers in the 1950s and 1960s illustrate a different approach. With a problem identified, he typically followed the advice he later offered to young theorists faced with a new phenomenon: "take a standard model for the system and look for higher-order corrections to the standard calculation for the model."[21] This is advice the heterodox Anderson would never offer or follow.

However they were achieved, Kondo's results stimulated many people, not least Bob Schrieffer. In 1965, Schrieffer was 34 years old, a full professor at the University of Pennsylvania,

---

[21] J. Kondo, "Sticking to My Bush," *Journal of the Physical Society of Japan* **74**, 1–3 (2005).

and eight years past his great success as the 'S' in BCS. His soon-to-be-classic monograph *Theory of Superconductivity* had just appeared and he was casting around for a new problem to attack and conquer.[22]

Schrieffer's choice for a new problem was consequential because he was already regarded as an informal leader of the solid-state physics community. That is, his reputation and past achievements alone were sufficient to induce others to work on problems that he worked on. This is one of the ways that social control operates within the community of scientists.[23]

One person who paid careful attention to Schrieffer's activities was Marvin Cohen, then a beginning Assistant Professor at the University of California, Berkeley and later a National Medal of Science winner.[24] Cohen paid similar attention to Phil Anderson, but this was hardly surprising as he had just completed a postdoc job in the theory group at Bell Labs.

Anderson's status as an informal leader was taken for granted. After all, he had just won the 1964 Oliver E. Buckley Prize of the American Physical Society for his "outstanding contributions to solid-state physics."[25] Nevertheless, Anderson knew he lacked Schrieffer's easy charm and skill with interpersonal relations. Indeed, Schrieffer was the only younger theorist Anderson regarded as a potential rival to his own stature and authority in their community. Schrieffer's subsequent achievements and influence fully justified this assessment until tragedy silenced his voice.[26]

[22] J. R. Schrieffer, *Theory of Superconductivity* (W.A. Benjamin, New York, 1964).

[23] Warren O. Hagstrom, *The Scientific Community* (Basic Books, New York, 1965), pp. 184–5.

[24] Interview of Marvin L. Cohen by the author, March 16, 2016.

[25] The 1964 Buckley Prize cited Anderson for "his contribution concerning many-body and superexchange interactions, which have led to new theoretical insights into superconductivity, liquid ^{3}He, plasmons, and magnetism." He and Joyce spent the Buckley Prize money on a drilling rig to sink a well on their New Vernon, NJ property.

[26] See the contributions to *Selected Papers of J. Robert Schrieffer, In Celebration of His 70th Birthday*, edited by N.E. Bonesteel and L.P. Gor'kov (World Scientific, New Jersey, 2002). Schrieffer (1931–2019) was a superb researcher, teacher, and

Returning to the Kondo effect, Schrieffer focused on the fact that Kondo's resistance calculation *assumed* that an immersed impurity atom retained its magnetic moment. This motivated him and a collaborator to analyze Anderson's impurity model using a method superior to the Hartree–Fock approximation. They concluded that there were situations when the immersed atom would *not* retain its magnetic moment even though Anderson's calculation said it would. This happened because a cloud of antiparallel conduction electron spins gathered near the impurity atom and exactly compensated its spin.[27] In other words, the conduction electron spins completely screened (or hid) the impurity spin moment from the rest of the metal (see Figure 11.3).

Schrieffer's method was not exact and, like all methods available to many-body theorists at the time, it ran into serious problems as the temperature approached zero. Identifying the ground state of a Kondo system became the Holy Grail of solid-state physics. Between 1965 and 1970, over one hundred different theoretical physicists published papers on the subject.[28] Unfortunately, not one of them could give a reliable answer to the question of whether the ground state possessed a local magnetic moment or not. Anderson wrote at the time that theorists were "wandering in the "wilderness.""[29]

research mentor. He served from 1984 to 1989 as the second Director of the Institute for Theoretical Physics at the University of California, Santa Barbara. He also waged a lifelong battle with bipolar disorder. In 2004, a fateful decision to stop taking his medications led to an automobile accident, a conviction for vehicular manslaughter, and a two-year prison sentence. Alan J. Heeger, *Never Lose Your Nerve* (World Scientific, Singapore, 2015), pp. 237–8.

[27] J.R. Schrieffer and D.C. Mattis, "Localized Magnetic Moments in Dilute Metallic Alloys: Correlation Effects," *Physical Review* **140**, A1412–A1419 (1965).

[28] Google Search. February 24, 2018. See also A.C. Hewson, *The Kondo Problem to Heavy Fermions* (Cambridge University Press, Cambridge, 1993), Chapter 3.

[29] P.W. Anderson, "Kondo Effect III: The Wilderness—Mainly Theoretical," *Comments on Solid-State Physics* **3** (6), 153–8 (1971).

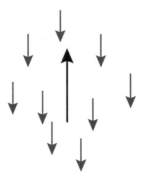

Figure 11.3 A cloud of antiparallel conduction electron spins (red) surround and cancel the magnetic moment produced by an impurity atom spin (black).

Out of the Wilderness

Anderson's path out of the Kondo wilderness was long and winding. It also exploited a sequence of mathematical models not obviously related to each other or to the problem he was trying to solve. This is not unusual when a creative physicist grapples with a difficult problem. Three people played important roles in his journey: John Hopfield, Donald Hamann, and Gideon Yuval.

Hopfield was a theorist ten years Anderson's junior who had been a postdoc at Bell Labs and a faculty member at Berkeley before moving to Princeton. His expertise was the optical properties of semiconductors and insulators. Years after the events described below, Anderson called Hopfield a "hidden collaborator" on the Kondo project and admitted he should have offered him co-authorship on several of the papers where he is only thanked.[30]

This admission twenty-five years after the fact is notable because Anderson was famously sensitive about receiving proper

---

[30] P.W. Anderson, *A Career in Theoretical Physics* (World Scientific, Singapore, 1994), p.281. Thanks to Hopfield appear at the end of four of Phil's papers on the Kondo effect.

credit for physics work he had done. More than once, he sent letters to the editors of scientific magazines to correct the record.[31] He was livid in a similar way when he was not mentioned in a series of posters depicting "A Century of Physics" prepared by the American Physical Society in 1999 to help celebrate its centennial.[32] On the other hand, a visitor to Anderson's home in the early 2000s was surprised when his host had to pull out a banker's box stashed in a closet to display the physical evidence of his scientific achievements (e.g., diplomas given for honorary degrees, memberships of learned societies, and even a facsimile of the Nobel Prize medal).

Don Hamann was hired as a postdoc at Bell Labs at the beginning of the wilderness period. He had trained as a many-body theorist, so it was natural for him to work on the Anderson and Kondo models. Nevertheless, he and Anderson did not collaborate actively until Hamann gained permanent staff status at the Labs. Years later, Hamann drew Phil's ire when he abandoned pencil-and-paper many-body theory in favor of large-scale computer calculations designed to understand the behavior of specific materials systems.[33]

Gideon Yuval was one of Anderson's first PhD students at Cambridge. A self-described "typically aggressive Israeli," Yuval was impressed by his advisor's intellectual arrogance—which he felt was fully justified—and worked to channel it to his advantage during the two years they worked together.[34] Yuval had a reputation among his fellow students as a hyperkinetic wild man who never minced words with anyone, including the professors. Much later, Anderson remembered that:

[31] Two examples of Anderson's strong interest in receiving credit for his work are: P. W. Anderson, Letter to the Editor, *Science* **308**, June 3, 2005, p. 1412, and P.W. Anderson, Letter to the Editor, *Mosaic* **19**(2), Summer 1988, p. 53.
[32] Interview of Brian Schwarz by the author, March 4, 2015.
[33] Interview of Donald Hamann by the author, March 17, 2016.
[34] Interview of Gideon Yuval by the author, September 16, 2015.

> I got Gideon and he was good, although difficult. He believed that
> his role in life was to have the ideas and my role in life was to write
> them down. I saw things differently.[35]

Soon after earning his PhD, Yuval switched from physics to computer science and he enjoyed a long career as an algorithm expert at Microsoft Corporation.

Anderson's Kondo odyssey began in June 1967 when Gerald Mahan, a former PhD student of John Hopfield, presented a seminar at Bell Labs about x-ray absorption by metals. This problem is a higher-energy cousin of the microwave absorption problem Anderson had worked on for his PhD thesis. Mahan limited himself to photons with just enough energy to excite an atomic inner shell electron to an empty conduction band state just above the Fermi energy.[36] The approximate absorption spectrum Mahan calculated was very peculiar and reflected the fact that a single absorption event perturbs an enormous number of conduction band electrons.

Anderson realized that Mahan's x-ray absorption scenario was an ideal setting for a theoretical deduction he had just published and applied to the Kondo problem.[37] The possibility of a connection between the two problems was tantalizing. Could one repurpose advances made in the theory of x-ray absorption to get closer to the solution of the Kondo problem?

The answer came in May 1968 when Anderson's old friend Philippe Nozières and a colleague produced an exact solution to the x-ray absorption problem. This difficult task was made even

[35] Interview of PWA by P. Chandra, P. Coleman, and S. Sondhi on October 15, October 29, and November 5, 1999, Niels Bohr Library & Archives, American Institute of Physics, College Park, MD.

[36] See Figure 9.1. Only x-ray photons supply enough energy to excite an electron from an atomic inner shell state to the Fermi level of a metal.

[37] P.W. Anderson, "Infrared Catastrophe in Fermi Gases with Local Scattering Potentials," *Physical Review Letters* **18**, 1049–51 (1967); P.W. Anderson, "Ground State of a Magnetic Impurity in a Metal," *Physical Review* **164**, 352–9 (1967).

more difficult by the angst the two scientists felt over the extreme civil unrest that was convulsing France at just at that moment.[38] Barely a month earlier, Phil experienced similar angst when riots broke out in African-American communities in the United States in the wake of the assassination of Martin Luther King, Jr. He and Joyce had admired Dr. King enormously and they sympathized whole-heartedly when he expanded his fight against racism to include issues of economic justice and opposition to the war in Vietnam.[39] The civil unrest continued in late August when police attacked anti-war protestors at the Democratic National Convention in Chicago. It was a relief to the Andersons when fall came and it was time to return to England.

Back in Cambridge, Anderson and Yuval studied the French solution of the x-ray absorption problem and realized that a Mahan-type disturbance of the sea of conduction electrons occurred every time a conduction electron and the Kondo impurity flipped each other's spins. As John Hopfield put it, the Kondo problem of a magnetic impurity in a metal was related to an *iterated* version of the x-ray absorption problem.[40]

A breakthrough occurred when Phil and Gideon recognized a formal correspondence between the mathematics that described the *quantum* behavior of the Kondo problem and the mathematics which described the *thermal* behavior of a one-dimensional array (or chain) of spins with ferromagnetic Heisenberg exchange between every pair of spins.[41]

[38] P. Nozières, This Week's Citation Classic, *Current Contents* **32**, 20 (1984). Nozières did his PhD at Princeton with David Pines and spent the summers of 1956 and 1957 at Bell Labs.

[39] Taylor Branch, *At Canaan's Edge: America in the King Years 1965–1968* (Simon & Schuster, New York, 2006).

[40] J.J. Hopfield, "Infrared Divergences, X-Ray Edges, and All That," *Comments on Solid State Physics* **2**, 40–9 (1969).

[41] G. Yuval and P.W. Anderson, "Exact Results for the Kondo Problem: One-Body Theory and Extension to Finite Temperature," *Physical Review B* **1**, 1522–8 (1970).

By now, Don Hamann was an active collaborator and the team of Anderson, Yuval, and Hamann (AYH) introduced a two-step iterative process to help them determine the ground state of the Kondo model.[42] The first step was to reduce the total energy scale of the spin chain as would necessarily be the case at low temperature. They did this by removing the most energetically costly pairs of spins from the chain and calculating the changes in the chain's properties produced by their removal.

Next, AYH showed that the same changes occurred if they scaled the original spin system by renormalizing (altering) the parameters which defined the chain. The relationship between the spin chain and the Kondo model implied that repeating these two steps over and over generated "scaling" equations which described the variation of the Kondo model parameters with decreasing energy scale and thus decreasing temperature.[43]

The equations derived by AYH showed that reducing the energy scale of the spin chain always renormalized the Kondo exchange parameter K to larger values. This led them to conjecture that the $T=0$ ground state of the Kondo model corresponded to the limit of infinite exchange. Since Kondo's model produced a resistance minimum for $K<0$ only, AYH interpreted an infinitely strong exchange to mean that a cloud of antiparallel conduction electron spins exactly compensated the impurity spin (see Figure 11.3).

In other words, the AYH scaling analysis confirmed the idea that a Kondo impurity was *always* non-magnetic at zero temperature. This conclusion did not contradict Matthias's experiments done at higher temperature because thermal energy could detach

[42] P.W. Anderson, G. Yuval, and D.R. Hamann, "Exact Results in the Kondo Problem. II. Scaling Theory, Qualitatively Correct Solution, and Some New Results on One-Dimensional Classical Statistical Models," *Physical Review B* 1, 4464–73 (1970).

[43] Interview of Donald Hamann by the author, March 24, 2015. P.W. Anderson, "Kondo Effect IV: Out of the Wilderness," *Comments on Solid State Physics* 5, 73–9 (1973).

the antiparallel spin cloud from the impurity and expose the magnetic moment. Phil's original Hartree–Fock theory of the impurity problem was simply not sophisticated enough to capture these subtleties.

The AYH paper is typical of the mature Anderson because it "ferrets out the physics qualitatively and then does some math to support his intuition."[44] Anderson was sensitive to the complexity of AYH and also to the fact that it mixed intuitive physical arguments with less-than-rigorous mathematics. On the other hand, the qualitative idea of successively reducing the energy scale to extract the zero temperature behavior appealed to him greatly. He resolved to recast this methodology directly in Kondo model language, without a detour through a spin chain.

One month before AYH appeared in print, Anderson delivered on this promise with the manuscript, *A Poor Man's Derivation of Scaling Laws for the Kondo Problem.*[45] The key idea was to progressively reduce the (energy) width of the conduction band and then renormalize the Kondo exchange parameter K in order to reproduce the effect of the missing states. The final scaling equations were the same as those found by AYH.

Anderson attended a low-temperature physics conference in Japan six months after he submitted his "Poor Man's" paper. A letter to Joyce praised the "infinitely improved" quality of the food since their visit to the country fifteen years earlier. Otherwise, he complained about spending the day at a long session devoted to the Kondo effect and remarked that "my work has been taken up by [others] and the problem is ready to be left."[46] As always, his interest was to identify the essentials. The details and implications could be left to others.

---

[44] Interview of John J. Hopfield by the author, October 13, 2015.

[45] P.W. Anderson, "A Poor Man's Derivation of Scaling Laws for the Kondo Problem," *Journal of Physics C: Solid State Physics* 3, 2436–41 (1970).

[46] Letter from PWA to Joyce Anderson, September 7, 1970. Courtesy of Philip W. Anderson.

The Renormalization Group

The scaling arguments presented in the AYH and Poor Man's papers were revolutionary and had no precedent in the literature of many-body physics. On the other hand, whether consciously or unconsciously, AYH made contact with ideas from particle physics and statistical physics, which later became part of a grand theoretical edifice called the *renormalization group*. That is why Michael Kosterlitz used the occasion of his 2016 Nobel Prize lecture to cite the "early version of the renormalization group" used by Anderson and his collaborators as very influential in his own research focused on phase transitions in two-dimensional systems.[47]

The renormalization group is a theoretical program designed to attack problems where many length scales and/or energy scales contribute to the physics. The program: (1) eliminates a group of variables from a theory; (2) renormalizes (alters) the theory's parameters to compensate for the loss of those variables; and (3) repeats the process. At each step, the renormalized theory has an increasingly restricted domain of applicability, e.g., very small or very large distances, or very low or very high energies or temperatures. The physics emerges from an analysis of the sequence of parameter values obtained by the renormalization steps.

Renormalization ideas entered physics in the late 1940s with the discovery that suitable redefinitions of the mass and charge of the electron eliminated some infinities that had plagued the quantum theory of electrons and photons.[48] Later work showed that a more general renormalization procedure made it possible to study the high-energy and short-distance behavior of quantum field theory without grappling with the entire theory. The key observation was that certain quantities in the theory did not

---

[47] J. Michael Kosterlitz, "Nobel Lecture: Topological Defects and Phase Transitions," *Reviews of Modern Physics* **89**, 040501:1–7 (2017).

[48] Silvan S. Schweber, *QED and the Men that Made it: Dyson, Feynman, Schwinger, and Tomonaga* (Princeton University Press, Princeton, NJ, 1994), pp. 595–605.

change if practitioners multiplied them by scale factors and simultaneously renormalized the parameters of the theory.[49]

Scaling ideas appeared independently in statistical physics in the 1960s. The motivation there was to understand the behavior of a fluid in the vicinity of its "critical point" where, among other peculiarities, its vapor and liquid phases became indistinguishable. Remarkably, many of the peculiarities became explicable with the realization that a scaling and renormalization procedure could be applied to an equation which connected the temperature of the fluid to its density.[50]

Motivated by the fluid problem, a young theorist named Leo Kadanoff presented a novel analysis of a lattice of atomic spins coupled by nearest-neighbor Heisenberg exchange.[51] He first reduced the number of spin variables by grouping neighboring spins together and replacing them by a single "block spin" to represent their average behavior. He then chose the exchange interaction between nearest-neighbor block spins to capture the same physics as the original spin model. From there, a bit of algebra permitted Kadanoff to draw quantitative conclusions about the thermodynamic behavior of his spin system when it passed from its disordered state to its ordered state.

The years 1970–1971 were a watershed when particle physicist Kenneth Wilson brilliantly generalized the scaling and renormalization ideas sketched above. In his hands, the renormalization group became a fully realized theory capable of addressing previously intractable problems in both quantum field theory and the statistical mechanics of phase transitions.[52] Several profound new

---

[49] M. Gell-Mann and F.E. Low, "Quantum Electrodynamics at Small Distances," *Physical Review* **95**, 1300–12 (1954).

[50] B. Widom, "Equation of State in the Neighborhood of the Critical Point," *Journal of Chemical Physics* **43**, 3898–905 (1965).

[51] Leo P. Kadanoff, "Scaling Laws for Ising Models Near T_c," *Physics* **2**, 263–72 (1966).

[52] See the contributions to *Ken Wilson Memorial Volume: Renormalization, Lattice Gauge Theory, the Operator Product Expansion, and Quantum Fields*, edited by Belal E. Baaquie,

ideas emerged, but a centerpiece remained the idea to progressively eliminate variables and renormalize model parameters to produce an "effective" theory that is valid at a particular scale and accurately describes phenomena occurring at that scale.

Wilson learned about the Kondo problem from a solid-state physics colleague and promptly developed a numerical version of his renormalization group method to attack it. He announced his solution at a June 1973 meeting of fifty (mostly) theorists from the worlds of solid-state and statistical physics.[53] Among other things, he demonstrated that the conclusion of AYH was correct. The magnetic moment of the Kondo impurity was zero at $T = 0$.

Phil Anderson attended this conference and, by pre-arrangement with the organizers, he offered a personal view of the meeting's highlights at its conclusion. His summary characterized Wilson's "magnificent solution" of the Kondo model as "an incredible intellectual feat."[54] He backed up this opinion a few years later by nominating Wilson for a Nobel Prize.[55] For his part, Wilson and two colleagues devoted 80 pages of a 1979 issue of the *Physical Review* to a renormalization group analysis of the Anderson impurity model.[56]

It is not always easy to identify the influences which lead a theoretical physicist to produce a particular result. However, when asked by a historian whether Anderson's work on the Kondo effect influenced him in any way, Wilson answered:

Kerson Huang, Michael E. Peshkin, and K.K. Phua (World Scientific, Singapore, 2015).

[53] K.G. Wilson, "Solution of the Spin-1/2 Kondo Hamiltonian," in *Collective Properties of Physical Systems*, edited by Bengt Lundqvist and Stig Lundqvist (Academic Press, New York, 1973), pp. 68–75.

[54] P.W. Anderson, "Conference Summary," in *Collective Properties of Physical Systems*, edited by Bengt Lundqvist and Stig Lundqvist, pp. 266–71.

[55] P.W. Anderson, Nomination for the Award of the 1979 Nobel Prize for Chemistry to Kenneth G. Wilson, Michael E. Fisher, and Leo P. Kadanoff, AT&T Archives, Warren, NJ. Wilson won an unshared Nobel Prize for Physics in 1982.

[56] H.R. Krishna-murthy, J.W. Wilkins, and K.G. Wilson "Renormalization Group Approach to the Anderson Model of Dilute Alloys. I & II," *Physical Review B* **21**, 1003–43 & 1044–83 (1980).

No. It came from my utter astonishment at the capabilities of the
Hewlett-Packard pocket calculator...I bought this thing and I
could not take my eyes off it. I had to figure out something
that...would somehow enable me to have fun with this calculator. At the same time...I learned about the Kondo problem and
discovered that it was very similar to the static model of the
nucleon that I had been working on for many years.[57]

Hewlett-Packard introduced its first pocket calculator, the HP-35,
on February 1, 1972. It is not known exactly when Wilson became
aware of the 1970 AYH and Poor Man's papers. But it seems that
their existence impelled him to push his method all the way
through to get explicit numerical results for the Kondo problem.[58] In the event, Anderson complained to a colleague that
Wilson gave short shrift to AYH in a major review article Wilson
wrote about the renormalization group in 1975.[59] A fair reading
does not support that complaint. Wilson refers to Anderson's
work a half-dozen times and he states (correctly) that "the
method developed here is more complex and more powerful
than Anderson's but the basic ideas are the same."

What about the influences that led Anderson to his pre-Wilson
renormalization group equations? The abstract of the Poor Man's
paper points out that the strategy to reduce the energy scale and
renormalize a parameter to capture the effect of the eliminated
high-energy processes appears already in his 1962 paper with
Morel on the "bad actor" superconductors. That idea originated
in early work by Bogoliubov's group, but neither they nor
Anderson and Morel thought to use renormalization to produce

[57] Interview of Kenneth G. Wilson conducted for the Physics of Scale project
by Babak Ashrafi, Karl Hall, and Sam Schweber, July 6, 2002, Gray, Maine.
Available on-line at https://authors. library.caltech.edu/5456/1/hrst.mit.edu/
hrs/renormalization/Wilson/index.htm.

[58] Interview of David R. Nelson by the author, October 22, 2015.

[59] Letter from PWA to Michael Fisher, August 4, 1998, Princeton University
Archives. The review article in question is Kenneth G. Wilson, "The
Renormalization Group: Critical Phenomena and the Kondo Problem," *Reviews
of Modern Physics* **47**, 773–840 (1975).

"a set of scaling laws connecting a given problem to ones with different parameter values" as AYH say in their paper. The place where this notion *does* appear is Leo Kadanoff's 1966 block spin scaling paper.

Anderson knew Kadanoff's paper well because he was the person who reviewed it and accepted it for publication in the journal *Physics* that he and Bernd Matthias edited. There is no reference to Kadanoff in Phil's Kondo papers, but it seems plausible that the idea to remove spin pairs and then renormalize the model parameters to take account for their absence was suggested, even if only subconsciously, by Kadanoff's spin removal and parameter renormalization methodology.

On the other hand, there is no suggestion in Kadanoff's paper to iterate this process and no notion of deriving scaling equations for the model parameters. Anderson and his coworkers deserve full credit for these innovations which, a bit later, Wilson developed independently and made important parts of his renormalization group program.[60] Today, the renormalization group is a standard technique learned by all advanced students of theoretical physics.

A Nobel Rumor

The Nobel selection committee broke seventy-five years of precedent when it named John Van Vleck, Philip Anderson, and Nevill Mott as the winners of their 1977 Prize for Physics. It was common to split the honor among two or three people. However, consistent with Alfred Nobel's specification of a prize for "an important discovery or invention", all previous co-winners had worked on

[60] Two pre-1970 papers that could have influenced AYH are A.M. Polyakov, "Microscopic Description of Critical Phenomena," *Soviet Physics JETP* **28**, 533–9 (1969) and C. Di Castro and G. Jona-Lasinio, "On the Microscopic Foundation of Scaling Laws," *Physics Letters A* **29**, 322–3 (1969). These were first available in March and May of 1969, respectively. However, neither seems to have influenced AYH, whose work was complete by September of 1969. Gideon Yuval and Donald Hamann, private communication.

the same physics problem. The people who shared the Prize were either collaborators, competitors, or individuals focused on limited parts of one well-defined larger problem. By contrast, the 1977 Prize went to three solid-state theorists for work done on two quite distinct problems. Anderson had worked on both problems, Van Vleck and Mott had each worked on one. As is their custom, the Nobel committee did not explain their decision.[61]

According to the selection committee's press release, Van Vleck was "the father of modern magnetism".[62] His former student Anderson "succeeded in explaining how magnetic moments can occur in metals." Anderson also "published a paper in which he showed under what conditions an electron in a disordered system can either move through the system as a whole, or be more or less tied to a specific position as a localized electron." Mott "called the attention of experimentalists to [Anderson's] paper" and "created a multitude of new concepts which have turned out to be central for the understanding of disordered materials."

Very soon, a rumor began to circulate that the selection committee had intended to break precedent even more dramatically.[63] Their plan was to award a one-third share of the Prize to each of three septuagenarians: John Van Vleck, Nevill Mott, and John Slater. All three were founders of solid-state physics and all three had made many important contributions. However, none of the three could claim a single, transformative achievement similar to those associated with theorists who had won unshared Nobel Prizes in the past. On the other hand, a strong case could be made for the group, particularly because Slater's work was complementary to the work of Van Vleck and Mott.

---

[61] Fifty years after the award of a Nobel Prize, the Royal Swedish Academy of Sciences releases information about the selection process for that Prize.

[62] Press Release: The Nobel Prize for Physics 1977. https://www.nobelprize.org/nobel_prizes/physics/laureates/1977/press.html

[63] Author interviews with Martin Blume, Marvin Cohen, James Phillips, David Pines, and Richard Zallen.

According to the rumor, the plan went awry when Slater died on July 25, 1976. This was two months after the selection committee's period of consultation had ended and two months before their short list was due. Posthumous awards of a Nobel Prize are forbidden by statute. Rather than split the prize equally between Van Vleck and Mott, the committee conceived the elegant solution of replacing Slater by Anderson. Not being a septuagenarian, they cited him for two unrelated and unequal pieces of work in order to connect him equally to his two co-winners.

It is unlikely that the selection committee had any qualms about their decision. At fifty-two years old, Anderson was on the cusp of winning a Nobel Prize anyway. With the advantage of hindsight, there is every reason to believe that the dramatic increase in interest in localization stimulated by the 1979 Gang of Four scaling theory of localization would have secured him at least a share in a future award.

12

From Emergence to Complexity

Phil Anderson's career divides into early, middle, and late periods (see Appendix). For the first twenty years (1947–1967), he worked mostly without collaborators and mostly on solid-state physics problems. Often, he began writing a paper long before he had worked out the mathematics needed to justify the conclusions he had reached intuitively. His subject matter during this period included spectral line shapes, ferroelectricity, magnetic resonance, superexchange, antiferromagnetism, symmetry breaking, localization, superconductivity, superfluid helium, the Josephson effect, and magnetic impurities in metals.

Anderson's middle period (1968–1986) began when he started splitting his time between industry and academia. He trusted his intuition more and more and he became less and less interested in supplying the mathematical details needed to convince others that his insights were correct. He turned that job over to students, postdocs, and senior collaborators who enjoyed doing calculations more than he did. The Gang of Four collaboration and his work on the Kondo problem, superfluid ^{3}He, and the spin glass (see later in this chapter) are characteristic of this period.

It was during this middle period that Anderson revealed another side of himself. He stepped out of his role as a solid-state physics researcher and sought to influence events and people on a larger scale. Personal politics motivated his first steps in this direction. He was driven to greater action by two beliefs, one old and one new. The old and festering belief was that federal funds for research were inequitably divided among the sciences. The new

and exciting belief was that a large class of non-physics problems—
both scientific and societal—could be attacked using methods
developed by physicists like himself.

This chapter and the next explore how Anderson responded to
the challenges posed by these beliefs. He came to understand (and
then to proselytize) that *emergence* and *complexity* were organizing
principles of great value for understanding problems where many
elements interact with one another in time and space. The force
of his arguments made him one of the leaders of an intellectual
mini-revolution focused on these questions.

Prestige Asymmetry

The social and political unrest that swept the United States in the
mid-to-late 1960s had a profound effect on nearly every adult
American. Joyce Anderson had been active in local Democratic
politics in New Jersey for years, but Phil chose to broadcast his
views widely for the first time only in 1966. Encouraged by the
Industrial and Research Scientists' Committee on Vietnam, he
was among the first signers of an open letter published in the *New
York Times* opposing the United States' involvement in that country.[1] Two years later, he added his name to a *Times* advertisement
supporting Senator Eugene McCarthy and his anti-war campaign
for President.[2]

As much as the war in Vietnam disturbed Anderson as a private
citizen, it had a greater effect on him as a card-carrying solid-state
physicist. The background is the post-World War II acclaim given
to physicists when their role in the development of the atomic
bomb and radar became widely known. New federal agencies
appeared expressly to provide financial support to basic physics

---

[1] Letter of March 17, 1966 from the Industrial and Research Scientists'
Committee on Vietnam, AT&T Archives, Warren, NJ. The open letter itself
(signed by 6500 persons) appeared as "On Vietnam," *New York Times*, June 5, 1966,
p. 207.

[2] "A Long List of McCarthy Supporters," *New York Times*, August 12, 1968, p. 26.

research (particularly at universities) and the level of that support grew every year.[3] The successful launch of the Sputnik satellite by the Soviet Union in 1957 and the attendant fear that the United States was falling behind in science opened the federal purse even more. Between 1955 and 1964, support for scientific research and development from Washington increased by 450%.[4]

The same period saw a slow increase in the involvement of the United States in the civil war in Vietnam. The cost of that involvement ballooned abruptly in 1965 when President Lyndon B. Johnson sent American combat troops to the region. Johnson did not want to disrupt the 1961 promise of President John F. Kennedy to put a man on the moon by the end of the decade and he did not want to cut back on his own Great Society social programs. As a result, scientists watched as the growth of federal spending for scientific research and development stagnated. Between 1965 and 1974, federal support for non-space related science increased by only 20%.

As an employee of Bell Labs and the University of Cambridge, Anderson did not suffer directly when the United States government put the brakes on its support of scientific research. That was not the case for many of his academic colleagues in solid-state physics. They felt badly squeezed and the reason was simple to identify. Over one-third of all federal funds for physics went to building and maintaining the high-energy accelerators used by experimental particle physicists to ply their trade. This was a source of friction because only ten percent of all American physicists worked in high-energy physics while twenty-five percent worked in solid-state physics.[5]

[3] Daniel J. Kevles, *The Physicists: The History of a Scientific Community in Modern America* (Harvard University Press, Cambridge, MA, 1995), Chapter 21.

[4] National Science Foundation, "Table A: Federal Obligations for Research and Development, by Character of Work, R&D Plant, and Major Agency: Fiscal Years 1951–2002."

[5] David C. Cassidy, *A Short History of Physics in the American Century* (Harvard University Press, Cambridge, MA, 2011), p.129.

Alvin M. Weinberg, the director of Oak Ridge National Laboratory, had seen a budget squeeze coming and publicly suggested criteria the government might use to distribute limited funds among the various subfields of science. One of Weinberg's criteria was "social value" and he probably surprised many of his former Manhattan Project colleagues when he used a 1964 article in the trade magazine *Physics Today* to question the wisdom of continuing to fund high-energy physics at the very high levels to which it had become accustomed.[6]

Sensing a threat to the funding of their activities, thirty theoretical high-energy physicists responded to Weinberg in a report designed to correct "the apparent existence of some misunderstanding of the objectives of high-energy physics."[7] In his Foreword to this report, J. Robert Oppenheimer emphasized the physics, but concluded by noting the "possibility of an unanticipated discovery of profound importance to technology and to human welfare."[8] Most of the contributed essays followed suit: a nod toward possible technological spin-offs, but always emphasizing the fundamental nature of the subject. The report circulated widely among members of Congress and the staff of the White House.[9] The prestigious magazine *Science* devoted seven full pages to excerpting it.

One of the essayists, Victor F. Weisskopf, the Director-General of CERN (the multi-national laboratory for particle physics in Europe), pointedly distinguished "intensive" research which

[6] Alvin M. Weinberg, "Criteria for Scientific Choice," *Physics Today* 17(3), 42–8 (1964). See also, Alvin M. Weinberg, "Impact of Large-Scale Science in the United States," *Science* **134**, 161–4 (1961).

[7] Luke C.L. Yuan (editor), *Nature of Matter: Purposes of High Energy Physics* (Brookhaven National Laboratory, Upton, NY, 1965).

[8] Oppenheimer was the highly respected former director of the Manhattan Project during World War II. He died two years after the report was issued.

[9] Silvan S. Schweber, "A Historical Perspective on the Rise of the Standard Model," in *The Rise of the Standard Model: Particle Physics in the 1960s and the 1970s*, edited by Laurie M. Brown, Lillian Hoddeson, Michael Riordan, and Max Dresden (Cambridge University Press, Cambridge, UK, 1997), p. 662.

"goes for the fundamental laws" from "extensive" research which "goes for the explanation of phenomena in terms of known fundamental laws."[10] According to Weisskopf, the "fundamental laws" are those that regulate the elementary particles and the interactions among them. Weisskopf granted that further progress in biology and solid-state physics was possible without any further research in subnuclear physics. Nevertheless, he asserted that "the study of science is based on the burning interest in fundamental problems" and any diminution in fundamental intensive research would produce an "over-emphasis on extensive research, and this would harm all fields of science."

There is no reason to believe that Weisskopf personally disrespected physicists working in solid-state physics. However, not a few of his colleagues in the high-energy physics community made no secret of their belief that a pecking order existed in the physics profession with particle physics at the top and solid-state physics at the bottom. There was plenty of precedent for this attitude.

The quantum pioneer Werner Heisenberg once remarked that he had "thought a little about the theory of ferromagnetism, conductivity, and similar filth."[11] Another of the revered founders, Wolfgang Pauli, believed that the approximations needed to make progress in solid-state physics consigned that activity squarely to the category of 'Schmutziger Physik' (dirty physics).[12]

[10] Victor F. Weisskopf, "In Defense of High-Energy Physics," in *Nature of Matter: Purposes of High Energy Physics*, edited by Luke C.L. Yuan (Brookhaven National Laboratory, Upton, NY, 1965).

[11] Heisenberg quote from Lillian Hoddeson, Gordon Baym, and Michael Eckert, "The Development of the Quantum Mechanical Electron Theory of Metals," in *Out of the Crystal, Chapters from the History of Solid-State Physics*, edited by Lillian Hoddeson, Ernest Braun, Jurgen Teichmann, and Spencer Weart (Oxford University Press, New York, 1992), p. 129.

[12] Wolfgang Pauli, *Scientific Correspondence with Bohr, Einstein, Heisenberg, a.o., Volume II: 1930–1939*. (Springer-Verlag, Berlin, 1985), p. XIV. For the prevalence of this attitude among particle physicists in later years, see David J. Gross, "Asymptotic Freedom, Confinement, and QCD," in *History of Ideas and Basic Discoveries in Particle*

The corrosive effect of this attitude on the self-image of solid-state physicists helps explain why members of this group generally agreed when a 1960s sociologist asked them if their field was considered less important than nuclear or particle physics.[13]

Public perceptions followed the lead of science journalists who had little trouble finding interview subjects to document the evolution of the frontier of physics from atomic physics to nuclear physics to high-energy physics, all in the pursuit of fundamentality. Newspapers wrote breathlessly about every newly discovered subatomic particle and reported that the job of particle physics was to reveal "the nature of matter, life, and the Universe."[14] Solid-state physics earned little or no reportage, a practice which largely continues to the present day, despite the fact that solid-state and nuclear-particle physicists have won roughly equal numbers of Nobel Prizes since World War II.

The historian of physics Joseph Martin has studied what he calls the "`prestige asymmetry" between high-energy and condensed matter physics.[15] Even when news outlets do discuss solid-state research, they routinely ignore its intellectual content and focus solely on its technological import. Often, they choose the most pedestrian technical applications to illustrate their point. A typical example cited by Martin comes from the *The New York Times*, which linked the theoretical work for which Brian Josephson and Philip Anderson won their Nobel Prizes to the workings of television sets and office copying machines.

Physics, edited by Harvey B. Newman and Thomas Ypsilantis (Plenum Press, New York, 1996), p. 81. Pauli's biographer suggests that his attitude about solid-state physics was ambiguous. See, Charles P. Enz, *No Time to be Brief: A Scientific Biography of Wolfgang Pauli* (Oxford University Press, Oxford, UK, 2002) p. 157, 207.

[13] Warren O. Hagstrom, *The Scientific Community* (Basic Books, New York, 1965), p. 53.

[14] Robert C. Toth, "Three Answers Sought: Scientists Pin Hopes on Giant A-Smasher," *Los Angeles Times*, August 15, 1965, p. A3.

[15] Joseph D. Martin, "Prestige Asymmetry in American Physics: Aspirations, Applications, and the Purloined Letter Effect," *Science in Context* **30**, 475–506 (2017).

All of this helps explain why the achievements of condensed matter science remain obscure to laypeople compared to their awareness of the latest news about black holes, dark matter, particle accelerators, and supersymmetry. It is a safe bet this will continue when Edward Witten of Princeton's Institute for Advanced Study can explain to a science journalist that:

> Generally speaking, all the really great ideas of physics are spin-offs of string theory. Some of them were discovered first, but I consider that a mere accident of the development of the planet Earth.[16]

The experimentalist Leon Lederman was simply harvesting well-tilled earth when he promoted the quasi-religious mystique of his subject by choosing the title *The God Particle* for his popular book extolling particle physics.[17] Quasi-religiosity is a difficult sell for a solid-state physicist who literally holds the samples of interest in his or her hands.

Are the Big Machines Necessary?

Phil Anderson had known Victor Weisskopf for nearly twenty years. A paper co-authored by Weisskopf and John Van Vleck had inspired his thesis work and Weisskopf (then a professor at MIT) had served on his PhD examination committee.[18] Anderson liked the man personally, but he was unhappy with the implication of Weisskopf's essay that he—as a solid-state physicist—was little more than an applied quantum mechanic servicing old theories

---

[16] Quoted in John Horgan, *The End of Science: Facing the Limits of Knowledge in the Twilight of the Scientific Age* (Addison-Wesley, Reading, MA, 1996), p. 69.

[17] Roy Gibbons, "Scientist Sees Awesome Force in Atom That Points to God," *Chicago Daily Tribune*, February 11, 1962, p. 10. Leon Lederman and Dick Teresi, *The God Particle: If the Universe is the Answer, What is the Question?* (Houghton-Mifflin, Boston, 1993).

[18] J.H. Van Vleck and V.F. Weisskopf, "On the shape of collision-broadened lines," *Reviews of Modern Physics* **17**, 227–36 (1945).

compared to the profound quantum artistry that engaged high-energy physicists when they designed entirely new theories.

Anderson did not react immediately to Weisskopf's essay. He was busy with superconductivity and superfluid helium and he ran into particle physicists only occasionally. Then, in 1967, he joined the faculty at the University of Cambridge and he was elected to the National Academy of Sciences (NAS). Both places teemed with particle physicists, particularly the NAS, which he soon discovered had very few solid-state scientists as members.

The trigger for Anderson to act was a 1970 panel discussion on the future of high-energy physics sponsored by the American Physical Society. Attendees learned that funding for smaller scientific projects in Britain and the United States might suffer due to the commitments made by those countries to two very large projects, the Super Proton Synchrotron at CERN and the National Accelerator Laboratory in Batavia, Illinois.[19]

This information, and a Science and Society course he co-taught at Cambridge, motivated Anderson to write a lengthy opinion piece (his first) for the London-based magazine *New Scientist*.[20] His goal was to refute four specific intellectual arguments used to justify the expensive particle accelerators used in high-energy physics:

1. The money spent on the field has produced remarkably exciting breakthroughs.
2. The next generation of accelerators will answer a set of very important questions.
3. Everything is composed of particles and fields and their laws govern all matter and energy.

[19] Session CB1, April 27, 1970, *Bulletin of the American Physical Society*, Series II **15**(4), 517 (1970). The National Accelerator Laboratory was renamed Fermilab in 1974.
[20] Philip Anderson, "Are the Big Machines Necessary?," *New Scientist* **51**, 510–13 (1971).

4. High-energy physics is more fundamental than anything
   else.

Against point #1, Anderson quoted the mathematical physicist
Freeman Dyson to the effect that the "remarkably exciting breakthroughs" could have been obtained from inexpensive cosmic ray
experiments rather than from costly particle accelerator experiments.[21] Against point #2, he attacked the "eternal assumption"
of the particle physicists that the answers to their questions were
"just around the next decade in energy and that all that needs to
be done is to spend another billion dollars."

Anderson added depth to his criticism by using ideas advanced
by his old Harvard classmate, Thomas S. Kuhn. The second edition of Kuhn's ground-breaking book, *The Structure of Scientific
Revolutions* had just appeared and Anderson admiringly summarized Kuhn's thesis that scientific progress is revolutionary rather
than evolutionary.[22] He suggested that particle physicists may
have reached what Kuhn calls a "crisis situation" and that the crisis would be resolved only by adopting a new understanding
based on a conceptual revision of their subject rather than by
building more big machines and collecting more data.

Anderson must have relished using Thomas Kuhn's theories
to attack the particle physics community. Kuhn had famously
distinguished the short but revolutionary periods in scientific history when scientists adopted a new paradigm for conceptualizing
their activities from the long periods of "normal science" when
scientists simply work out the consequences of an existing paradigm. To many particle physicists, normal (boring) science was all
that solid-state physicists ever did. Anderson now accused them
of mindlessly pursuing their own long-standing normal science
paradigm that bigger machines were always justifiable, no matter
the cost.

[21] Freeman J. Dyson, "The Future of Physics," *Physics Today* **23** (9), 23–8 (1970).
[22] Thomas S. Kuhn, *The Structure of Scientific Revolutions*, 2nd edition (University
of Chicago Press, Chicago, 1970).

Anderson did not dispute point #3 except to point out its irrelevancy. As far as he knew, nothing learned from the big accelerator machines had any bearing on any field of science except particle physics itself. The basic microscopic laws that constrained solid-state physics, chemistry, and biology had been known since the mid-1930s. Finally, Anderson dismissed point #4 out of hand. It was "based on an outmoded philosophy of science" which ignored the fact that "complex aggregates of matter generate their own new laws." For example,

> the laws of economics and the fascination with that subject in no way depended on the size, color, and shape of the elementary particles (the money in use in a given country).

Anderson's responses to points #1 and #2 criticized the methodologies of high-energy physics, a subject for which he had no training and no expertise. For that reason, particle physicists who took the time to discover who he was were probably outraged at the arrogance of a solid-state physicist passing judgment on their field. If so, it matched the arrogance Anderson assigned to them when they dismissed *his* field as lacking fundamentality. Predictably, his broadside produced no response, at least in print.[23]

More is Different

A year later, Anderson responded to Victor Weisskopf in a subtler and more interesting way. His new essay, "More is Different," did not discuss the big machines of particle physics at all. Instead it focused entirely on the intellectual achievements of solid-state and many-body physics.[24] To that end, he ignored points #1 and

[23] The November 11, 1971 issue of the *Brookhaven Bulletin* published by Brookhaven National Laboratory noted Anderson's article in a sidebar titled "Selected Reading." There *was* a response in the Russian literature: M.A. Markov, "The Future of Science—Is it Really Necessary to Build Accelerators for Large Energies?" (in Russian), Joint Institute for Nuclear Research, P2-7079 (1973).

[24] P.W. Anderson, "More is Different," *Science* 177, 393–6 (1972). For a detailed analysis of the origin and later uses of the ideas in this paper, see Núria Munoz Garganté,

#2 above and rephrased point #3 as "all matter obeys the same fundamental laws." He called the latter the "reductionist hypothesis" while admitting that philosophers might disagree about the exact meaning of that phrase.[25]

With his definition (the same set of laws control all animate and inanimate matter), Anderson regarded reductionism as a undeniable truth with little practical value because:

> the more the elementary particle physicists tell us about the nature of fundamental laws, the less relevance they seem to have to the very real problems of the rest of science, much less to those of society.

The existence or non-existence of the Higgs boson or of quarks or neutrinos simply did not matter to chemists, solid-state physicists, or biologists. Some other truth about the subnuclear world would be just as good as long as the existence of atoms and the laws governing them remained intact.

Anderson then turned to his most important observation:

> The reductionist hypothesis does not by any means imply a *constructionist* one: The ability to reduce everything to simple fundamental laws does *not* imply the ability to start with those laws and reconstruct the Universe.

In other words, perfect knowledge of the laws which govern the subnuclear constituents of cabbages and kings does <u>not</u> imply that one can gain an understanding of either cabbages or kings beginning with just those laws, at least not in a finite amount of time.

Anderson went on to remind his readers that it is straightforward to organize the sciences into a hierarchy based on size and complexity. Nevertheless,

---

The Development of Emergence in Modern Physics: Revisiting the Story of "More Is Different", PhD thesis, Max Planck Institute for the History of Science, in preparation.

[25] See Raphael van Riel and Robert Van Gulick, "Scientific Reduction," in *The Stanford Encyclopedia of Philosophy*, edited by Edward N. Zalta (Spring 2019 Edition), https://plato.stanford.edu/archives/spr2019/entries/scientific-reduction/.

at each stage, entirely new laws, concepts, and generalizations are necessary, requiring inspiration and creativity to just a great a degree as the previous stage...The relationship between a system and its parts is intellectually a one-way street. Synthesis is expected to be all but impossible; analysis, on the other hand, is not only possible but fruitful.

This was his rejoinder to point #4. Despite what the particle physicists may claim, it was a mighty intellectual achievement to uncover, say, the laws of hydrodynamics and a deep and profound task to understand how fluid turbulence is a consequence of those laws. As Anderson saw it, scientific activity directed to this aspect of hydrodynamics was every bit as fundamental as scientific activity directed to understanding how charmed quarks interact with each other.

There are two messages here. First, the laws at each scale must be consistent with the laws at all the smaller scales. The latter *constrains* the former but it is essentially impossible to *derive* from it. That is why biology is not applied chemistry, solid-state physics is not applied atomic physics, and nuclear physics is not applied particle physics.[26] Second, the appearance of new laws and concepts virtually guarantees that truly fundamental research (often using new language) is required at each level of the hierarchy.

The deeper roots of "More is Different" lay in a public lecture Anderson had presented five years earlier at the University of California at San Diego.[27] The theme of this 1967 lecture was how

---

[26] "More is Different" also states that "psychology is not applied biology." The British mathematician Alan Turing held similar ideas in 1944 when he was inventing the idea of the electronic computer. According to his biographer, in contrast to "behaviorist psychology, which spoke of reducing psychology to physics, Turing did not seek to explain one kind of phenomenon, that of mind, in terms of another...His thesis was that mind or psychology could properly be described in terms of Turing machines because they both lay at the *same* level of description of the world." Andrew Hodges, *Alan Turing the Enigma* (Princeton University Press, Princeton, NJ, 1983), pp. 365–6.

[27] Probably the earliest statement of the theme of "More is Different" appeared in a 1962 talk Anderson gave to theoretical graduate students at the

systems composed of a large number of atoms (like solids) differ in fundamental ways from systems composed of a small number of atoms (like molecules). He dwelt on the importance of symmetry and how a macroscopic system often does not possess all the symmetry of the laws that govern it. In other words, he gave a qualitative account of the concept of symmetry breaking and its physical consequences.

Fifty years after the event, the power (if not the details) of Anderson's message came quickly to mind for people present at the original lecture.[28] Particularly memorable was the idea that a macroscopic system with broken symmetry invariably possessed novel properties that could not be predicted knowing only the properties of its constituent elements.

"More is Different" elevated this statement to a fundamental research result in condensed matter physics and offered as an example the property of *rigidity*. According to Anderson, all crystalline solids, magnets, antiferromagnets, superconductors, and superfluids exhibit rigidity as a consequence of broken symmetry. To illustrate this, return to Figure 6.8 and imagine a crystal formed by condensation from the gas phase. This process destroys the continuous translational symmetry of the gas, but it simultaneously creates a set of long-ranged *spatial correlations* which fix the positions of the atoms in a crystal relative to each other.

In other words, if you know the position of *one* atom in a crystal, you know the position of *every* atom of the crystal. This is a consequence of the relatively short-ranged, microscopic forces between atoms which ensure that the solid is indeed a crystal, i.e.,

---

University of Cambridge. Anderson used the opportunity to disagree violently with his Cambridge colleague Brian Pippard [A. B. Pippard, "The Cat and the Cream," *Physics Today* 14(11), 38–41 (1961)] who had suggested that little beyond applied work was left to do for solid-state physicists. Interview of T. Maurice Rice by the author, July 11, 2015.

[28] Author correspondence with Richard More, Zachary Fisk, Gregory Benford, and Christiane Caroli. At the time of Anderson's lecture, More, Fisk, and Benford were graduate students. Caroli was a postdoc.

a many-atom system where the minimum energy is achieved by a repeated stacking of identical unit cells of atoms.[29] Rigidity follows immediately from this requirement that the atoms of a crystal maintain their relative positions.[30] Or as Anderson put it later, "when we move one end of a ruler, the other end moves the same distance."[31]

Anderson goes on to tell his readers that similar spatial correlations fix the relative orientations of the spins in a magnet and the local phases in a superconductor. Hence, a torque tending to rotate one spin (or one local phase angle) causes all the spins (or all the phase angles) to rotate together as rigid objects. Later, Anderson likened this kind of generalized rigidity to the transmission of rotation over a long distance by the many-geared crankshaft shown in Figure 12.1.

"More is Different" appeared in the weekly magazine *Science*, a publication physicists did not read routinely in 1972. An exception was Frederick Seitz, whose book *The Modern Theory of Solids* had been Anderson's early bible on the subject. Seitz was then president of Rockefeller University, and he praised Anderson in a private letter for having performed a "great service to the scientific community by formalizing the central issue in such an elegant way."[32]

The closest any particle physicist came to responding to Anderson was a *Science* article published eight months later by the theorist Steven Weinberg.[33] In that essay, Weinberg reviewed

---

[29] There is no mathematically rigorous proof of this minimum-energy statement about crystals. It is a folk-theorem universally believed by condensed matter physicists.

[30] A related argument establishes rigidity for non-crystalline solids, see Grzegorz Szamel and Elijah Flenner, "Emergence of Long-Range Correlations and Rigidity at the Dynamic Glass Transition," *Physical Review Letters* **107**, 105505(1)–105505(5) (2011).

[31] P.W. Anderson, *Basic Notions of Condensed Matter Physics* (Benjamin-Cummings, Menlo Park, CA, 1984), p. 49.

[32] Letter from Frederick Seitz to Philip W. Anderson, August 16, 1972, AT&T Archives, Warren, NJ.

[33] Steven Weinberg, "Where We Are Now," *Science* **180**, 276–8 (1973).

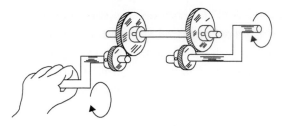

Figure 12.1 A crankshaft transmits rotation over a macroscopic distance in a manner similar to the way spatial correlations transmit a translation or a rotation in a broken symmetry system. Figure from P.W. Anderson, "Some General Thoughts About Broken Symmetry," in *Symmetries and Broken Symmetries in Condensed Matter Physics*, edited by N. Boccara (IDSET, Paris, 1981), pp. 11–20.

"the ultimate laws of nature," which he defined as what quantum field theory tells us about "the few simple general principles which determine why all of Nature is the way it is." He addressed Anderson indirectly when he allowed that "I am not under any illusions that any discoveries in elementary particle physics are going to make life any easier for the biologist or the solid-state physicist." Otherwise, he and his colleagues continued their campaign to brand themselves as voyagers embarked on a grand cultural enterprise to discover what was most profound about the cosmos.[34] A decade later, Anderson and Weinberg would spar in a very public way about fundamentality and funding priorities in physics.

Emergence

Shortly before its publication, Anderson discussed "More is Different" with William Homan Thorpe, the Professor of Animal Behavior at the University of Cambridge. Thorpe outlined Anderson's thinking at a small meeting of biologists, physiologists,

[34] Hallam Stevens, "Fundamental Physics and Its Justifications, 1945–1993," *Historical Studies in the Physical and Biological Sciences* **34** , 151–97 (2003).

and philosophers.[35] The reaction was very positive. Anderson's claim that new laws appeared at each level of the science hierarchy resonated with the physiologists and organism-scale biologists who felt disrespected by the claims of molecular biologists who asserted the unique fundamentality of their own work.[36] This may be why many of the citations to "More is Different" in the first decade after its publication came from those two types of scientist.

Anderson's point of view also resonated with the philosophers in Thorpe's audience because they recognized his arguments as an updated and improved version of the concept of *emergence*. This was a late-nineteenth century idea that reached the peak of its popularity in British philosophical circles of the 1920s.[37] A representative emergentist was C.D. Broad, a philosopher of science who wanted to understand how life, the mind, and consciousness evolved from inanimate matter. In his view:

> Emergence is the theory that the characteristic behavior of the whole *could* not, even in theory, be deduced from the most complete knowledge of the behavior of its components.[38]

In short, *the whole can be much more than the sum of its parts*. As an example, Broad pointed to molecules, whose properties seemed to have nothing to do with the properties of their constituent atoms.

[35] W.H. Thorpe, "Reductionism in Biology," in *Studies in the Philosophy of Biology*, edited by Francisco Jose Ayala and Theodosius Dobzhanshy (University of California Press, Berkeley, 1974), pp. 109–38.

[36] See, for example, the discussion remarks of the biochemist and Nobel Prize winner J. Monod following V.F. Weisskopf, "The Connection Between Physics and Other Branches of Science," *Il Nuovo Cimento. Supplemento* 4(1), 465–91 (1966). See also Carl R. Woese, "A New Biology for a New Century," *Microbiology and Molecular Biology Reviews* 68(2), 173–86 (2004).

[37] Brian P. McLaughlin, "The Rise and Fall of British Emergentism," in *Emergence: Contemporary Readings in Philosophy and Science*, edited by Mark A. Bedau and Paul Humphreys (MIT Press, Cambridge Press, 2008), pp. 19–58.

[38] C.D. Broad, *The Mind and Its Place in Nature* (Kegan Paul, Trench, Trubner & Co. Ltd., 1925), p. 59.

The problem with emergence was that none of its advocates could explain exactly how it worked. Worse, the rise of genetics and quantum mechanics provided a way to make sense of the behavior of molecules and other supposedly emergent phenomena from more basic ingredients. This drove emergence into a prolonged eclipse from which it was just emerging when Anderson's article appeared.[39]

The word 'emergence' does not appear in "More is Different" and Anderson was unaware of its provenance when his essay appeared. He nevertheless contributed significantly to its renaissance. An important reason was that many of the diverse readers of *Science* magazine thought creatively about the hierarchical structure of science and its consequences for the first time after reading his piece.[40]

Unlike all previous commentators on the subject, Anderson offered a specific *mechanism* that was capable of producing emergence. That is, producing unforeseen (perhaps even unimaginable) properties of a macroscopic system that were unrelated to the properties of the system's constituents. His suggested mechanism was symmetry breaking and we sketched just above how the property of rigidity emerges from that process. The novelty of this kind of specificity led one philosopher to rate "More is Different" as one of the most influential articles on emergence written in the twentieth century.[41]

The temperature of a gas, the color of a gold earring, and the magnetism of a piece of iron are all examples of emergent properties. So is the wetness of water, a property that is unfathomable to anyone who has not experienced it or who knows only the properties of a single water molecule. In that case, the mechanism of

[39] Jaegwon Kim, "Making Sense of Emergence," *Philosophical Studies* 95, 3–36 (1999); Peter A. Corning, "The Re-Emergence of Emergence and the Causal Role of Synergy in Emergent Evolution," *Synthese* **185**(2), 295–317 (2012);.

[40] See, for example, R.J. Huggett, "A Schema for the Science of Geography, its Systems, Laws, and Models," *Area* **8**, 25–30 (1976).

[41] Paul Humphreys, *Emergence* (Oxford University Press, New York, 2016) p. 180.

emergence is the phenomenon of condensation. Similarly, an off-duty scientist at a cocktail party may define temperature as a measure of the average speed of the particles in a gas. But to pass from a collection of microscopic particles to the concept of temperature in thermodynamics requires a mechanism for the system of particles to reach a quiescent state where macroscopic changes no longer occur.[42]

Anderson did not use the word 'emergence' himself until 1981, when he and his former PhD student Dan Stein wrote a paper about the origin of life.[43] An interesting sequence of events led him to this topic, which was so far from his usual beat. First, at the 1977 Nobel Prize award ceremony, Anderson stood next to the Russian-Belgian theoretician Ilya Prigogine. Prigogine received the Nobel Prize for Chemistry that year for, among other things, his use of symmetry-breaking ideas to analyze situations (he suggested living systems) where continuous energy input balances energy loss due to dissipation. Anderson did not believe Prigogine's theory, primarily because his old Harvard classmate Rolf Landauer had demonstrated that Prigogine's reasoning failed in at least one specific case.[44]

Immediately after the Nobel ceremony, Anderson attended a large biology conference to which he had been invited by the physiologist F. Eugene Yates. Yates was an enthusiastic fan of "More is Different" and he was anxious to meet its author and

[42] Robert C. Bishop, *The Physics of Emergence* (Morgan & Claypool, San Raphael, CA, 2019), Section 4.2.

[43] P.W. Anderson and D.L. Stein, "Broken Symmetry, Emergent Properties, Dissipative Structures, Life: Are They Related?," preprint received by the Defense Technical Information Center, February 12, 1981. First published in P.W. Anderson, *Basic Notions of Condensed Matter Physics* (Benjamin-Cummings, Menlo Park, CA, 1984).

[44] R. Landauer, "Stability and Entropy Production in Electrical Circuits," *Journal of Statistical Physics* **13**, 1–16 (1975). See also Joel Keizer and Ronald Forrest Fox, "Qualms Regarding the Range of Validity of the Glansdorf–Prigogine Criterion for the Stability of Non-Equilibrium States," *Proceedings of the National Academy of Sciences USA* **71**, 192–6 (1974).

exchange ideas.[45] Anderson did so and was delighted to discover that Yates and a few of his like-minded colleagues had serious and specific scientific ideas about the origins of life and the organizing principles of the brain.[46]

It bothered Anderson's sense of scientific integrity that Prigogine had discussed his theory in general terms without any mention of its limitations. The Yates meeting energized him to start thinking about dissipation so he could join Landauer as a critic. He elaborated his perspective in a letter to a physiologist:

> I have always made it my specialty to confine myself to the concrete and specific in research. I prefer problems in search of answers rather than answers in search of problems. This was my reason for entering the [biological] field. I felt Ilya Prigogine was presenting generalities which have repeatedly not stood up to the test of specific models.[47]

An invitation to speak at the 17th International Solvay Conference on Physics in Brussels at the end of 1978 gave Anderson a forum to try out his criticism. Prigogine was present and Phil did not mention him by name. However, the identity of his antagonist was obvious from the content of his lecture, which (in its published form) does not use the word "life" until its last sentence.[48] Dan Stein had contributed to the research reported at the Solvay meeting and over the next few years he and Anderson kept

[45] F. Eugene Yates, "Reductionist versus Organismic Biology," *American Journal of Physiology* **233**, R73–R74 (1977); F. Eugene Yates, "Complexity and the Limits to Knowledge, " *American Journal of Physiology* **235**, R201–R204 (1978).

[46] P.W. Anderson, *More and Different: Notes from a Thoughtful Curmudgeon* (World Scientific, Hackensack, NJ, 2011), p.351.

[47] Letter from PWA to Dr. Sang Chul Ji (Rutgers University), December 18, 1984, Princeton University Archives, Department of Rare Books and Special Collections, Princeton University Library.

[48] P.W. Anderson, "Can Broken Symmetry Occur in Driven Systems," in *Order and Fluctuations in Equilibrium and Non-equilibrium Statistical Mechanics*, edited by G. Nicolis, G. Dewel, and J.W. Turner (John Wiley & Sons, New York, 1981), pp. 289–97.

returning to the question that had motivated the original British emergentists: how can life arise from inanimate matter?

Emergence was still not in Anderson's lexicon in May 1980 when he presented the Cherwell-Simon Memorial Lecture at the University of Oxford. On the other hand, he was ready and willing to debunk Prigogine's ideas about the origin of life.[49] He and Stein had convinced themselves that, when life developed from inanimate matter, it could *not* have happened by a sequence of symmetry-breaking processes as Prigogine imagined. Some other mechanism(s) must be involved. Only in the 1981 written version of that lecture did Anderson and Stein finally use the word emergent, which they did ten times.

Emergence became a more and more popular idea over the years, even as commentators disagreed over its exact meaning. In "More is Different," Anderson is unclear (or coy) about whether he regarded the constructivist program (e.g., to derive the laws of biology from the laws of chemistry) as impossible in principle or merely impossible in practice.[50] As a result, some viewed emergence as a common phenomenon while others judged it as rare. Anderson wrote a paper asking whether measurement itself was an emergent property.[51] Robert Laughlin, the 1998 Nobel Prize winner in Physics went so far as to suggest that *all* physical laws were emergent.[52]

Anderson's emergence ideas resonated with many different kinds of scientists. But what about the particle physicists? It was their reductionist claims of fundamentality that launched

[49] Letter from Dr. Anthony Michaelis (Editor of *Interdisciplinary Science Reviews*) to PWA, May 28, 1980, Nicholas Kurti letters, Bodleian Library, Oxford University Library.

[50] Compare Paul Mainwood, "Is More Different? Emergent Properties in Physics," PhD Thesis, Merton College, University of Oxford, 2006 to Robert C. Bishop, *The Physics of Emergence* (Morgan & Claypool, San Raphael, CA, 2019).

[51] P.W. Anderson, "Is Measurement Itself an Emergent Property?," *Complexity* 3, 14–16 (1997).

[52] Robert B. Laughlin, *A Different Universe (Reinventing Physics from the Bottom Up)*, (Basic Books, New York, 2005), p. xv.

Anderson on his crusade in the first place. Steven Weinberg, for one, never backed down. Thirty years after "More is Different," he wrote:

> After all, emergent phenomena do *emerge*, ultimately from the physics of elementary particles, and if you want to understand why the world is the way it is, you have to understand why elementary particles are the way they are.[53]

On the other hand, the eminent historian of physics (and former quantum field theorist) Silvan Schweber used an essay in *Physics Today* to warn of a "crisis in physical theory" driven by shrinking budgets for science, changing political agendas, and changes in the perception of science driven by a growing acceptance of Anderson's fundamental premise. As Schweber rephrased it:

> A hierarchical arraying of parts of the physical universe has been *stabilized*, each part with its quasi-stable ontology and quasi-stable effective theory, and the partitioning is fairly well understood. For the energy scales that are experimentally probed in atomic, molecular, and condensed matter physics the irrelevance (to a very high degree of accuracy) of the domains at much shorter wavelengths has been justified. Effectively a kind of "finalization" has taken place in these domains.[54]

Finally, a younger generation of fundamental physicists seems willing to grant that emergence plays an important role in their own endeavors. Thus the string theorist Edward Witten (quoted earlier) approaches Robert Laughlin's point of view when he suggests "gauge symmetry may be emergent" and muses that "maybe the space-time we experience and the particles and the fields in it are all emergent from something deeper."[55]

[53] Steven Weinberg, *Facing Up: Science and Its Cultural Adversaries* (Harvard University Press, Cambridge, MA, 2001), p. 58

[54] Silvan S. Schweber, "Physics, Community, and the Crisis in Physical Theory," *Physics Today* **46**(11), 34–40 (1993).

[55] Edward Witten, "Symmetry and Emergence," *Nature Physics* **14**(2), 116–19 (2017).

Near the end of "More is Different," Anderson advises his readers that a point inevitably comes when one must stop talking about decreasing *symmetry* and start talking about increasing *complexity*. For him, that point came in the early 1980s when he connected complexity to an exotic magnetic system called a *spin glass*. His understanding of the unusual characteristics of the spin glass, in turn, helped facilitate the substantial contribution he made to the creation and early activities of a private research organization called the Santa Fe Institute.

The Spin Glass

Phil Anderson and Sam Edwards did not have complexity in mind in 1974 when they began meeting on Saturday mornings for coffee in the break room of Cambridge's Theory of Condensed Matter group. Edwards was on unpaid leave of absence from the university and he traveled by train every day to London to serve as Chairman of the Science Research Council of the United Kingdom.[56] His weekend chats with Anderson provided food for thought and calculation during his two-hour commute.

One weekend, Anderson told Edwards about a class of disordered metal alloys he had been thinking about for several years. In these alloys, magnetic atoms substitute randomly at the lattice sites of a non-magnetic host crystal.[57] Figure 7.7 represents such a system if the randomly distributed blue dots are magnetic atoms and the grey dots are non-magnetic atoms. This system intrigued Anderson because different experiments did not agree about whether a transition to a low-temperature magnetic phase occurred or not.

Solid-state physicists at the time understood that the exchange interaction between two magnetic atoms embedded in a metal

---

[56] Mark Warner, "Sir Sam Edwards, 1 February 1928–7 July 2015," *Biographical Memoirs of Fellows of the Royal Society* **63**, 243–71 (2017).

[57] These are the same alloys used to study the Kondo effect except that the concentration of magnetic atoms is higher.

oscillated in sign as a function of the distance between the atoms. At some distances, the two atomic spins preferred to be parallel and at other distances, they preferred to be antiparallel. Because the magnetic atoms were distributed at random in the metal alloys of interest, a coin flip decided whether the spins of any two magnetic atoms preferred parallel alignment or antiparallel alignment.

Edwards and Anderson constructed a Heisenberg-type model for this situation and used their complementary styles of doing theoretical physics (Phil's deep intuition and Sam's nose for just the right mathematical tool) to analyze it approximately.[58] They concluded that a phase transition took place between a high-temperature disordered state where the direction of every spin fluctuated wildly and a low-temperature disordered state where the direction of each spin "froze" into a fixed but random direction.[59] This conclusion justifies the name *spin glass* given to these alloys because the disorder in the spin directions is similar to the disorder in atomic positions found in ordinary window glass.[60]

The approximate nature of their solution encouraged others to use computers to search numerically for the exact ground state spin configuration of the Edwards–Anderson model. By about 1980, it became clear that this was a fool's errand. The problem defined by their model was "NP-complete" and computationally

[58] David Sherrington, "Edwards–Anderson: Opening up the World of Complexity," in *Stealing the Gold: A Celebration of the Pioneering Physics of Sam Edwards*, edited by Paul M. Goldbart, Nigel Goldenfeld, and David Sherrington, (Clarendon Press, Oxford, UK, 2004), pp. 179–191.

[59] S.F. Edwards and P.W. Anderson, "Theory of Spin Glasses," *Journal of Physics F: Metal Physics* **5**, 965–974 (1975).

[60] P.W. Anderson, "Localization Theory and the Cu-Mn Problem: Spin Glasses," *Materials Research Society Bulletin* **5**, 549–54 (1970). In this paper, Anderson takes credit for the name "spin glass" and credits the Welsh experimentalist Bryan Coles for inventing the related term "magnetic glass." In a later conference paper, Anderson credits Coles alone for coining the term "spin glass." See P.W. Anderson, "Topics in Spin Glasses," in *Amorphous Magnetism*, edited by Henry O. Hooper and Adriaan M. De Graaf (Plenum Press, New York, 1973), pp. 1–11.

complex. This meant that the time required to find a solution increased *exponentially* with the number of spins in the system.[61] Another example of an NP-complete problem arises when a traveling sales person seeks the shortest possible path to visit every city on his/her route exactly once before returning home.

The spin glass and the traveling sales person problems both exhibit *disorder* and *frustration*, by which we mean conflicting constraints or desires.[62] For the sales person, the locations of the cities are random and the frustration comes from the conflicting desires to make the total path as short as possible while still visiting every city. For the spin glass, the pairwise exchange interactions are randomly ferromagnetic or antiferromagnetic and frustration arises because every spin cannot always satisfy the energetic needs of all of its neighbors simultaneously.

To see this, imagine a square lattice with a spin at every site that can point up or down only. Assume that a coin flip determines if the link between any pair of nearest-neighbor spins prefers ferromagnetic (FM) alignment or antiferromagnetic (AF) alignment. Figure 12.2 shows four sites of such a lattice where, by random chance, one link favors FM spin coupling and three links favor AF spin coupling.

Suppose we traverse the square in Figure 12.2 counterclockwise beginning with the upper left spin. Choose this spin to point up and note that the link immediately below the spin is AF. Therefore, the spin on the lower left must point down. The link to the right of this spin is also AF, so the spin on the lower right must point up. Finally, the link above this spin is AF, so the spin on the upper right should point down, like the blue spin shown. However, the link to the left of this spin is FM which is not satisfied by the up spin at the upper left. Starting from the beginning and moving clockwise around the square, the red spin would

[61] M.L. Garey and D.S. Johnson, *Computers and Intractability: A Guide to the Theory of NP Completeness* (W.H. Freeman, San Francisco, CA, 1979).

[62] Scott Kirkpatrick, "Models of Disordered Systems," in *Disordered Systems and Localization*, edited by C. Castellani, C. Di Castro, and L. Peliti (Springer, Berlin, 1981).

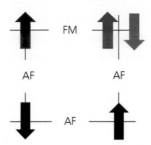

Figure 12.2 Frustration in a system of (up/down only) spins on a square lattice. By random chance, three pairs of nearest-neighbor spins prefer anti-ferromagnetic (AF) exchange and one pair of nearest-neighbor spins prefers ferromagnetic (F) exchange. The spin on the site at the upper right is frustrated. It cannot decide whether to point up or down because, in either case, three of the four exchange interactions are satisfied and one is not.

occupy the upper right site and the spin on the lower left would point up. This would not satisfy the AF link on the far left. This system is frustrated because no single choice of spin orientations satisfies all the links simultaneously.

The Edwards–Anderson model has enjoyed steady popularity over many years. This is because, with a change of variables, the model applies to many *non-physics* problems like airplane scheduling, mail delivery, pattern recognition, integrated circuit wiring, and message encoding.[63] The features common to all of these are the presence of many agents (spins, packages, people, automobiles, etc.) and a competition between disorder and frustration to achieve a common goal.

Besides the model itself, a whole set of theoretical methods invented to study spin glasses transfer easily to study NP-complete problems in other fields. Among the most successful is a technique called "simulated annealing" where a computer mimics a heat-treatment process used in metallurgy.[64] Simulated annealing

---

[63] Daniel L. Stein and Charles M. Newman, *Spin Glasses and Complexity* (Princeton University Press, Princeton, NJ, 2013), p.3.

[64] S. Kirkpatrick, C.D. Gelatt, and M.P. Vecchi, "Optimization by Simulated Annealing," *Science* **220**, 671–80 (1983). This paper had earned 45,000 citations by the end of 2019.

has been applied with great success to all the tasks mentioned above and many more. It was precisely this extraordinary flexibility of the spin glass model that fired Anderson's imagination to push its use even further.

Pausing for a Breath

Anderson was uniquely receptive to new problems in the early 1980s when he began to contribute significantly to a subject that is today called *complexity*. His Nobel Prize and the Gang of Four paper brought satisfying closure to the localization problem. The many applications of the spin glass, both inside and outside of physics, were eye-opening and the Nobel Prize conveniently generated invitations to attend conferences outside his expertise.

A relevant fact is that Anderson's PhD thesis advisor died at the end of October 1980. John Van Vleck was the first person in the United States to complete a PhD based entirely on quantum mechanics and his subsequent work can be seen as symbolic of the early, heroic phase of the quantum era.[65] Perhaps that is why, around the time of Van's death, it seemed to Anderson that "after the quantum revolution solved so much, physics drew a breath and looked around for new fields to conquer"[66] More accurately, Phil drew a breath, even as he continued to publish in areas he knew well.

If solid-state physics drew a breath at that time, the caesura lasted no more than a year or so because, early in 1982, three experimenters at Bell Labs announced that a two-dimensional electron gas exposed to a strong magnetic field exhibited remarkable properties.[67] Former Bell Labs postdoc Robert Laughlin

---

[65] Brebis Bleaney, "John Hasbrouck Van Vleck, March 13, 1899–October 27, 1980," *Biographical Memoirs of the Fellows of the Royal Society* **28**, 627–65 (1982).

[66] P.W. Anderson, "My Brief Life as an Economist," August 25, 2016, unpublished.

[67] D.C. Tsui, H.L. Stormer, and A.C. Gossard, "Two-Dimensional Magnetotransport in the Extreme Quantum Limit," *Physical Review Letters* **48**, 1559–662 (1982). The phenomenon they observed was later named the fractional quantum Hall effect.

guessed the many-body wave function for this system and the phenomenon it explained—the fractional quantum Hall effect—touched off an avalanche of work that continues to the present day.[68] Anderson played no role in any of this.[69] As usual, if he could not lead, he would not follow.

Instead, Anderson looked outside his comfort zone for a new field to conquer. An unusual part of this process was his attendance at a physics conference facilitated by Werner Erhard, the founder of a self-awareness training program called *est* that thrived in the 1970s and 1980s. Erhard was a physics enthusiast (some would say groupie) who, for over ten years, used his *est*-generated wealth to host a yearly conference for A-list theoretical physicists at his San Francisco mansion.[70] The physicists' only obligation was to discuss the latest and most exciting developments in their subfields.

An Erhard conference in 1982 devoted to "Complex Systems" was an important event for Anderson because it was there that he met Norman Packard and Stuart Kauffman. These two scientists (and most of the others present) were deeply interested in the behavior of nonlinear systems. [71] This was a subject Anderson had dipped his toe into years earlier.[72] Packard was about to complete his PhD thesis on the effect of random perturbations on nonlinear systems. Earlier, he had contributed to an effort to use nonlinear science (and miniature computers hidden in shoes) to beat

[68] R.B. Laughlin, "Anomalous Quantum Hall Effect: An Incompressible Quantum Fluid with Fractionally Charged Excitations," *Physical Review Letters* **50**, 1395–8 (1983).

[69] Anderson published one short and quickly forgotten paper on this subject: P.W. Anderson, "Remarks on the Laughlin Theory of the Fractionally Quantized Hall Effect," *Physical Review* B**28**, 2264–5 (1983).

[70] David Kaiser, *How the Hippies Saved Physics* (W.W. Norton, New York, 2011), pp. 179–193. See also, Kitty Ferguson, *Stephen Hawking: An Unfettered Mind* (St. Martin's Press, New York, 2012), pp. 95–9.

[71] James Gleick, *Chaos: Making a New Science* (Viking Penguin, New York, 1987).

[72] P.W. Anderson, "The Reaction Field and Its Use in Some Solid-State Amplifiers," *Journal of Applied Physics* **28**, 1049–53 (1957).

the odds at the roulette tables in Las Vegas.[73] Kauffman was one of the first people to exploit nonlinear methods to address problems in cellular biology and genetics.

Like all physicists, Anderson had been trained to analyze *linear* situations where the response of a system is directly proportional to the stimulus it receives. A hanging pendulum bob displaced by a small amount from the vertical responds by swinging downward with an acceleration that is proportional to its displacement. However, if the initial displacement away from the vertical is large, the acceleration is <u>not</u> proportional to the displacement, the standard theoretical analysis fails, and there is no simple analytic way to proceed. This is typical of a *nonlinear* system.

Physics has been a successful predictive science for centuries because most classical and quantum systems respond linearly if the stimulus they receive is gentle enough. However, as the pendulum demonstrates, the very same physical systems often respond nonlinearly when the stimulus is large. The problem of understanding fluid turbulence has always been excruciatingly difficult precisely because it is highly nonlinear. Most physicists ignored nonlinear problems for years because they lacked the tools to attack them.

Help arrived in the form of computers, which can be programmed to tackle nonlinear problems. Bell Labs theorist Don Hamann (one of Anderson's collaborators for the Kondo problem) made this point when he discussed the uses of computers in physics in a 1983 survey article.[74] Hamann wrote particularly about "simulations" where a computer evaluates nonlinearities directly so the practitioner is free to look for qualitative behaviors that appear in the simulations. At the Erhard meeting, Packard presented novel simulations of this kind. Anderson was impressed, but he was in no way qualified to take a leadership position in nonlinear science.

---

[73] Thomas A. Bass, *The Eudaemonic Pie* (Houghton Mifflin, New York, 1985).

[74] Donald R. Hamann, "Computers in Physics: An Overview," *Physics Today* **24**(5), 24–33 (1983).

The Santa Fe Institute

Anderson's entrée to complexity theory came in June 1984 when his old friend David Pines asked him to attend a workshop on "Emerging Syntheses in Science." The purpose of the workshop was to help define the goals of a proposed Institute focused on interdisciplinary studies.[75] The Institute was the brainchild of George Cowan, a Senior Fellow and former Associate Director for Research at Los Alamos National Laboratory (LANL). Cowan was also a sitting member of the White House Science Council. The committee Cowan put together to brainstorm about a new Institute consisted of LANL Senior Fellows and a few Visiting Senior Fellows like Pines and the Nobel Prize winner, Murray Gell-Mann.

Anderson was suspicious. Gell-Mann had invented the idea of the quark and he and his Caltech colleague Richard Feynman had dominated the field of theoretical particle physics in the 1950s and early 1960s. Gell-Mann was a terrifyingly brilliant polymath but, as far as Anderson knew, he had never worked on an interdisciplinary problem in his life. Was this new Institute for real or was it just a soft-landing spot for aging scientists?[76]

Pines assured Anderson that both Cowan and Gell-Mann had serious intent. Cowan's experience with science policy at the White House had confirmed an observation he had made after a forty-year career at Los Alamos. Contemporary scientists appeared to value disciplinary specialization over the highly interdisciplinary skills he and others of his generation had developed to design and build the atomic bomb during World War II and later to tackle the very complex problem of nuclear weapons forensics (his personal expertise).[77]

---

[75] David Pines, *Emerging Syntheses in Science: The Founding Papers of the Santa Fe Institute* (Santa Fe Institute, Santa Fe, NM, 2015). Archives of the Santa Fe Institute.

[76] M. Mitchell Waldrop, *Complexity: The Emerging Science at the Edge of Order and Chaos* (Simon & Schuster, New York, 1992), pp. 79–89.

[77] George A. Cowan, *Manhattan Project to the Santa Fe Institute* (University of New Mexico Press, Albuquerque, NM, 2010).

Problems of this sort were highly nonlinear and the govern-ment's need to make progress with them was one reason Los Alamos had always boasted one of the largest concentrations of computer power anywhere in the world. Moreover, LANL had created a Center for Nonlinear Studies in 1980 to capitalize on their local expertise. Cowan's dream was much more expansive. He hoped to build a new kind of learning institution that would train young scholars to embrace and attack interdisciplinary problems.

Gell-Mann's story was different. Pines had been his close friend for thirty years and he was well-acquainted with Gell-Mann's knowledge of subjects like archaeology, arms control, ecology, ornithology, linguistics, and numismatics. With such broad interests, the mere idea of an entire Institute devoted to interdis-ciplinary problems excited him very much. As it happened, as a board member of the John D. and Catherine T. MacArthur Foundation, Gell-Mann had worked hard to convince his col-leagues to support interdisciplinary research on the psychobiol-ogy of depression.

Pines secured Anderson's participation in the workshop by assuring him he could influence the Institute's final shape. Therefore, on November 10, 1984, Phil sat in an elegant board room in Santa Fe, New Mexico and listened to Gell-Mann open the workshop with a broad vision statement. Anderson followed with a talk about "Spin Glass Hamiltonians: A Bridge between Biology, Statistical Mechanics, and Computer Science." Other participants discussed linguistics, computer science, primate behavior, materials science, unconscious thought, and macro-molecular evolution.[78]

By the end of the workshop, it was clear that the initial focus of the Institute would be on complex, nonlinear problems that bridged (or fell inside) the gaps between traditional academic disciplines. Anderson was excited about what he heard and agreed to serve on

[78] *Emerging Syntheses in Science*, edited by David Pines (Addison-Wesley, Redwood City, CA, 1987).

the Science Board of what would soon be called the Santa Fe Institute (SFI).

His serious involvement began a year later when George Cowan responded to an unusual request from John Reed, the new CEO of Citicorp and a friend of a member of the SFI Board of Directors. Citicorp stood to suffer enormous losses because, on the advice of their in-house economists, they had lent billions of dollars to emerging countries and now those countries could not repay their loans. Reed was willing to come to Santa Fe to hear what the brilliant folks Cowan had assembled had to say about how economists went about their business.[79]

Anderson had an amateur's interest in economics going back to his Cambridge days. For that reason, he agreed to join a group of SFI people to meet with Reed and a few of his people. He and Joyce hitched a ride on Reed's corporate jet and flew with him from New Jersey to Santa Fe. The meeting went well.[80] Cowan now tasked Anderson to organize a full-scale SFI workshop devoted to economic theory.

Anderson needed a professional as a co-organizer and he thought immediately of James Tobin, the Nobel Prize-winning economist who had preceded him by a few years at Uni High School. Tobin did not want to participate, but he knew someone who might: Kenneth J. Arrow, another winner of the Nobel Prize for Economics. Anderson arranged to meet Arrow in person and the affable economist agreed to co-organize a workshop in Santa Fe devoted to "Evolutionary Paths of the Global Economy." Arrow would invite ten economists and Anderson would invite ten physical scientists.

[79] M. Mitchell Waldrop, *Complexity: The Emerging Science at the Edge of Order and Chaos* (Simon & Schuster, New York, 1992), pp. 91–96.

[80] George Cowan and Robert McCormick Adams, "Summary of Meeting on 'International Finance as a Complex System' at the Rancho Encantado, Tesuque, New Mexico, August 6–7, 1986," in *The Economy as an Evolving Complex System*, edited by Philip W. Anderson, Kenneth J. Arrow, and David Pines (Addison-Wesley, Redwood City, CA, 1988), pp.307–12.

Arrow was widely regarded as a truly outstanding economist. He had made seminal contributions to the theory and application of mainstream (so-called neoclassical) economics but he had also published several papers questioning some of its fundamental assumptions.[81] This made him a nearly ideal partner for Anderson for a program designed to rethink the subject.

The first economics workshop of the Santa Fe Institute convened in early September of 1987 in the chapel of the Christo Rey Convent. George Cowan had leased the unused convent to be the Institute's first home. The official charge to the participants was to:

> expand the horizons of conventional economic theory so that it might eventually be able to deal with such complex macroeconomic problems as the global economy, rather than in applying pre-existing economic theories to this problem and so becoming a forum for conflicting views of causation or cure, based on manifestly incomplete theories.[82]

By all accounts, the ten days of all-day discussions were lively and exhausting. The fact that economists and physicists share a belief they are always the smartest people in any room produced some interesting initial judgments. The scientists were puzzled by the economists' emphasis on "equilibrium" situations where economic forces like supply and demand balance so that the values of economic variables do not change as a function of time. Anderson reportedly blurted out, "Do you guys really *believe* that?"[83] The economists were skeptical that the scientists' models of complex many-particle systems could be of much value to

---

[81] George R. Feiwel, "The Many Dimensions of Kenneth J. Arrow," in *Arrow and the Foundations of the Theory of Economic Policy*, edited by George R. Feiwel (Palgrave Macmillan, London, 1987), pp. 1–115.

[82] David Pines, "An Introduction to the Workshop," in *The Economy as an Evolving Complex System*, edited by Philip W. Anderson, Kenneth J. Arrow, and David Pines (Addison-Wesley, Redwood City, CA, 1988), p.4.

[83] M. Mitchell Waldrop, *Complexity: The Emerging Science at the Edge of Order and Chaos* (Simon & Schuster, New York, 1992), p. 142.

them when their agents had no expectations, no goals, no foresight, and no memory.

There were also differences in style and modeling philosophy.[84] Typically, the scientists built models designed to capture the most important features of relevant experimental data and then analyzed those models using just enough mathematics to get a sensible answer. By contrast, the economist's lack of reproducible empirical data (as opposed to non-reproducible observations of functioning economies) led them to build mathematical models based on the simplest assumptions, even if they were unrealistic. They then used quite rigorous mathematics to analyze them. In this way, neoclassical economics aimed to build up its subject from its foundations. It must have amused Anderson that this was exactly the constructionist approach to a complex problem that he had attacked in "More is Different."

Attitudes evolved over the course of the workshop. Anderson's summary conceded that mainstream economic theories "which appeal to the concept of equilibrium do not necessarily avoid the apparently random fluctuations in the course of time which are characteristic of driven dynamical systems."[85] On the other hand, everything he heard only reinforced his belief that the economy had many similarities to a spin glass.

Ken Arrow's summary acknowledged that economists needed to eliminate the unrealistic assumptions they typically made in their standard models.[86] One topic that captured his imagination was the colorfully named phenomenon of *chaos*, a behavior observed in nonlinear systems where practically undetectable

---

[84] Interview of economist John Miller by the author, August 10, 2018.

[85] P.W. Anderson, "A Physicist Looks at Economics: An Overview of the Workshop," in *The Economy as an Evolving Complex System*, edited by Philip W. Anderson, Kenneth J. Arrow, and David Pines (Addison-Wesley, Redwood City, CA, 1988), pp. 265–73.

[86] Kenneth J. Arrow, "Workshop on the Economy as an Evolving Complex System: Summary," in *The Economy as an Evolving Complex System*, edited by Philip W. Anderson, Kenneth J. Arrow, and David Pines (Addison-Wesley, Redwood City, CA, 1988), pp. 275–81.

changes in initial conditions can lead to extremely different out-comes. Chaos did not bode well for economists paid to make pre-dictions about the future.

Arrow also shared Anderson's enthusiasm for the ideas of John Holland, a computer scientist who later suggested that it is pre-cisely the presence or absence of emergence that distinguishes complexity from the merely complicated.[87] Thus, a watch is a complicated object composed of many gears and springs, but its ability to keep time differs little from other devices of entirely dif-ferent design. By contrast, a financial market is a truly complex system that exhibits emergent phenomena like the South Sea bubble of 1720 and the 1929 US stock market crash.

The consensus view when the workshop ended was that insights and methods drawn from many-body and nonlinear physical science had real promise for transforming economic modeling. The ghost of the philosopher and economist Friedrich von Hayek must have sat up and recalled his 1974 Nobel Prize for Economics lecture where he asserted that "unlike the position that exists in the physical sciences, [we] in economics deal with essentially complex phenomena...whose characteristic proper-ties can be exhibited only by models made up of relatively large numbers of variables."[88] In the event, Cowan and his senior advi-sors promptly created the first residency program of the Santa Fe Institute and dedicated it to economics. Anderson was involved for a year or so until the siren song of high-temperature super-conductivity became irresistible.

At the second SFI workshop devoted to economics, Anderson made it his business to educate the economists about *self-organized criticality*, a phenomenon which had only recently been discovered by physicists. To an SFI audience, he described the self-organized

[87] John H. Holland, *Complexity, A Very Short Introduction* (Oxford University Press, Oxford, UK, 2014).

[88] Friedrich von Hayek, "The Pretense of Knowledge," Nobel Lecture, December 11, 1974.

critical state as relevant to situations where many small-scale events drive an aggregate system to a unique large-scale structure.[89]

The formal SFI program in economics ran for fifteen years.[90] Two decades into the twenty-first century, it is not difficult to find both enthusiasts and skeptics of its unique approach.[91] But even if SFI ideas never affected the foundations of economic theory, it is undeniable that several financial forecasting companies founded on its principles have had great success analyzing fluctuating financial markets and advising clients on investment strategies.[92]

Complexity and its associated concepts rapidly took hold as a touchstone for practically every program sponsored by the SFI. A short list of program topics plucked at random from a long list includes mental processes, the nature of time, intelligence, globalization, molecular evolution, immunology, measurement theory, the origin of life, technology transfer, the limits to growth, chemical signaling, and urban crime. If imitation is the sincerest form of flattery, the fact that there are now more than fifty institutions around the world modelled on the SFI suggests that the ideas promoted by its founders had real merit.[93]

Their service on the SFI Science Board put Phil Anderson and Murray Gell-Mann together on a regular basis. The two had not met when Gell-Mann responded to Phil's blast solicitation letter of 1964 and submitted a paper to his fledgling journal, *Physics*.

---

[89] P.W. Anderson, "How to Follow a Great First Act," *Santa Fe Bulletin*, Winter–Spring 1989.

[90] Magda Fontana, "The Santa Fe Perspective on Economics: Emerging Patterns in the Science of Complexity," *History of Economic Ideas* 18(2), 167–96 (2010).

[91] Compare the latest edition of the neoclassical textbook by Paul A. Samuelson and William D. Nordhaus, *Economics*, 19th edition (McGraw-Hill, New York, 2009) to the SFI-influenced treatment by the CORE Team, *The Economy: Economics for A Changing World* (Oxford University Press, Oxford, UK, 2017). Free online at https://www.core-econ.org/.

[92] James Owen Weatherall, *The Physics of Wall Street: A Brief History of Predicting the Unpredictable* (Houghton Mifflin Harcourt, Boston, 2013), Chapter 6.

[93] Private communication with Laurence Gonzales.

They got better acquainted years later when Anderson spent a semester at Caltech. Hiking turned out to be a common passion, as was a disdain for pseudoscience and insincere politicians.

In part, David Pines had recruited Anderson to the SFI to provide a counterbalance to Gell-Mann. Everyone admired Murray's intellectual leadership, fund-raising skill, and eloquence in promoting the Institute. But there was a down side:

> By sheer intellectual power and force of personality, Murray [tended] to displace every other point of view. The danger everyone saw was that the Institute would just become a vehicle for Gell-Mann's personal enthusiasms.[94]

Anderson's relationship with Gell-Mann was always a bit fraught. In print, Murray generously credited Phil with anticipating what he called the Anderson–Higgs mechanism for mass generation in particle physics.[95] On the other hand, Murray often needled Phil for choosing to work in "squalid-state physics."[96] Gell-Mann neglected to tell Anderson that he had dipped his toe into solid-state physics research himself precisely when the SFI was getting off the ground.[97] According to one observer who enjoyed Gell-Mann's company:

> Murray looked up to Anderson because he feared that Anderson was better than he was. But he also looked down on him because Anderson was not Murray.[98]

For his part, Anderson thought Gell-Mann knew more about almost everything than any living person. But he grew tired of

---

[94] M. Mitchell Waldrop, *Complexity: The Emerging Science at the Edge of Order and Chaos* (Simon & Schuster, New York, 1992), p. 347.

[95] Murray Gell-Mann, *The Quark and the Jaguar: Adventures in the Simple and the Complex* (W.H. Freeman, New York, 1994), p. 193.

[96] George Johnson, *Strange Beauty: Murray Gell-Mann and the Revolution in Twentieth-Century Physics* (Alfred A. Knopf, Inc., New York, 1999), p. 323.

[97] Lester De Raad, Murray Gell-Mann, and Richard Latter, "Developing New Materials at Elevated Temperatures and Pressures," Report RDA-TR-124200–001, Defense Advanced Research Projects Agency, April 1984.

[98] Author correspondence with Laurence Gonzales, September 29, 2019.

Murray's need to show off that knowledge. It also galled him that Gell-Mann's autobiography gave an inaccurate explanation of the flow properties of superfluid helium, a subject Anderson had explained years earlier as an elegant example of emergence by symmetry breaking.[99]

Anderson was honest enough to admit that:

> I was always a little afraid of Murray. He was both very fast and very profound. One time, I was about to speak up in a group and explain the concept of emergence when he began speaking. He explained it as well as I could, if not better.[100]

In the final analysis, Anderson realized he could never have as large an influence on the Santa Fe Institute as Gell-Mann. That would require his physical presence in Santa Fe and he and Joyce were unwilling to give up the time they spent in Port Isaac. Gell-Mann had already built a second home in nearby Tesuque, NM.

Anderson resigned from the SFI Science Board in 1989 but he remained on its Steering Committee for another decade and made many short visits. He particularly enjoyed dropping into sessions of programs focused on a topic he knew nothing about— just to learn. Overall, he regarded his involvement with the Institute as the single most satisfying intellectual experience of his life. It was exciting to help shape a new institution the likes of which no one had ever seen before. In a reminiscence published on the occasion of the Institute's thirtieth birthday, David Pines remarked that, without planning to be, the SFI had developed rather quickly into a center devoted to the study of emergent behavior.[101] More was indeed different.

---

[99] P.W. Anderson, "Coffee-Table Complexities," *Physics World*, August 1964, p. 47–8.

[100] Interview of PWA by the author.

[101] David Pines, "Emergence: A Unifying Theme for 21ˢᵗ Century Science," *Santa Fe Institute Bulletin*, Fall 2014. Available at https://medium.com/sfi-30-founda-tions-frontiers/emergence-a-unifying-theme-for-21st-century-science-4324ac0f951e. Accessed September 2, 2019.

13

The Pope of Condensed Matter Physics

The French physicist Pierre-Gilles de Gennes won the Nobel Prize in 1991 for his groundbreaking work on the theory of liquid crystals, polymers, and other 'soft' condensed matter systems. Nevertheless, the urbane De Gennes always regarded Phil Anderson (ten years his senior) as the leading solid-state physicist of his generation, going so far as to call him the "pope of solid-state physics."[1]

This nickname is apt. Anderson consciously tried to establish doctrine in many of his writings, the faithful paid close attention to his every utterance, and many made special efforts to seek his views and approval. The preceding chapters aimed to demonstrate that, by the beginning of what we called his "middle period" (1968–1986), Anderson had become a dominant figure in the worldwide community of physicists. This chapter looks back over that period and explores some of the less technical aspects of the life such a person leads.

Bell Labs Redux

Phil Anderson's career at Bell Labs (1949–1984) coincided with the glory years of that institution.[2] His stature there grew over time until, by the mid-1960s, he occupied a Director-sized office and he

[1] Laurence Plévert, *Pierre-Gilles de Gennes, A Life in Science* (World Scientific, Hackensack, NJ, 2011), pp. 75, 287.

[2] John Gertner, *The Idea Factory: Bell Labs and the Great Age of American Innovation* (Penguin Books, New York, 2012); A. Michael Noll, *Memories: A Personal History of Bell Telephone Laboratories* 2015. Available at http://quello.msu.edu/wp-content/uploads/2015/08/Memories-Noll.pdf. Accessed September 26, 2019.

was paid a higher salary than anyone else in the theory group. No one was hired or retained in the group without his agreement. This was significant because *eleven* theorists joined the theory group between 1962 and 1973, most of whom remained at AT&T for many years.[3] This was an ideal situation for Anderson who, during those years, developed a *modus operandi* of ceaselessly bouncing ideas off his colleagues. Lively debate was the norm at a morning coffee klatch which rotated among several offices. Eventually, he published papers with all but a handful of the group's permanent members.

Anderson felt a particular kinship with William McMillan and William Brinkman. McMillan shared Phil's need to challenge conventionality. He wore graphic T-shirts to work at a time when most everyone else wore a coat and tie. He also grew and shaved off his beard periodically so his ID badge never matched his actual appearance. Anderson and McMillan worked to make the BCS theory of superconductivity quantitative and Bill showed Phil that computers could be used in creative ways to do physics. McMillan died tragically in 1984 and Anderson was the obvious choice to write his friend's biography for the National Academy of Sciences.

In 1966, Bill Brinkman was poised to make the step from postdoc to assistant professor. He did not because his PhD mentor advised him to accept an offer of a second postdoc at Bell Labs. He should not pass up the opportunity to work with the best theoretical solid-state physics group in the country.[4] Thirty-five years later, Brinkman retired from Lucent Technologies (the successor to AT&T) as Vice-President for Research. Throughout the decade

---

[3] The theorists hired by the Bell Labs Theory Group between 1962 and 1973 were Richard Werthamer, William McMillan, Pierre Hohenberg, Don Hamann, Bertrand Halperin, T. Maurice Rice, William Brinkman, Joel Appelbaum, James C. Phillips, Chandra Varma, and Patrick Lee. McMillan stayed the shortest time (eight years) and Hamann stayed the longest time (thirty-six years).

[4] Interview of William F. Brinkman by the author, March 14, 2016. Brinkman's PhD supervisor was Bernard Goodman.

of the 1970s, he and Anderson co-authored ten papers on subjects ranging from semiconductors to superfluid ^{3}He (see Chapter 10) to liquid crystals. Only Brinkman's ascent into upper management brought their technical work together to an end.

In January 1975, Anderson's manager asked him to make a presentation to the Bell Labs Research & Development Council (its senior R&D governing body) about possible future directions for its Physical Sciences Division. Anderson's talk began with a short review of the Division's activities and then focused on two subjects he thought were ripe for study by the Division's scientists: non-crystalline solids and solid surfaces.[5] Both later became rich subjects for physicists.[6]

Anderson told the top administrators on the Council that:

> I feel it is all too seldom that the great value of formal and informal collaborations across organizational lines is emphasized at the higher levels of management and I feel that all of us should do everything we can to foster it.

Looking farther into the future, Anderson identified fluid turbulence and wave propagation in nonlinear systems as exciting and appropriate topics for the Division to investigate. Indeed,

> since we have a great interest in understanding the organization of complex communications systems, it seems suitable for us to take an interest in research in cell and organism development and the nervous system and the brain. We should try to foster communication between artificial intelligence and other computer work, and the biological approach.

This statement preceded by three years Anderson's own entry into research in this class of problems.

---

[5] P.W. Anderson, transcript of oral presentation to the Bell Labs Research & Development Council, April 1975, AT&T Archives, Warren, NJ.

[6] See, e.g., Richard Zallen, *The Physics of Amorphous Solids* (Wiley, New York, 1983) and Andrew Zangwill, *Physics at Surfaces* (Cambridge University Press, Cambridge, 1988).

Anderson spent the years 1976–1984 as Consulting Director of the Physical Research Laboratory. This job—created just for him—required a certain amount of administrative work, but it mostly gave him the ear of the Director. His last boss, Arno Penzias, used a weekly breakfast meeting to chat with his 'consultant' whom he regarded as having "high character, iron-clad integrity, tremendous curiosity, and a razor-sharp ability to get to the heart of any matter."[7] Penzias took seriously the advice he got from Anderson. Among other things, he shut down a substantial effort devoted to Josephson effect logic elements based, in part, on Phil's description of the relevant physics and his analysis of the project's prospects.

Anderson self-identified strongly with Bell Labs. He retired soon after the court-ordered break-up of AT&T caused Bell Labs to fission and change mission.[8] It pained him to watch successive owners slowly dismantle the magnificent institution he loved so much. The final straw occurred in September 2002 when Bell released a report revealing that one of its young scientific stars, Jan Hendrik Schön, had committed scientific fraud on multiple occasions.[9] Many of Schön's most spectacular results, it turned out, were the result of data-faking only a little less crude than the medical researcher who had used a magic marker to discolor the transplanted skin tissue of a mouse.[10]

Schön was able to hide in plain sight because his Bell Labs supervisor and co-author, Bertram Batlogg, had a very good reputation inside and outside the Labs. Batlogg's initial response was "when I am a passenger in a car and the driver drives through a red light, then I am not responsible."[11] This statement offended

[7] Interview of Arno Penzias by the author, January 12, 2016.

[8] Gloria B. Lubkin, "Bell Labs Fissions, Yielding AT&T Bell Labs and Bellcore," *Physics Today* **37**(5), 77 (1984).

[9] Eugenie Samuel Reich, "The Rise and Fall of a Physics Fraudster," *Physics World* **22**(5), 24–9 (2009).

[10] Joseph Hixson, *The Patchwork Mouse* (Anchor Press, New York, 1976).

[11] Quoted in Kenneth Chang, "On Scientific Fakery and the Systems to Catch It," *New York Times*, October 15, 2002, F1.

Phil's sense of personal and scientific integrity. Anderson had assisted the Bell Labs team looking into Schön's activities and he knew Batlogg personally. Eventually, Batlogg took responsibility and Anderson drew the final conclusion that "some combination of temptation, pressure, and enthusiasm overcame any natural caution that Batlogg may have had."[12]

Princeton

Two years before the celebrity of the 1977 Nobel Prize descended on him, Anderson feared that shuttling back and forth between Britain and the United States would prevent him from having any influence on the scientific establishment and culture of either country. A desire to do this had begun to form in his mind a few years earlier. However, as he wrote in his official Nobel Prize biography, "the sense of being tourists in two cultures with no really satisfactory role in either, led Joyce and me to reluctantly return to the United States."[13] Lobbying initiated by his friend John Hopfield paid off and he exchanged his half-time professorship at the University of Cambridge for a similar position at Princeton University.

Back in New Jersey, the Andersons sold the handsome colonial-style house Joyce had designed for them twelve years earlier and moved into a modern Scandinavian house two miles away. She designed and supervised the construction of that house also. The new house sat on six acres of unspoiled land which catered wonderfully to their passion for horticulture. Bell Labs was a bit closer than before and Princeton was only an hour away.

Princeton had built its physics faculty around two eminent theorists, John Wheeler and Eugene Wigner. This produced a department focused on general relativity, astrophysics/cosmology, plasma physics, particle physics, and mathematical physics.

---

[12] P.W. Anderson, "When Scientists Go Astray," in *More and Different: Notes from a Thoughtful Curmudgeon* (World Scientific, Hackensack, NJ, 2011), pp. 204–17.

[13] P.W. Anderson, Nobel autobiography, 1977.

Similar topics dominated the research at the nearby (two miles) Institute for Advanced Study which J. Robert Oppenheimer had overseen for twenty years after World War II. When Anderson arrived on the Princeton campus in the fall of 1975, John Hopfield was the only other tenured solid-state physicist on the faculty. Also present was Phil's former PhD student, Richard Palmer, who was working as a Lecturer.

The condensed matter cupboard was even barer two years later when Princeton celebrated Anderson's Nobel Prize. Palmer had left, Hopfield had switched completely to theoretical biophysics, and the tiny experimental solid-state effort at Princeton had all but disintegrated. Phil had only his graduate students, a new postdoc, and (he hoped) the clout to build a strong condensed matter group reasonably quickly. This was not to be.

The Princeton physics department operated on a system where half the salary of every Assistant Professor came from the department's teaching budget and the other half came from the research grant of a senior faculty member. There was little or no expectation that any Assistant Professor would survive and be granted tenure. Anderson did not always judge junior talent accurately and when he did, he was not very adept at the politics of the promotion and tenure process. It did not help that he dismissed mathematical physics, one of the strengths of the department, as "a subject where people try to prove things that everybody already knows."[14] Whatever the reasons, each of the five junior faculty members he supported over the next fifteen years came and went; some by their own choice, some not.

The strategy of the physics department was to induce already-successful scientists to leave their home institutions and come to Princeton as tenured professors. Some of Phil's faculty colleagues had little knowledge of, and even less interest in, solid-state physics. As a result, he frequently found himself outvoted when

---

[14] Interviews of James Sethna (January 30, 2015) and Raphael Benguria (October 10, 2016) by the author.

the faculty considered candidates for senior faculty positions.[15] Anderson's personal friendship with the experimentalist Val Fitch was one of his few bridges to Fitch's particle physics colleagues.

An implacable opponent was Eugene Wigner. In the 1930s, Wigner had supervised the solid-state physics PhD theses of Frederick Seitz, John Bardeen, and Conyers Herring. Unfortunately, Wigner appeared to believe that nothing fundamental was left to do in the field he helped create.[16] It frustrated Anderson that Wigner thought that the BCS approach to superconductivity was incorrect because it supposedly violated a theorem Wigner had proved.[17]

Anderson had to wait ten years before the Princeton physics department hired its first senior condensed matter experimentalist.[18] A senior theorist followed two years later and only then did hiring in condensed matter physics begin to occur more frequently. Observers cite two circumstances for influencing the department to make a commitment to condensed matter physics in the mid-to-late 1980s. One was the steadily increasing clout of younger particle physicists who, like Peter Higgs in Edinburgh, had benefitted personally from the mathematical connections between their field and condensed matter physics.[19] The other was the fear of the train leaving the station without them when high-temperature superconductivity and soft-condensed matter physics suddenly became fashionable.[20]

[15] Interview of Nai Phuan Ong by the author, May 3, 2016.

[16] Interview of Ravindra Bhatt by the author, May 3, 2016. Wigner refers to Conyers Herring's "brilliant career as an applied physicist" in Eugene P. Wigner and Andrew Stanton, *The Recollections of Eugene P. Wigner* (Plenum Press, New York, 1992), p. 167.

[17] Magdolna Hargittai and István Hargittai, *Candid Science IV. Conversations with Famous Physicists* (Imperial College Press, London, 2004), Chapter 29.

[18] Nai Phuan Ong moved to Princeton from the University of Southern California in 1985. The Electrical Engineering department had hired the Bell Labs experimentalist Daniel Tsui three years earlier.

[19] Interview of Thomas Banks by the author, October 31, 2019.

[20] Interview of Nai Phuan Ong by the author, May 3, 2016.

Anderson supervised a dozen senior thesis students and twenty-seven doctoral students at Princeton between 1975 and 1993.[21] The experience of his Princeton PhD students was similar to that of the Cambridge students who had preceded them. He would suggest interesting problems and then wait for them to solve them. He would tell a student if he or she was on the right or wrong track, but he did not offer much advice or explain his own methods of problem solving.[22] Anderson's nickname among his graduate students was "the Delphic Oracle" because they routinely left private meetings with him at a complete loss to understand what he had told them, only to have his remarks make sense to them days, weeks, or even months later.[23]

This helps explain why there is no "Anderson school" of physicists analogous to, say, the Landau or Slater schools where a particular style of physics dominated and then was transmitted to the next generation. On the other hand, Anderson encouraged his students to attend the weekly solid-state seminar at Bell Labs and to make connections with the scientists they met there. Introducing oneself as a PhD student of Phil Anderson opened every door in Murray Hill and several students wound up interacting with (and even collaborating with) Bell Labs personnel.

When it came to teaching, Princeton required somewhat less from Anderson, the Joseph Henry Professor of Physics from 1975 to 1997, than it required from his chair's namesake in 1833. Joseph Henry's duties included six hours of teaching per week plus daily

[21] Anderson's final PhD student graduated in 2010. He supervised a senior thesis student in 2006 and a final one in 2013.

[22] Interviews of James Sethna (January 20, 2015), Piers Coleman (May 2, 2016), and Gabriel Kotliar (May 3, 2016). Author correspondence with Henry Greenside, Clare Yu, and Ross McKenzie.

[23] Daniel Stein, "Round-Table Discussion about the Life and Science of Philip W. Anderson", 2020 Global Summit of the Institute for Complex Adaptive Matter, July 15, 2020. Available online at https://www.youtube.com/watch?v=ifwXo-6VOrs.

attendance at 5:00 am chapel.[24] The mostly graduate-level classes Anderson taught focused on many-body theory, solid-state physics, statistical mechanics, disordered matter, two-dimensional magnetism, and other special topics. By all accounts, he often seemed ill-prepared and was not a good lecturer. He co-taught freshman physics once and was annoyed to learn that the students did not know he was a Nobel Prize winner.[25]

Anderson brought ten years of writing and re-writing to a conclusion with the 1984 publication of *Basic Notions of Condensed Matter Physics*. This is his magisterial identification and exposition of the fundamental principles of condensed matter physics. It is an unusual book, half original text and half reprints from the research literature. The latter (twenty-two papers in total, nine by Anderson himself) includes both "classics" and others chosen to illustrate points brought out in the text.

Basic Notions makes no attempt to explain the vast phenomenology of condensed matter physics. Instead, it aims to "present the logical core of the discipline of many-body physics as it is practiced [and to] supplement and deepen the [reader's] understanding of what he first learned elsewhere."[26] Anderson is explicit about the philosophy that drove him to write the book:

> Most people entering research find that by far the most difficult question is where to start, especially when confronted with something that is actually new. This, it seems to me, is the kind of question a book like this should be designed to answer. Many books are simply compendia of methods that have already been used or techniques for calculating a little better something that is already understood....[When] faced with a genuinely novel problem, it is far more important to have some idea of what the

[24] Albert E. Moyer, *Joseph Henry, The Rise of an American Scientist* (Smithsonian Institution Press, Washington, DC, 1997), p. 139.

[25] Correspondence with V.N. Muthukumar, April 7, 2016. Interview of Kirk McDonald by the author, May 4, 2016.

[26] P.W. Anderson, *Basic Notions of Condensed Matter Physics* (Addison-Wesley, Reading, MA, 1984), pp. 4–5.

relevant questions are than it is to do any one calculation with great accuracy or rigor.

Anderson identifies broken symmetry, adiabatic continuity/ model building, and renormalization as the three basic principles of his subject.[27] His discussions of broken symmetry and renormalization begin with the basics as presented earlier in this book, but he then explores many applications and elaborations. He defines "adiabatic continuity" as the process of replacing a complex many-body system by a simple one which, in an average way, has already accounted for much of the system's many-body character. Anderson offers Landau's Fermi liquid theory of quasiparticles (Chapter 9) and his own analysis of the Kondo model as examples of the general method.

Basic Notions discusses a wide range of many-body problems, but never in a conventional manner. Anderson's interest is always to demonstrate that his handful of basic notions reappear over and over in many-body systems, albeit in different guises for different problems. The ten entries in a two-page "Master Table of Broken Symmetry Phenomena" serve this purpose and illuminate the depth, subtlety, and coherence of his subject. On the other hand, the book can be a challenge to read, mostly because it assumes a knowledge of condensed matter physics more typical of a professional than of a graduate student. It also suffers from an overly elliptical prose style. This is so despite a heroic effort by his Princeton PhD student Clare Yu, who edited the entire manuscript and rewrote parts of the earliest chapters.[28]

For someone of Anderson's stature, *Basic Notions* was not widely reviewed. The most useful review (by one of his former Cambridge PhD students) remarks that it is neither a textbook nor a research monograph but "more of an extended original work of the type

[27] Anderson actually states only two basic principles: (1) broken symmetry and (2) adiabatic continuation and renormalization. I separate the latter into two principles to more accurately reflect the discussion he gives in his text.

[28] Author correspondence with Clare Yu and Susan Coppersmith.

which is now very rare in science."[29] This may explain why an unscientific survey of a dozen condensed matter physicists produced reactions ranging from "it blew me away" and "very deep" to "undisciplined and idiosyncratic" and "I returned it for a refund."

The true impact of *Basic Notions* shows up clearly in the twenty-first century pedagogy of condensed matter physics. Lecture notes on the web speak often and in depth about the importance of broken symmetry, disorder, and renormalization. These topics are particularly prominent in textbooks of the subject published in the last twenty years or so. Like their predecessors, these books work hard to explain the most important phenomenology of the subject. Unlike their predecessors, the new books spend time making the connection between individual phenomena and the basic notions that Anderson felt were so important. The preface of one graduate-level textbook makes this explicit:

> We acknowledge our indebtedness to *Basic Notions of Condensed Matter Physics*, an inspiring book that is full of creative ideas. We try to make concrete some of these ideas, illustrating them with examples and placing them in suitable contexts.[30]

On the Road

Phil's status as a scientist ensured that, for many decades, he traveled about once per month to lecture or consult. A long-standing priority was to attend the one-week March Meeting of the American Physical Society. This was (and remains) the most important annual conference where condensed matter physicists gather to learn about each other's work.

Anderson gave many talks at March Meetings over the years, often at the invitation of its organizers. But it was in the hallways

[29] John C. Inkson, Review of "Basic Notions of Condensed Matter Physics" by Philip W. Anderson, *Foundations of Physics* **15** (8), 915–17 (1985).

[30] Feng Duan and Jin Guojun, *Introduction to Condensed Matter Physics, Volume 1* (World Scientific, Singapore, 2005), p. v.

outside the presentation rooms and at meals that he did his most important work: learning from experimenters about their latest results and trading insights with theorists whose opinions he respected. Unfortunately, the sheer size of the March Meeting (about 5000 attendees through most of the 1990s) and its hectic schedule is not conducive to extended discussions. For that purpose, Anderson relied on Gordon Conferences, special topic workshops, and summer schools.

The Gordon Research Conferences began in 1931 as occasions for small groups of scientists to discuss their work in a leisurely manner. Anderson attended his first Gordon Conference in the summer of 1961 at the Kimball Union Academy, a boarding school in the New Hampshire village of Meriden. The tradition at all Gordon Conferences is to leave afternoons unscheduled to maximize informal discussions among the participants. Phil did this, but he also walked the isolated countryside and challenged all comers to compete on the first-rate tennis courts. No other Gordon Conference venue ever appealed to him more than the isolated and bucolic Kimball Union Academy.

Invitations to lecture at workshops and summer schools gave Anderson the opportunity to travel the world. The fact that these events are staged at stunningly beautiful places like Banff (Canada), Cargèse (Corsica), Lake Como (Switzerland), Lerum (Sweden), Les Houches (France), Naples (Italy), St. Andrews (Scotland), and Trieste (Italy) is one of the perquisites of the physics business at the highest level. The quality of the food and the accommodations could be variable, but the price was always right. If the organizers could not cover all of his expenses, Bell Labs or his US government research grants made up the difference. Joyce only occasionally accompanied Phil on these trips. She had little interest in the activities that conference organizers typically prepared to entertain spouses.

In 1973, Anderson attended a symposium sponsored by the Nobel Foundation at a country villa near Göteborg, Sweden. The subject of the symposium was many-body theory and most of the movers and shakers in theoretical condensed matter physics were present.[31] By remarkable luck (or remarkable prescience by the organizers), the conferees listened to six future Nobel winners present their latest work, four for the first time at any length in public.[32] Anderson presented two technical talks and the conference summary. He later characterized the entire experience as "magical" and he was not alone in thinking so.[33] Fine tennis courts, opera and ballet performances, and a troupe of Swedish folk dancers who performed at a sumptuous conference banquet rounded out the kind of amenities that A-list physicists enjoy on a fairly regular basis.[34]

Anderson inevitably made many visits to the two most important centers for physics in the United States: the Aspen Center for Physics and the Institute for Theoretical Physics in Santa Barbara.[35] His experiences at these two were very different, at least as regards his influence on the development of American theoretical physics. The Aspen Center for Physics (ACP) began in 1962 as part of the Aspen Institute for Humanistic Studies. It split off as an independent organization in 1968 to provide:

> a place for physicists to work on their own problems during the summer, in a stimulating physics atmosphere, and in a location

[31] *Collective Properties of Physical Systems*, edited by Bengt Lundqvist and Stig Lundqvist (Academic Press, New York, 1973).

[32] The future Nobel Prize winners who spoke were Ken Wilson (the renormalization group and its use to solve the Kondo problem), Robert Richardson and David Lee (experimental discovery of superfluid ^{3}He), Walter Kohn (density functional theory of electronic structure), Alan Heeger (one-dimensional metals), and Anthony Leggett (theory of superfluid ^{3}He).

[33] P.W. Anderson, "Stig Olov Lundqvist, 9 August 1925–6 April 2000," *Proceedings of the American Philosophical Society* **147**, 287–91 (2003); Interview of William Brinkman by the author, March 14, 2016.

[34] Bengt Lundqvist, "Walter Kohn—Points of Contact," in *Walter Kohn: Personal Stories and Anecdotes Told by Friends and Collaborators*, edited by Matthias Scheffler and Peter Weinberger (Springer, Heidelberg, 2003), p. 142.

[35] Anderson believed that the organization of the Institute for Advanced Studies in Princeton, NJ into distinct departments severely hampered its effectiveness. That is one reason he and Murray Gell-Mann made sure this structure was not used at the Santa Fe Institute.

with pleasant surroundings and natural beauty....Unlike most summer institutes, which are organized around an intensive program of lecture courses, physicists pursue their work [here] with minimal distractions. The emphasis [is on] individual research and informal interchange of ideas.[36]

Murray Gell-Mann and David Pines were founding Trustees of the ACP and most of its visitors during its first years were particle physicists and astrophysicists. David Pines developed an interest in neutron stars (the interiors of which were believed to be in a superfluid state) and Anderson got interested in them also after hearing Pines speak about them at Bell Labs.[37] Pines did his neutron star work at Aspen and he invited Phil to join him there for part of the summer of 1974.

Elihu Abrahams (later of Gang of Four fame) was the only other solid-state visitor to Aspen at the time and he joined Pines in encouraging Anderson to visit again the following summer. Later, Phil learned that part of their motivation was that the National Science Foundation (which began funding the ACP in 1972) wanted to see more solid-state physics at the Center to prevent it from becoming "a mountain climbing club for the particle physics elite."[38]

Anderson loved the informality of Aspen that made it possible for the supernovae expert Stirling Colgate to present a seminar about fluid turbulence on an outdoor patio wearing only a pair of

[36] Excerpts from a letter sent to physicists publicizing the new Institute. Reproduced at the end of Chapter II of "The First 35 Years of the Aspen Center for Physics" by Jeremy Bernstein. Available at https://aspenphys.org/aboutus/history/first35years/chp2.html. Accessed October 3, 2019.

[37] Eventually, two of Anderson's PhD students at Cambridge worked on neutron stars. With a visitor, Anderson published a paper on the subject himself which won an award from the Observer newspaper of London for the scientific paper with the most opaque title. Namely, P.W. Anderson and N. Itoh, "Pulsar Glitches and Restlessness as a Hard Superfluidity Phenomenon," *Nature* **256**, 23–25 (1975).

[38] Interview of PWA by P. Chandra, P. Coleman, and S. Sondhi, March 8, 2002, Niels Bohr Library & Archives, American Institute of Physics, College Park, MD.

faded shorts.[39] It was an equal pleasure to climb the local mountains, hike to the nearby hot springs, and play tennis or volleyball whenever he liked. Anderson made a decision to return often and, in just a few years, he and Pines and Abrahams succeeded in making condensed matter physics a significant part of the center's activities.[40] Discussions about disordered systems, ³He, spin glasses, and phase transitions mixed well with the sounds of classical music that drifted across a meadow from the adjacent Aspen Music Festival and School. The year 1977 was a watershed when 40 condensed matter physicists converged on the center, including a group of eight Russian physicists led by Lev Gor'kov.

As the 1980s began, Anderson found that the flat administrative structure of the ACP made it easy for him to play an increasing role in shaping the Center's programs. He and a few others concentrated on a few key areas while remaining sensitive to the need to foster workshops on topics of more general interest. One success story was a paper about the quantum Hall effect authored by six theorists who hammered out the basic idea for the paper while sitting around an ACP picnic table.[41]

Anderson served as Chair of the ACP Board of Trustees from 1982 to 1986. He instituted winter conferences to supplement the summer programs, secured stable funding for free public lectures, and initiated a plan to bring a new group of Russian physicists to the Center.[42] But his real influence for about twenty years was raising the profile of condensed matter physics in the more general theoretical physics community and playing a major role

[39] Interview of PWA by Ravindra N. Bhatt, March 12, 2012. Videotape of the interview provided to the author courtesy of R.N. Bhatt.

[40] Ravindra N. Bhatt, "Condensed Matter Physics at the Aspen Center for Physics during the First Fifty Years." Available at https://www.aspenphys.org/science/sciencehistory/cm.html. Accessed October 4, 2019.

[41] R.B. Laughlin, Marvin L. Cohen, J.M. Kosterlitz, Herbert Levine, Stephen R. Libby, and Adrianus M.M. Pruisken, "Scaling of Conductivities in the Fractional Quantum Hall Effect," *Physical Review B* **32**, 1311–14 (1975).

[42] Author correspondence with Jane A. Kelly, Administrative Vice President, Aspen Center for Physics, January 11, 2016.

in defining the problems the best people chose to work on. Before
he knew it, he had spent at least one summer week in Aspen
25–30 times over a period of forty years.

Another key center for physics in the United States, the
Institute for Theoretical Physics (ITP) at the University of
California at Santa Barbara, was the brainchild of particle physicist Boris Kayser, the manager at the time of the Theory Program
at the National Science Foundation.[43] His goal, realized in 1979,
was to create a year-round center where 25–30 physicists (a few
permanent staff plus junior and senior visitors) would tackle
problems that spanned more than one subfield of physics.

Phil's friend and rival Walter Kohn served as the Institute's first
Director (1979–1984). He was assisted by a 17-member advisory
board, each of whom was replaced after three years and not permitted to serve again. The ubiquitous Murray Gell-Mann and
David Pines were inaugural advisory board members. Anderson
was part of the second wave of advisors. The ITP typically sponsored six-month programs devoted to a single subject with a
week-long intensive workshop to which experimenters were
invited to share their results.

Near the end of Kohn's term, the Institute approached
Anderson about becoming its second Director. The negotiations
got to the point where he and Joyce began looking at property.
Their real estate agent found several acres on a mountain top
overlooking Santa Barbara where Joyce thought she could build
her dream house. Unfortunately, a local history book the agent
gave to Joyce for the flight home happened to mention a famous
inn nearby that had burned down four times. Joyce's fire phobia
had derailed their earlier move to Stanford and now it killed the
ITP deal.

Anderson was disappointed, but he admitted later he probably
would have made a poor Director. He did not like dealing with

[43] Arthur L. Robinson, "Theoretical Physics Institute Gets Go Ahead,"
Science **203**, 1229–30 (1979).

problems filled with gray areas and he lacked Walter Kohn's skill dealing with people whose ideas differed from his own. In the end, Bob Schrieffer became the second Director of the ITP.

This incident notwithstanding, Anderson made a dozen extended visits to Santa Barbara over the years. And even if the ITP did not become quite the interdisciplinary physics think tank imagined by Kayser, Phil found many of its topical programs intellectually stimulating and well worth the time he spent there. A mild disappointment remained only because the rigid three-year rule about advisory board membership meant that his (or anyone else's) influence on setting directions for the Institute was very limited. Of course, that was the purpose for the rule, which continues to this day.

Private Communications

Anderson's behavior around others and their perception of him depended on whether his relationship with them was personal, professional and non-competitive, or professional and competitive. Close friends saw him as a warm, generous, and broad-minded human being with a genuine concern for social justice, economic justice, and environmental protection.[44] He and Joyce gave their time freely to friends and acquaintances who were in distress, but they rarely talked about it. On several occasions, Phil arranged for a troubled former apprentice to return to Princeton for a period of time so he could help that person regain their bearings.[45]

At the professional level, theoretical physicists do not compete with experimentalists and Anderson developed strong personal bonds with several experimental physicists. One recalled his delight at watching Anderson crawl around on the floor to entertain a very

[44] Author correspondence with Ingrid Kreissig, August 31, 2109, and interviews of David Pines (March 16, 2016), Nai Phuan Ong (May 3, 2016), and Claire and David Jacobus (May 5, 2016) by the author.

[45] Interview of Nai Phuan Ong by the author, May 3, 2016.

young child.[46] Another praised his patience while teaching a woman and her children to ice skate on a frozen lake.[47] On the other hand, experimenters had a duty to the profession and Anderson had little patience if they did not carry out that duty.

One experimenter remembers Phil yelling at her for keeping some puzzling Kondo effect data in her drawer rather than publishing it for all to see and ponder.[48] Another recalls Anderson bursting into his office to complain about the falseness of a theoretical interpretation the experimenter had supposedly applied to his data in a paper. Phil turned around sheepishly and left after being shown that the paper in question made no such statement.[49]

Theorists were another matter. Anderson was supremely self-confident and a fierce competitor. He was loyal and supportive if you earned his respect and brusque and impatient if you didn't. He could be withering and merciless when criticizing senior colleagues. But he was also capable of showing professional and personal kindness to younger theorists, including some he neither supervised nor employed.[50]

Anderson could be an enjoyable companion and a charming storyteller, if not a joke-teller. He enjoyed good food and good drink and, at least once at a summer school, he led the singing and drinking at a pub well into the wee hours.[51] Seven years after winning the Nobel Prize, he used an assumed name and wore a crude disguise (glasses and a fake moustache) to present a poster (an activity usually reserved for students and postdocs) at a four-day

[46] Interview of Albert Libchaber (May 9, 2015) by the author and correspondence with Hans Ott (August 12, 2016).

[47] Author correspondence with Theodore Geballe, December 26, 2015.

[48] Author correspondence with Myriam Sarachik, February 7, 2018.

[49] Interview of E. Ward Plummer by the author, November 20, 2015.

[50] Private communications with Premala Chandra, Martha Redi, Paul Soven, and Wayne Saslow.

[51] Interview of Milton Cole by the author, March 19, 2015.

international conference where fully ten percent of the talks presented had the phrase "Anderson model" in their titles.[52]

An indirect way for Phil to communicate with other physicists was by refereeing manuscripts sent to him by the editors of scientific journals. This task—performed by almost all working scientists—involves advising the editor whether the manuscript is suitable (or not) for publication in the editor's journal.

When he was positive about a paper, Anderson usually wrote a short report advising the editor to proceed with publication. When he was negative about a paper, he often wrote at length and combined detailed technical objections with advice for the author(s). One example among many is:

> I think the only honorable procedure is that the authors state clearly that their results, if correct, completely invalidate the explicit claims of Anderson and his coworkers, as well as disagreeing with the slightly less rigorous approaches of X, Y, and Z. After all, X has been wrong on such substantive matters before, and I suppose Anderson may have been; but before doing so it might be wise to inquire whether they have ever both been so disastrously wrong at the same time.[53]

With other negative reports, Anderson felt compelled to offer a *bon mot*:

> This paper will add immeasurably to the confusion on this subject and should not be published. It is a pity that the author's earlier paper cited as Reference 1 cannot be "unpublished".
>
> Like many of his other papers, this work has a pedantic character that is the author's greatest weakness.
>
> I don't know whether to be amused or sad that no one in the amazingly long list of individuals thanked by the author in his

---

[52] Author correspondence with Hans-Rudolf Ott (August 12, 2016) and Piers Coleman (November 23, 2019). The venue was the International Conference on Valence Fluctuations at Cologne, Germany, August 27–30, 1984.
[53] I have replaced the names of the physicists mentioned in this referee report by X, Y, and Z.

acknowledgements failed to see the basic physical fallacies of this paper or at least failed to convince the author of them.

This paper sets up a straw man and then knocks it down with great fanfare, arriving in the end at precisely Anderson's conclusions but very poorly understood and stated.

Anderson sometimes felt that a just-published paper, or a preprint sent to him by its author, was seriously flawed. If so, he occasionally felt compelled to write a short paper (called a Comment) to point out the paper's problems. He always sent a copy of his Comment to the author at the time he submitted it for publication.

In 1970, Anderson did exactly that after receiving a preprint from his friend and former postdoc, Marvin Cohen.[54] Cohen and a PhD student had used a computer-based method to calculate approximately the electron wave functions for six different crystalline semiconductors. Using these, they computed the charge density of the electrons and discussed what chemists called the "bond charge" in the region of space between nearest-neighbor atoms.

Anderson had worked on related issues with some of his Cambridge PhD students, but their quantitative results were limited to molecules. That experience and his intuition told him that Cohen's results for the bond charge could not be correct. He guessed that numerical truncation was to blame, a suggestion which supported his almost primal distrust of using computers to make quantitatively accurate statements about solids. He detailed his objections in a manuscript and sent a copy to Cohen.[55]

Unlike Anderson, Cohen had extensive personal experience with numerical methods and their pitfalls for solid-state calculations. Indeed, one of his reasons to specialize in using computers to study crystals was because he did not have to worry about

[54] John P. Walter and Marvin L. Cohen, "Electronic Charge Densities of Semiconductors," *Physical Review Letters* **26**, 17–19 (1971).

[55] P.W. Anderson, "Bond Charges in Semiconductors: A Comment on 'Electronic Charge Densities of Semiconductors'," unpublished manuscript, AT&T Archives, Warren, NJ.

stretching his intuition beyond its limits or rely on perturbation theory to get physically meaningful results. He implored his former mentor not to submit his Comment, saying:

> Phil, 99.99% of the time you win, but you are going to lose on this one. I guarantee it's going to be embarrassing for you because they are going to do the experiment and it's going to come out the way I say. I've done this for other things and it's correct. Please don't do this. Not for my sake, but for yours.[56]

For whatever reason, Anderson relented and withdrew his Comment from consideration for publication. It was good that he did so. Experiments a few years later showed that Cohen's description of the bond charge was correct.[57]

Anderson reacted angrily to a manuscript only when he suspected that an author had compromised the integrity of the scientific enterprise. This occurred when he and the theoretical chemist William Goddard III (a fellow member of the National Academy of Sciences) participated in a 1988 symposium devoted to chemical valence. Earlier that year, Goddard had published a theory to rationalize the existence of a class of recently discovered superconductors whose transition temperatures were much higher than those observed for any previously known superconductor.[58] Phil had seen a version of Goddard's symposium talk ahead of time and he prepared a set of remarks for publication in its proceedings.[59]

Anderson charged that Goddard had successfully avoided the normal referee process, dismissed criticism of his work provided

[56] Interview of Marvin L. Cohen by the author, March 2016.

[57] Y.W. Yang and P. Coppens, "On the Experimental Electron Distribution in Silicon," *Solid State Communications* 15, 1555–9 (1974).

[58] Yuejin Guo, Jean-Marc Langlois, and Willliam A. Goddard III, "Electronic Structure and Valence-Bond Band Structure of Cuprate Superconducting Materials," *Science* 239, 896–902 (1988). See Chapter 14.

[59] P.W. Anderson, "Discussion Remarks," *Proceedings of the Robert A. Welch Foundation Conference on Chemical Research. XXXII. Valency* (Robert A. Welch Foundation, Houston, TX, 1988), pp. 36–8.

by physicists who had heard him speak about it, and ignored all previous theoretical work on the problem. Worse,

> Dr. Goddard claims to calculate the transition temperatures for five or more superconductors to three-figure accuracy using a single, simple formula.... For the older [conventional] superconductors, McMillan and others got good one-figure accuracy for T_c using a much more sophisticated theory and several empirical inputs.
>
> Goddard computes a number of other [measurable quantities] with equally implausible accuracy. Equally serious is the fact that at least one of the numbers used in Goddard's formula for T_c is known [experimentally] and the value he gives is wrong by a factor of five, even though it enters very sensitively into the calculation of T_c. This kind of disregard of experimental fact permeates the theory.
>
> Science in the US is under considerable stress from restricted funding levels and general public distrust. It does science harm to promulgate shoddy work and raise unrealistic expectations, and to mislead the public as to the actual level of thought and effort necessary to solve an important problem like this one.

Goddard tried to respond, but Anderson made it clear he was not interested in a dialog. The symposium continued without further drama.

The Nobel Prize

To non-scientists, a Nobel Prize is the ultimate recognition that the scientific world can bestow on its practitioners. Scientists know that winning a Nobel Prize is a bit of a crap shoot and that politics and lobbying play a role.[60] Those who win are usually deserving, even if many who are deserving do not win. Our conclusion (Chapter 11) is that Anderson was deserving, whatever the circumstances that led to the decision.

[60] See James R. Bartholomew, "One Hundred Years of Nobel Science Prizes," *Isis* **96**, 625–32 (2005).

Phil was transplanting a pussy willow in his garden when the Associated Press called with the news that he, John Van Vleck, and Nevill Mott had been awarded equal shares of the 1977 Nobel Prize for Physics. Joyce was delighted and flustered. The phone never stopped ringing with inquiries from news outlets. Anderson scrapped his original plan to spend the day at Princeton and went to Bell Labs instead. Not only had he done his prize-winning work at Bell Labs, its publicity department was vastly more efficient than Princeton's.

The Labs hosted a reception, facilitated several telephone interviews, and arranged a news conference attended by twenty representatives of major media.[61] Anderson told *United Press International* that his work on localizing electrons in non-crystalline materials was novel and important because so many important materials were not crystals. The *New York Times* wanted to know what experimental work Anderson had done. The *New York Daily News* asked whether his theories might be applied to telephones or toasters.

The *Daily News* continued the low intellectual tone set by its electrical appliance question in its article about the award. The headline and photograph on the front page of its October 12, 1977 issue virtually guaranteed that its readers would confuse the new Nobel Prize winner Anderson and his wife with a local couple apprehended by police for running a sex-for-cash operation (see Figure 13.1). The headline of the Nobel Prize story itself also managed to err by suggesting that Anderson did experimental work.

After the news conference, the Labs arranged a convocation at which several of the Labs's top executives spoke. The vice-president of research, N.B. Hannay, emphasized the significance of Anderson's work to Bell Labs. The president of the entire organization, William O. Baker, stated that Anderson's influence would eventually be seen to reach far beyond what either he or the Swedish Academy had yet recognized.

[61] *Bell Labs News* 17 (42), October 17, 1977, pp. 1–5.

Figure 13.1 Photograph of Phil and Joyce Anderson on the front page of the *New York Daily News* for October 12, 1977. The accompanying story about the Nobel Prize begins at the lower right corner of the page. The main headline at the top of the page refers to different people. Sources: the Associated Press and the *New York Daily News*.

Princeton responded to Anderson's Nobel Prize the next day with a cocktail party and a one-gun salute. At that time, the Princeton physics department required every graduate student to complete a machine shop course. That year, many students elected to make toy cannons out of brass. At the cocktail party, one graduate student filled his cannon with gun powder and fired it off.[62]

Science magazine published a long article about the Prize written by Marvin Cohen and his Berkeley colleague Leo Falicov. They began by reporting that it was "generally believed" that the awards to the three men were for lifetime achievement and that their influence had reached practically every area of condensed matter physics.[63] The article discussed Anderson's prize-winning work, but then continued with a concise evaluation of his entire career to that point:

> He has greatly influenced the development of our understanding of magnetism, superconductivity, quantum properties of helium, tunneling, ferroelectricity, and the electronic structure of crystalline and amorphous solids. He has developed and applied new concepts and techniques, such as broken symmetry, in such a unique and inventive manner that his ideas have found their way into many other branches of physics. He is probably the broadest contributor to the forefront of solid-state theory, and his work almost always carries his particular stamp of originality. Much of Anderson's work has been motivated by the observation of something new or puzzling in experimental data.

In early December, an airplane carrying Phil, Joyce, and Susan Anderson touched down into the semi-darkness that envelopes Stockholm most of the day at that time of the year. They had been coached to expect a week of Nobel Prize pomp and circumstance including lectures, press conferences, TV interviews, receptions, seminars, and even a 7:00 am wake-up serenade by candle-bearing children on St. Lucia's Day.

---

[62] Interview of Raphael Benguria by the author, October 10, 2016.
[63] Marvin L. Cohen and Leo M. Falicov, "The 1977 Nobel Prize in Physics," *Science* **198**, November 18 1977, pp. 713–15.

There were also formal dinners. Before each, Anderson received a card printed with the name of a woman he was to locate, escort to the dining table, and converse with throughout the meal. Joyce was on the receiving end of this custom and, one evening, the elderly ambassador to Sweden from Spain escorted her to her seat. Breaking protocol, the ambassador directed all his attention to the attractive wife of a Belgian diplomat who was seated on his other side. This induced Joyce to break decorum by pouring champagne down his collar to get his attention. The Swedes forced him to offer *her* a formal apology the following day.

Anderson was obliged at one point to make a welcoming speech to a group of about 150 dinner guests. Joyce was amazed at how composed he was and guessed that "he was so high on the whole Nobel thing that he became very verbal."[64] This is consistent with the description of him as a "modest, even shy man" published at the time.[65] Even when speaking before physicists, Anderson was never particularly demonstrative and recordings of his seminars over the years reveal a soft-spoken delivery that borders on mumbling.

A few months after he returned home, Phil's Bell Labs colleagues filled the largest auditorium at the Murray Hill complex and waited for him to repeat his Nobel Prize lecture for their benefit. However, claiming "mental exhaustion" at "having given all or part of my Nobel Lecture eleven times in the last week," he proceeded to talk for an hour about spontaneous symmetry breaking instead.[66] It is not known how many in the audience were disappointed.

Like most laureates in the sciences, the Nobel Prize had relatively little effect on Anderson's life. We have seen that he expanded his technical interests to include complexity, but he mostly continued to work on many-particle problems in condensed matter physics.

[64] Letter from Joyce Anderson to her mother, January 21, 1978. Courtesy of Philip W. Anderson.

[65] "Profiles of 4 Nobel Prize Winners," *New York Times*, October 12, 1977, p. 92.

[66] P.W. Anderson, Transcript of a talk, "Broken Symmetry from Higgson to Boojums," Nobel Symposium at Bell Labs. March 21, 1978.

Autograph-seekers had to ply their trade elsewhere and, except for a few political issues close to his heart (see below), he mostly deflected requests to pontificate. One change in behavior concerned George Feher, the experimenter whose work led to his theory of localization. Feher was a poker player who competed (and won) at the national level.[67] Anderson had always refused to play with him, but Feher's teasing about the amount of money that came with a Nobel Prize forced him to finally agree. The result was predictable.

Political Activism

Anderson never forsook the progressive politics he learned from the University of Illinois Saturday Hikers. Joyce shared his views, having run around in high school with a gang of kids whose parents were faculty members at the University of Chicago. In fact, it was the smart, funny, and formidable Joyce (she reminded some of Katherine Hepburn) who always brought up politics and controversial subjects in private discussions with friends.[68]

The Andersons enthusiastically supported the intellectual liberal Democrat Adlai Stevenson II when he ran for President of the United States in 1952 and 1956. They sensed in Stevenson the same impulses that had made Phil refuse to answer the Bell Labs security questionnaire and later to sign the petition opposing the War in Vietnam. These political expressions grew in time and reached their apogee when Anderson became the most visible scientist to publicly oppose the construction of the six billion dollar Superconducting Super Collider.

Nobel laureates gain social power through their scientific celebrity and Anderson sometimes exploited that power after he joined the club in 1977.[69] He signed public letters to protest the imprisonment

[67] G. Feher, "Playing Poker," *EPR Newsletter* **2** (1–2) 10–12 (2003).

[68] Interview of David Pines (March 16, 2016) by the author and correspondence with Joyce McMillan (January 2, 2016) and T.V. Ramakrishnan (May 10, 2016).

[69] Declan Fahy, "The Laureate as Celebrity Genius: How *Scientific American's* John Horgan profiles Nobel Prize winners," *Public Understanding of Science* **27**, 433–45 (2018).

of the Russian physicist and dissident Yuri Orlov (1978), to call for a nuclear freeze (1984), to reject a proposed oil pipeline as a threat to the environment (1992), to endorse a Comprehensive Test Ban Treaty (1999), and to oppose the war in Iraq (2003).

On three occasions in the twentieth century, US scientists joined together publicly to express their feelings about a weapons system. The first was an attempt in the late 1950s to influence the government to prohibit the testing of nuclear weapons in the atmosphere.[70] The Partial Test Ban Treaty of 1963 forced all future tests of nuclear weapons to be conducted underground.

Anderson involved himself in the other two, both of which amounted to large-scale lobbying efforts to convince Congress to oppose Defense Department plans to develop countermeasures against ballistic missile attacks. In 1969, he opposed a proposed anti-ballistic missile (ABM) system designed to protect missile launch sites. In 1985, he opposed a plan to protect the entire country called the Strategic Defense Initiative.

Anderson was not a leader of those opposed to the ABM, but he and eighty-seven other Bell Labs researchers signed a "Stop ABM" petition that appeared in the *Washington Post*. By and large, they trusted the technical objections to the system voiced in *Scientific American* by their physics colleagues Richard Garwin and Hans Bethe.[71] Phil's action stopped with his signature, but the fact that he identified himself on the petition as a Bell Labs employee was almost enough to cost him his job.[72]

AT&T was the prime contractor for the ABM project and the company's Vice-President for Military Systems was livid when he saw the title above the list of names in the *Washington Post*: "Who Opposes the ABM? The Scientists Who May Have to Build it...That's Who!!"[73] This was unfortunate because almost none of

[70] Paul Robinson, "Crucified on a Cross of Atoms: Scientists, Politics, and the Test Ban Treaty," *Diplomatic History* 35, 283–319 (2011).

[71] Richard L. Garwin and Hans A. Bethe, "Anti-Ballistic Missile Systems," *Scientific American* 218(3), 21–31 (1968).

[72] Interview of William Brinkman by the author, March 19, 2016.

[73] *Washington Post*, July 23, 1969.

the Bell Labs signees were involved in the project or were privy to any of its details. Unbeknownst to the petition organizers, the person who had volunteered to hand-deliver the petition to the *Post* had replaced the "Stop ABM" title with the inflammatory and misleading title printed by the newspaper.[74]

The Military Systems VP demanded that the signees be fired. Luckily for them, he backed down because Al Clogston, the Director of the Physical Research Laboratory at the time (and a five-time co-author with Phil Anderson), threatened to quit if any of his people were fired. In the end, the opponents of ABM were unable to stop the project, but it wound up severely limited by the 1972 Anti-Ballistic Missile Treaty.

President Ronald Reagan created the Strategic Defense Initiative Organization (SDIO) in 1984 and tasked it to develop a nationwide defensive shield. The SDIO solicited the participation of university researchers and many responded by signing a pledge not to accept government money to work on SDIO projects. Phil Anderson, by then a half-time employee at both Princeton University and Bell Labs, got publicity for signing this academic pledge, as well as for sponsoring an anti-SDI petition written for physicists who worked at national and industrial laboratories.[75]

Anderson's outspoken opposition to the SDI and his Nobel Prize celebrity gave him various opportunities to influence others on the subject. For example, he argued against the project in the pages of the *Princeton Alumni Weekly*. An alumnus of the university, the sitting Secretary of State, George P. Shultz, responded and defended the project in the same publication.[76] Phil wrote that

[74] The petition courier and title-switch culprit was Richard Tuck, a notorious Democratic Party operative who made a career out of supposedly harmless political pranks. Correspondence with Anthony Tyson, October 20, 2019.

[75] William Sweet, "Star Wars Petitions Attract Strong Support at Some Schools," *Physics Today* **38**(11), 95–6 (1985); William Sweet, "APS and Academy Members Polled on SDI; Physicists Mobilize," *Physics Today* **39**(11), 81–3 (1986).

[76] The Anderson and Shultz essays are reprinted in P.W. Anderson, *More and Different: Notes from a Thoughtful Curmudgeon* (World Scientific, Hackensack, NJ, 2011), pp. 300–313.

the technical objections to the SDI were even more compelling than those that dogged the ABM. He also asserted that the contemporary political situation made the SDI more politically destabilizing than the ABM had ever been.

Twice, Anderson debated Edward Teller, the Hungarian-American veteran of Los Alamos who was probably the most vocal proponent of SDI among theoretical physicists. One dialog with Teller occurred at a Washington, DC hotel where the audience consisted of several hundred high school seniors and a diverse collection of others, including his Bell Labs boss Arno Penzias. According to Penzias:

> Phil never really had a chance. Teller brushed aside [Phil's] technical objections. With so many nay-sayers proved wrong in the past, Teller said, why should we believe this fellow now? Assuring the audience that science would find a way out of the threat of nuclear destruction, Teller won a standing ovation from the assembled students.[77]

Not long after, the non-profit media organization National Public Radio hosted and recorded a debate between Anderson and Teller at their Washington studios. The debate never aired and Anderson told an interviewer much later that "I lost."[78]

The Superconducting Super Collider

Anderson's interest in national defense policy was heartfelt but ultimately frustrating because he had no real ability to affect decision-making. He hoped the situation would be different when he turned his attention to national *science* policy.

In late 1981, David Shirley, the Director of the Lawrence Berkeley Laboratory, pitched a multi-million dollar project to George

---

[77] A.A. Penzias, "Sakharov and SDI," in *Andrei Sakharov. Facets of a Life* (Editions Frontières, Gif-sur-Yvette, 1991), pp. 507–16.
[78] Interview of PWA by István Hargittai in *The Chemical Intelligencer* 6(3), 26–32 (2000).

Keyworth, the Science Advisor to President Ronald Reagan.[79] Shirley convinced Keyworth to support the construction of what he called the Advanced Light Source (ALS). Electrons injected into the ALS would circulate at nearly the speed of light around a 1/8-mile circumference ring and radiate an intense beam of x-rays that could be used to probe the properties of materials. Shirley embedded his ALS request in a much larger request to fund a National Center for Advanced Materials (NCAM) at Berkeley.

Many condensed matter and materials scientists were unhappy when they heard about NCAM. Shirley had circumvented the standard review process and gone directly to Keyworth. The National Academy of Sciences convened a forum of its Solid-State Science Committee in February 1983 to discuss the matter and Anderson was among those who participated. Afterward, Anderson circulated a much-discussed letter where he remarked that "this is the first case at this level of funding in the history of American science where consensus has not preceded a budget request. I see this as a very dangerous precedent."[80] The precise impact of this letter is not known, but it is a fact that Shirley was forced to separate the ALS request from the NCAM request. The ALS was built eventually, but only after it passed through a series of review committees and was made part of a much larger Department of Energy spending plan for facilities.

Three months after the NAS forum on the ALS, *Science* magazine reported that the US particle physics community was thinking seriously about an unprecedentedly large proton accelerator.[81] A lead editorial in the *New York Times* told readers that recent discoveries at the CERN particle physics accelerator in Geneva provided convincing

[79] Gloria B. Lubkin, "NCAM not Peer-Reviewed, Critics Say; Review Panel Named," *Physics Today* **36**(6), 17–19 (1983).

[80] Quoted in Catherine Westfall, "Retooling the Future: Launching the Advanced Light Source at Lawrence's Laboratory, 1980–1986," *Historical Studies of the Natural Sciences* **38**, 569–609 (2008).

[81] M. Mitchell Waldrop, "High Energy Physics Looks to the Future," *Science* **220**, 809–11 (1983).

evidence for "a theory that unites electromagnetism and the 'weak' nuclear force seen in radioactivity."[82] The editorial quoted George Keyworth that "our world leadership in high energy physics has been dissipated" and added that any future American accelerator "should be designed to win or not be built at all."

By that summer (1983), *Science* readers knew that American particle physicists planned to put all their eggs in one basket and build an accelerator to be called the Superconducting Super Collider (SSC).[83] This mammoth machine would require more than 10,000 magnets (wound using superconducting wire) to guide protons around the 54-mile circumference of the device.[84] The estimated cost was $2 billion spread over a decade. For comparison, the total federal expenditure for research in all fields of science and engineering in 1983 was $14 billion.[85]

Anderson had questioned the need for large particle accelerators a decade earlier. He did so again now, but not until December 1986 when he wrote a short letter to *Physics Today* to support his friend Pedro Echenique, a former member of the Cambridge TCM group.[86] Echenique had marshalled Phil's more-is-different ideas to rebut claims about the exclusive fundamentality of particle physics that figured prominently in a long article promoting the SSC written by the particle physicists Sheldon Glashow and Leon Lederman.[87]

---

[82] "Europe 3, US Not Even Z-Zero," *New York Times*, June 6, 1983, p. A16.

[83] M. Mitchell Waldrop, "Physicists Nix ISABELLE, Endorse Super Machine," *Science* **221**, 344 (1983).

[84] A current-carrying wire wound like a corkscrew over a cylindrical surface produces a magnetic field which points mostly along the axis of the cylinder.

[85] National Science Foundation, Division of Science Resources Statistics, *Federal Obligations for Research by Agency and Detailed Field of Science and Engineering: Fiscal Years 1970–2002*. Table 1A.

[86] Letters to the Editor by P.W. Anderson and reply by Sheldon L. Glashow, *Physics Today* **40**(8), 90–1 (1987).

[87] Letters to the Editor by Pedro Echenique and Sheldon L. Glashow, *Physics Today* **39**(12), 12–14 (1986); Sheldon L. Glashow and Leon M. Lederman, "The SSC: A Machine for the Nineties," *Physics Today* **38**(3), 28–37 (1985).

Glashow and Lederman wrote their cheerleading article because the SSC had attracted criticism from both the scientific community and from members of Congress.[88] A particularly fierce opponent was Donald L. Ritter, a Republican congressman from Pennsylvania who had earned a ScD degree in metallurgy from MIT.[89] The initial cost estimate for the project had tripled and to Ritter, the SSC was "a $6 billion assault on the nation's science priorities and science budget." He ridiculed the "quark-barrel politics" involved in the debates about where to build the machine.

In April 1987, the Committee on Science, Space, and Technology of the US House of Representatives held hearings on the SSC.[90] The first two witnesses were the particle theorist Steven Weinberg and the condensed matter theorist J. Robert Schrieffer, both winners of a Nobel Prize in Physics. Weinberg was bullish about the SSC, as might be expected from one of the inventors of the standard model of particle physics, the theory the SSC had been designed to test. Schrieffer was also supportive of the SSC and he went to some trouble to address a common criticism of the project:

> A negative point of view frequently expressed is that a single large project usually gains a great deal of visibility and therefore can be moved forward at the expense of a collection of smaller projects, regardless of the relative merit of the case. I personally take exception to this view point. Rather, I would view the funding of the SSC as a benchmark for a renewed federal commitment to dramatically improve the level of funding of science as a whole.[91]

Phil Anderson did not attend the hearing, but he submitted testimony in the form of a prepared statement. He knew Schrieffer's

[88] The definitive history of the SSC is Michael Riordan, Lillian Hoddeson, and Adrienne W. Kolb, *Tunnel Visions: The Rise and Fall of the Superconducting Super Collider* (University of Chicago, Chicago, 2015).

[89] A Doctor of Science (ScD) degree at MIT is equivalent to a Doctor of Philosophy (PhD) degree elsewhere.

[90] *Superconducting Super Collider.* Hearings before the Committee on Science, Space, and Technology of the House of Representatives. April 7–9, 1987.

[91] Testimony before the Committee on Science, Space, and Technology of the House of Representatives. April 7–9, 1987, pp. 258–260.

views and was amazed that his colleague did not appreciate the potential of the SSC to drain funds from other areas of science or the unlikelihood that the federal government would increase funds for science across the board. Accordingly, his testimony broke ranks with almost every other technical witness and vigorously opposed the SSC.[92] Representative Ritter drew attention to Anderson's testimony repeatedly and thereby ensured that his virtual presence did not go unnoticed.

Anderson wrote to bury Caesar, not to praise him, and he proceeded to rebut each of the four reasons to support the SSC given by Glashow and Lederman (GL) in their *Physics Today* article: scientific challenge, potential spin-offs, national pride, and sense of duty.[93] Because GL distinguished the "fundamental" challenges of particle physics from the "relevant" challenges of condensed matter physics, Anderson reprised his earlier rejoinder to Victor Weisskopf that the phenomenon of emergence demonstrates that fundamentality and relevance can easily co-exist in complex, many-particle systems.

GL noted that the intense study of superconducting magnets by particle physicists could spin-off "socially relevant technologies" like medical diagnostics. Anderson cited his intimate knowledge of the invention of superconducting magnets at Bell Labs in 1961 and blandly remarked that "the truly innovative part of this technology owes nothing to particle physics."[94]

GL linked the loss of leadership in experimental particle physics to the broader issue of national pride and technological self-confidence. "When we were children," they wrote, "America did most things best. So it should again." Finally, GL asserted that it

[92] One other witness criticized the SSC at this hearing: James A. Krumhansl, a condensed matter theorist and the sitting Vice-President of the American Physical Society.

[93] Sheldon L. Glashow and Leon M. Lederman, "The SSC: A Machine for the Nineties," *Physics Today* **38**(3), 28–37 (1985).

[94] Prepared statement to the Committee on Science, Space, and Technology of the House of Representatives. April 7–9, 1987, Appendix 1, pp. 903–911. See also the final paragraph of this chapter.

was the "sacred duty" of the particle physicist to seek the deepest secrets of the Universe. Not to build the SSC would "terminate the 3000-year-old quest for a comprehension of the architecture of the sub-nuclear world." To these sentiments, Anderson simply stated that:

> It disturbs me to see accelerator physics seen as a nationalistic, competitive race; science is too serious a matter for that. And if the lack of the right accelerator here at exactly the right time is really going to kill high-energy physics, I must say it is better off dead, if only for the crippling lack of imagination it reveals.

By sheer coincidence, these events followed by only a few months the tremendously exciting discovery of a new class of materials that exhibited superconductivity at a much higher temperature than anyone had thought possible (Chapter 14). This had obvious implications for the SSC: should one delay the project to see if superior and/or cheaper superconducting magnets could be made from the new materials?

Anticipating this question, SSC proponents had arranged for a special witness to be present at the hearing. John Hulm was one of the most qualified people in the world to speak to this question. Hulm had introduced Anderson's colleague Bernd Matthias to superconductivity in 1951 and he had been intimately involved in the study, development, and manufacture of superconducting magnets at the Westinghouse Electric Corporation for twenty-five years. Hulm was explicit:

> It would be irresponsible to stop the development of the existing SSC magnets because of the recent discovery of the new high-temperature superconductors....A great deal of scientific and engineering development work lies between the discovery of a new superconductor and the achievement of a sophisticated magnet cable. The development we are facing may take several years, possibly even a decade, [and] it may not be possible at all due to the limitations of the materials, which are brittle ceramics.[95]

---

[95] Testimony of John K. Hulm, Ref. 632, pp. 533–540.

One week later, the *New York Times* explored the connection between the SSC and the new superconductors in a substantial above-the-fold article. However, instead of Hulm, the writer of the article sought a quote from the Nobel laureate Phil Anderson, who accommodated him: "It is important to wait a while. I had not thought this technology was going to move fast enough to make a difference to the Super Collider, [but] I've changed my mind."[96] The most charitable interpretation of this comment is that Anderson was summarizing the view of technologists of his acquaintance who were more optimistic than Hulm.[97]

SSC management chose to stick with conventional superconducting magnets and work on the project continued, even as estimates of its cost increased. By the spring of 1990, the *Times* had come to the view that "intellectual leadership [in particle physics] does not necessarily require owning the biggest and most expensive accelerator" and proposed "buying into the [European] Large Hadron Collider" as a cost-effective alternative.[98]

The apparently ballooning cost to build the SSC emboldened the project's critics in Congress and induced an increasing number of physical scientists to voice their opposition in public. Anderson's early willingness to do this, and his colorful way of expressing himself, led to six separate requests from Congressional subcommittees for his testimony. His arguments evolved somewhat over time, but three themes recur.[99]

In 1991 testimony, Paul Fleury (then the Director of Phil's old Physical Research Laboratory at Bell Labs) agreed with Anderson

---

[96] James Gleick, "Advances Pose Obstacle to Atom Smasher Plan," *New York Times*, April 14, 1987, p. C1.

[97] At this writing, high-field magnets wound using wires made from high-temperature superconductors have yet to progress past the prototype stage. See, e.g., David Larbalestier, "Wires and Tapes," in *100 Years of Superconductivity*, edited by Horst Rogalla and Peter H. Kes (CRC Press, Boca Raton, FL, 2012), Section 11.1.7.

[98] *New York Times*, March 21, 1990, p. A26.

[99] Joseph D. Martin, *Solid State Insurrection: How the Science of Substance Made American Physics Matter* (University of Pittsburgh Press, Pittsburgh, PA, 2018), p. 189.

and dubbed these themes the myth of trickle-down technology, the myth of the intellectual frontier, and the myth of the no-zero-sum game.[100] In other words, few significant spin-offs result from a giant project designed for a specific technical purpose, particle physics presents problems that are no deeper than those presented by complex many-particle systems, and science funding *is* a zero-sum game because future decision-makers feel compelled to feed a megaproject whatever it needs to survive to justify the enormous initial investment.

The eminent particle physicist Steven Weinberg was probably the most eloquent spokesman for the pro-SSC forces. Reading the text of his testimonies before Congress (and his subsequent articles for the general reader), one senses that he was often conducting a debate with Anderson across space and time. Both theorists drew on a lifetime of thinking about their subject and argued their cases with economy, precision, and wit.

Anderson and Weinberg did not appear together at a Congressional hearing until August 4, 1993. The portions of their oral testimony reproduced in the Prologue give a sense of their positions. As a reductionist par excellence, Weinberg framed the battle over the SSC as a struggle over the intellectual importance of fundamental particle physics to the entire scientific enterprise and its legacy to future generations. He concluded with a dire warning:

> If the Super Collider is killed this year, it is killed for good. And what is killed with it is high energy physics in America. And that may be the beginning of the killing of support for basic science in this country.[101]

Anderson denied Weinberg's privileging of particle physics and its implication that the United States had a special responsibility to

---

[100] Testimony before the Subcommittee on Energy Research and Development of the Committee on Energy and Natural Resources of the United States Senate, April 16, 1991, pp. 37–40.

[101] Testimony before the Committee on Energy and Natural Resources and the Subcommittee on Energy and Water Development of the Committee on Appropriations of the Senate of the United States, August 4, 1993, p. 54.

fund the SSC to uncover the mysteries of that subject. To Phil, the SSC was the Advanced Light Source problem writ large because American particle physicists:

> did not choose to consult with [their] international colleagues or with the community of physicists as a whole before deciding on the particular course of action that they took.... What has first priority on the public purse should all along have involved the whole of science.[102]

The House of Representatives voted to terminate the SSC on October 19, 1993. By that time, it was a "clear fact" that the total project cost would exceed $10 billion, with some worrying that $15 billion was more likely.[103] A Congressional oversight committee concluded that physicists' testimony played little role in the decision to cancel the SSC. The most important factors were inaccurate cost estimates and poor project management.[104] Historians of science have reached the same general conclusion, to which they add that particle physicists were simply not prepared to construct an enormous new machine at an empty site where everything had to be assembled from scratch—including the management team—and simultaneously defend the project from public attacks before Congress in times of increasing fiscal austerity. The SSC was "a bridge too far" for them to cross.[105]

---

[102] Ibid., p. 59.

[103] Michael Riordan, Lillian Hoddeson, and Adrienne W. Kolb, *Tunnel Visions: The Rise and Fall of the Superconducting Super Collider* (University of Chicago, Chicago, 2015), p. 235.

[104] *Out of Control: Lessons Learned from the Superconducting Super Collider*, A Staff Report for the use of the Subcommittee on Oversight and Investigations of the Committee on Energy and Commerce of the US House of Representatives, December 1994.

[105] Michael Riordan, "A Bridge Too Far: the Demise of the Superconducting Super Collider," *Physics Today* **69**(10), 48–54 (2016). For the opinion of an experimental particle physicist who was involved in the SSC in its early design phase, but not thereafter, see Stanley Wojcicki, "The Supercollider: the Texas Days. A Personal Recollection of Its Short Life and Demise," in *Reviews of Accelerator Science and Technology: Medical Applications of Accelerators*, Volume 2, edited by Weiren Chou (World Scientific, Singapore, 2009), pp. 265–301.

By contrast, none of these problems dogged the Europeans at CERN who succeeded in building the Large Hadron Collider and discovering the Higgs particle. That project did suffer from setbacks and cost overruns, but its stable management, highly experienced team of physicists and engineers used to working together, and success in securing funds from countries across the globe gave them a huge advantage the SSC did not enjoy.

Some particle physicists—then and now—blame the demise of the SSC on the testimony and lobbying skill of the outspoken Anderson.[106] It is true that Anderson was one of the first scientists to go public with his criticism of a scientific project funded by the federal government. However, the most likely effect of his testimony was merely to provide intellectual cover for lawmakers who had political and parochial reasons to vote as they did.

In 2013, *APS News* noted the twenty-year anniversary of the cancellation of the SSC and correctly named Anderson as "among the most vocal detractors" of the project.[107] Phil responded angrily that he was "quite tired of being the only physicist quoted in assigning the blame for the demise of the SSC." He also recalled his unhappiness with politicians who somehow concluded that magnetic resonance imaging (MRI) spun-off from particle physics rather than from condensed matter physics.[108]

Anderson knew that the basic technology exploited by MRI flowed directly from the discovery of nuclear magnetic resonance by solid-state physicists right after World War II. He also knew that other solid-state physicists invented the superconducting magnet needed to make MRI practical. They also developed reliable designs and manufacturing processes for superconducting wire and magnets for small-scale research.

---

[106] Private communication with John Preskill, April 15, 2019.

[107] "October 1993: Congress Cancels Funding for the SSC," *APS News* **22**(9), 2 (2013).

[108] P.W. Anderson, "Clarifying the Record on the SSC," *APS News* **22**(10), 4 (2013).

On the other hand, it was particle physicists at Fermilab in Illinois who spent a decade developing and constructing the more than 1000 superconducting magnets needed for its Tevatron proton accelerator.[109] The Intermagnetics General Corporation supplied 90% of the superconducting wire for the Tevatron and, because of that demand, the company established large-scale manufacturing and quality assurance protocols which were critical to their subsequent entry into the commercial MRI magnet market.[110] Hence, while there is no question that it was solid-state physics research that drove the key enabling technologies behind MRI, one can say with fairness that particle physicists played an important role in making it cost effective.[111]

[109] Lillian Hoddeson, "The First Large-Scale Application of Superconductivity: the Fermilab Energy Doubler, 1972–1983," *Historical Studies in the Physical and Biological Sciences* **18** (1), 25–54 (1987). See also Steve A. Gourlay and Lucio Rossi, "The History of Superconductivity in High Energy Physics," in *100 Years of Superconductivity*, edited by Horst Rogalla and Peter H. Kes (CRC Press, Boca Raton, FL, 2012), Section 12.2.

[110] Carl H. Rosner, "Intermagnetics Remembered: From Superconductor-Based GE Spin-Off to Billion Dollar Valuation," *IEEE/CSC & ESAS European Superconductivity News Forum* **19**, January 2012.

[111] Private communication with James Bray, Chief Scientist, General Electric Corporation.

14

The Problem of a Lifetime

In 1986, the MGM Grand Hotel in Las Vegas hosted the annual March Meeting of the American Physical Society (APS). This is the meeting that most condensed matter physicists attend as a matter of tradition. When the meeting ended, the hotel casino discovered that its one-week profit was the lowest in its history.[1] As one pit boss reportedly put it, "The physicists came to Las Vegas with one twenty dollar bill and one shirt—and changed neither."[2] The March Meeting never returned to Las Vegas, supposedly at the request of the Chamber of Commerce.

The staid behavior on display in Las Vegas vanished one year later when the same group of physicists gathered for the 1987 March Meeting at the Hilton hotel in New York City. Raucous cheering and wild applause broke out repeatedly on the evening of Wednesday, March 18 in an 1140-seat ballroom packed to overflowing during a marathon session which lasted from 7:30 pm to 3:15 am. Another 2000 people watched the session on television monitors set up throughout the hotel. The subject matter of the session was superconductivity at high temperature.

Only a few months earlier, most physicists thought of superconductivity as a respected but no longer very exciting topic for research. Everyone knew that the zero resistance state was achievable only for certain metals and only if they were cooled to a

---

[1] Harry Lustig, "To Advance and Diffuse the Knowledge of Physics: An Account of the One-Hundred Year History of the American Physical Society," *American Journal of Physics* **68**, 595–636 (2000), Note 71.

[2] Marvin L. Cohen, Letter to the Editor, "More on Physicists and Their Shirts," *APS News* **22**, August/September 2013.

temperature so far below room temperature that practical applications were rare. Coincidently, the night before the marathon
session, Phil Anderson presented a long-planned historical talk
about the early, exciting days of superconductivity in the 1960s.

The cause for the unrestrained enthusiasm at the marathon
session was that speaker after speaker reported new observations
about the occurrence of superconductivity in a class of ceramic
materials called *cuprates* at temperatures vastly higher than had
ever been seen before. A press release from the American Institute
of Physics broadcast this news and listed the possible implications:
cheap, lossless electrical power lines, super-powerful computers,
electronic circuits with fantastic speeds, magnetically levitated
trains, and more.[3]

The press release also quoted Phil Anderson. In light of the discovery of high-temperature superconductivity and the importance
of superconducting magnet technology to the Superconducting
Super Collider, Anderson opined that "perhaps the first billion
dollars of the $5 billion budget of the SSC might profitably be
spent on materials research." He took this message to Congress
just one month later.

The morning after the marathon session, a front-page article
in the *New York Times* referred to the event as the "Woodstock of
Physics," a name chosen to evoke the legendary Woodstock rock
music festival that had attracted nearly half a million people in
1969. The American Physical Society staged a press conference
where a panel of seven bleary-eyed physicists answered questions.
Six of the seven panelists were experimentalists who ran laboratories devoted to measurements of superconductivity in various
materials.

Anderson was the only theoretical physicist on the panel
(Figure 14.1). He told the assembled journalists that the phonon

---

[3] News Release of the American Institute of Physics, "Highlights of the
Papers Presented at the American Physical Society Meeting in New York City,"
March 16–20, 1987.

Figure 14.1 Panel members at the Woodstock of Physics press conference, March 19, 1987. Left to right: K. Alex Müller, Paul C.W. Chu, Philip W. Anderson, M. Brian Maple. Source: Paul M. Grant.

mechanism responsible for superconductivity in metals could not possibly be operative in the new materials. In fact, a week before the March Meeting, he had beaten all his competition to the punch by publishing a novel theoretical explanation for the existence of superconductivity at high temperature in the new materials.

Anderson's outward demeanor during the press conference was calm and collected. But his interior state was close to elation. He was certain his intuition about the origin of superconductivity in the cuprates was correct. It was not only one of his best ideas, it combined elements from a variety of other problems he had attacked over the years. It was as if Nature had served up the cuprate problem to him on a platter. He had only to work out the details and a (nearly unprecedented) second Nobel Prize for Physics would be his.

High-Temperature Superconductivity

The 1911 discovery of superconductivity in solid mercury at temperatures below $T_C = 4\,\mathrm{K}$ was totally unexpected. Motivated by the dream of zero-resistance conductors at room temperature, experimenters worked for years to identify materials with higher and higher values of T_C. This effort made slow but steady progress before stalling out after the 1973 discovery that the compound Nb_3Ge was a superconductor with $T_C = 23K$.

Unfortunately, even with its relatively high transition temperature, Nb_3Ge became superconducting only if one used expensive liquid helium (similar in cost to good Scotch whiskey) as a refrigerant. That cost was the main reason superconductivity had not produced a truly transformative technology to that time. Specialists ached to find a superconductor with $T_C > 77K$. If they could, cooling into the superconducting state would require only liquid nitrogen (cheaper than beer) as a refrigerant.

Theorists also dreamed about increasing T_C. Conventional BCS theory exploits the subtle charge re-arrangements associated with small ion motions (phonons) to induce an attractive force between pairs of electrons. As early as 1964, William Little of Stanford University suggested that Nature might proceed differently and replace the ion motion with some other source of charge re-arrangement (e.g., excitations of localized electrons) to produce the "glue" needed to form Cooper pairs. Using this idea, he speculated that superconductivity near room temperature (300 K) might occur in quasi-one dimensional organic molecular crystals.[4]

Vitaly Ginzburg, the co-developer of the pre-BCS Ginzburg–Landau phenomenological theory of superconductivity, investigated a number of other situations where Little's mechanism might favor high-temperature superconductivity (HTS). One possibility studied by his group at the Lebedev Institute in Moscow

---

[4] W.A. Little, "Possibility of Synthesizing an Organic Superconductor," *Physical Review* **134**, A1416–A1424 (1964). A handful of organic molecular crystals do become superconductors, but only at very low temperature. See, e.g., A. Ardavan, S. Brown, S. Kagoshima, K. Kanoda, K. Kuroki, H. Mori, M. Ogata, S. Uji, and J. Wosnitza, "Recent Topics of Organic Superconductors," *Journal of the Physical Society of Japan* **81**, 011004.

was the "ginzburger," a theoretical structure composed of alternating thin layers of metal and semiconductor in intimate contact.[5] This geometry also intrigued John Bardeen as a possible
setting for non-phonon superconductivity.[6]

Phil Anderson disagreed strenuously with Little, Ginzburg,
and Bardeen on this issue. He co-authored two papers critical of
their work and made it a point to debunk them in talks he gave at
the time.[7] Anderson's skepticism of HTS was not unfounded. By the
late 1970s, funding agencies around the world had grown weary of
waiting for a high-temperature superconductor and eliminated
or severely cut back funding to many research programs devoted
to the subject.[8] Nevertheless, physicists would occasionally
receive preprints reporting indications of superconductivity in
some material at some elevated temperature. None of these
reports panned out. As one science blog put it:

> The history of superconductivity is littered with dubious claims
> of high-temperature activity that later turn out to be impossible
> to reproduce. Physicists have a name for them: USOs, or uniden-
> tified superconducting objects.[9]

[5] V.L. Ginzburg, "The Problem of High-Temperature Superconductivity,"
Soviet Physics Uspekhi **13**, 335–52 (1970). See also *High-Temperature Superconductivity*,
edited by V.L. Ginzburg and D.A. Kirzhnits (Consultants Bureau, New York,
1982).

[6] David Allender, James Bray, and John Bardeen, "Model for an Exciton
Mechanism for Superconductivity," *Physical Review B* **7**, 1020–9(1973).

[7] Marvin L. Cohen and P.W. Anderson, "Comments on the Maximum
Superconducting Transition Temperature," in *Superconductivity in d- and f-Band
Metals*, edited by Hugh C. Wolfe and D. H. Douglass, *AIP Conference Proceedings* **4**,
17–27 (1973); J.C. Inkson and P.W. Anderson, "Comment on 'Model for an
Exciton Mechanism for Superconductivity'," *Physical Review B* **8**, 4429–32 (1973).

[8] Helga Nowotny and Ulrike Felt, *After the Breakthrough. The Emergence of High-
Temperature Superconductivity as a Research Field* (Cambridge University Press,
Cambridge, UK, 1997), Chapter 2.

[9] "The Record for High-Temperature Superconductivity has Been Smashed
Again," *Emerging Technology From the ArXiv*, MIT Technology Review, December 10,
2018.   https://www.technologyreview.com/s/612559/the-record-for-high-tem-
perature-superconductivity-has-been-smashed-again/Accessed December 4,
2019. According to the physicist who coined the acronym, "If pronounced

The Cuprates

By 1986, Anderson was fully retired from Bell Labs and working full-time as a professor at Princeton University. The fall semester ended in mid-December and he took advantage of the break to visit old friends at the Labs. Another Bell Labs alumnus back for a visit was Stanford University professor Ted Geballe, an Anderson contemporary and a longtime collaborator of the now-deceased Bernd Matthias. Geballe had important news to share from a Materials Research Society conference he had just attended in Boston.

The Japanese physicist Koichi Kitazawa had shocked the conference by announcing that a lanthanum-barium copper oxide (LaBaCuO) compound was a superconductor with a transition temperature greater than 30 K. This was a bombshell rather than a USO because it confirmed an earlier report by Georg Bednorz and Alex Müller, two scientists at the Zürich Research Laboratory of the IBM Corporation.[10] One year later, these two shared a Nobel Prize in Physics for their discovery of LaBaCuO, the first *cuprate superconductor.*[11]

Bednorz and Müller defied conventional wisdom for several reasons.[12] First, both were outsiders to the superconductivity business. Müller was a senior IBM Fellow, which meant he could work on any problem he wished. He had an idea about superconductivity

incorrectly as 'woo-saw', [USO] means 'telling a lie' in Japanese. But I named it with much affection hoping that a USO—like a UFO—might someday really land on earth." Koichi Kitazawa, "Superconducting Materials: History and the Future," in *Superconducting Materials: Advances in Technology and Applications*, edited by A. Tampieri and G. Celotti (World Scientific, Singapore, 2000), pp. 3–24.

[10] J.G. Bednorz and K.A. Muller, "Possible High T_C Superconductivity in the Ba-La-Cu-O System," *Zeitschrift fur Physik B* **64**, 189–93 (1986).

[11] A cuprate is a material which contains negatively charged complexes of copper.

[12] Ulrike Felt and Helga Nowotny, "Striking Gold in the 1990s: The Discovery of High-Temperature Superconductivity and Its Impact on the Science System," *Science, Technology & Human Values* **17**, 506–31 (1992).

and recruited Bednorz because the younger man's knowledge of crystallography nicely complemented his own experience with ceramic oxides. Second, the Zürich laboratory where Bednorz and Müller worked was small and modestly equipped compared to the two large laboratories IBM maintained in the United States. Finally, the oxides they focused on violated the empirical rules for identifying potential candidates for high-temperature super-conductivity that Bernd Matthias had established after nearly thirty years of effort.

Resonating Valence Bonds

Koichi Kitazawa visited Bell Labs after his bombshell announce-ment in Boston. He gave a seminar about LaBaCuO and outlined the simple synthesis recipe: mix together the ingredients for the insulating compound La_2CuO_4, dope with barium atoms, and heat in a furnace. *Voila*, a superconductor results. Several audi-ence members left his talk with plans to begin working on the cuprates immediately.[13]

Anderson's many friends at Bell Labs were happy to share with him the progress being made there with high-temperature super-conductivity. He was particularly interested to learn that the arrangement of atoms in La_2CuO_4 bore a strong family resem-blance to the arrangement of atoms in $BaTiO_3$, the insulator that Bill Shockley had asked him to think about almost forty years earlier. By no later than the last week of December 1986, Anderson had formed the key ideas that would dominate his thinking about the cuprates for years to come.

He began by dismissing the relevance of lattice vibrations to superconductivity in the cuprates. To Anderson, it was obvious that the energy scale for ion motion in LaBaCuO was simply too low to account for the high value of T_C. Instead, the crystal chem-istry of La_2CuO_4 sketched below convinced him that the observed high transition temperature derived from the large energy

---

[13] See, e.g., R.J. Cava, "Disorder in the Ranks," *Nature* **394**, 126–7 (1998).

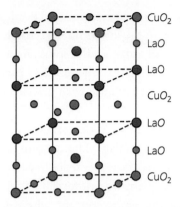

CuO$_2$
LaO
LaO
CuO$_2$
LaO
LaO
CuO$_2$

Figure 14.2 The crystal structure of La$_2$CuO$_4$. Each CuO$_2$ layer is a square lattice of copper atoms (orange) with oxygen atoms (green) interposed between nearest neighbor Cu atoms. Each layer of LaO consists of two interpenetrating square lattices of lanthanum atoms (purple) and oxygen atoms.

associated with the Coulomb repulsive force between two electrons localized at the same atomic site. Thus, his premise from the very beginning was that high-temperature superconductivity involved many-electron physics profoundly different from the physics relevant to the familiar low-temperature superconductors.

Anderson focused first on the non-superconducting parent compound, La$_2$CuO$_4$. It was crucial to him that the parent material had the layered structure shown in Figure 14.2: a repeating motif of one layer of CuO$_2$ followed by two layers of LaO. He guessed that the latter acted like barriers to electron motion. This meant that all the interesting physics occurred independently in the two-dimensional CuO$_2$ layers. Each of these consisted of a square lattice of copper atoms with an oxygen atom interposed between nearest-neighbor Cu atoms. The O electron shells were filled but one electron per copper atom was available to hop to a nearest-neighbor Cu atom in the same layer.

Anderson estimated some of the energies relevant to La$_2$CuO$_4$ using his extensive knowledge about transition metal oxides. He concluded that this compound was an electrical insulator as a consequence of Nevill Mott's Coulomb localization mechanism.

In short, the just-mentioned electrons do not hop from Cu atom to Cu atom because to do so would cost an electric energy U if any two of them (with anti-parallel spins) happened to wind up on the same Cu atom.[14] Thus, he identified each CuO_2 layer as a two-dimensional Mott insulator with one electron localized on every Cu atom.

Chapter 11 described how Anderson connected Mott's mechanism with antiferromagnetism. This led him to predict the likely orientations of the spins of the electrons in each CuO_2 layer. Since his 1952 work on the quantum antiferromagnet, Phil understood the importance of the *singlet* concept when discussing ground state spin configurations (Chapter 6). A three-dimensional (3D) antiferromagnet could exhibit the ordered Néel-type pattern shown in Figure 6.5, yet still achieve singlet status by glacial rotation of the spin pattern. A one-dimensional (1D) antiferromagnet was explicitly a singlet state with an essentially *disordered* pattern of spins.

To describe each 2D CuO_2 layer, Anderson invoked a "middle ground" suggestion he had made over a decade earlier.[15] One should expect the ground state of a 2D antiferromagnet to adopt a quantum state composed of singlet objects he called *resonating valence bonds*. The inspiration for this suggestion was a 1949 paper by the theoretical chemist Linus Pauling.[16] He knew this work because, soon after his arrival at Bell Labs, Bill Shockley asked him to explain it to a small group of Bell Labs colleagues.

The group learned that *resonance* was a term introduced by quantum pioneer Werner Heisenberg to describe situations where the wave function of a system was best written as a sum of terms,

[14] Recall that quantum mechanics forbids two *parallel* spin electrons to occupy the same orbital.

[15] P.W. Anderson, "Resonating Valence Bonds: A New Kind of Insulator?," *Materials Research Bulletin* 8, 153–60 (1973).

[16] L. Pauling, "A Resonating-Valence-Bond Description of Metals and Intermetallic Compounds," *Proceedings of the Royal Society (London) A* 196, 343–62 (1949).

each describing one possible configuration the system might adopt.[17] Anderson had not found Pauling's application of resonance to ordinary metals persuasive, but the basic idea (and the name) never left him.

Anderson guessed that the many-electron wave function of each copper layer of La_2CuO_4 was a resonating valence bond state constructed as a sum of spin configurations, each one similar to the configuration illustrated in Figure 14.3(a). There, each ribbon connects a pair of oppositely pointing spins on different atoms into a quantum mechanical 'valence bond' singlet. There are very many such configurations in the RVB sum because there are many different antiparallel spin pairs available in the 2D layer to connect into valence bonds.

Now, what about doping and the superconductivity which attends it? Already, in their breakthrough paper, Bednorz and Müller had pointed out that replacing lanthanum atoms with barium atoms in La_2CuO_4 removes electrons from the copper layer.[18] The presence of these "holes" [see Figure 14.3(b)] transforms the insulating layer into a metal because there are now empty sites in the copper lattice to which spins might hop. The previously grid-locked valence bonds are now mobile and therefore, Anderson imagined, they must somehow be indistinguishable from the singlet Cooper pairs characteristic of the BCS many-body state.

Bangalore and Beyond

As 1986 melted into 1987, Anderson challenged himself to establish quantitatively what he believed to be true qualitatively. His prospective resonating valence bond (RVB) wave function described

---

[17] Valeria Mosini, "A Brief History of the Theory of Resonance and of its Interpretation," *Studies in the History and Philosophy of Modern Physics* **31**(4), 569–81 (2000).
[18] Doping can *add* electrons if one chooses suitable elements for the impurity species.

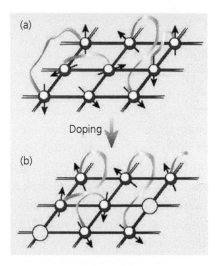

Figure 14.3 Resonating Valence Bond (RVB) model of a square lattice of copper atoms, i.e., a CuO_2 layer with the oxygen atoms not shown. (a) The Mott mechanism localizes one electron per copper atom. Ribbons connect antiparallel electron spins that are correlated into "valence bonds." (b) Same as (a) except that chemical doping has removed the spins from two of the atoms. Figure adapted from P. Coleman, *Nature* 424, 625–6 (2003).

electrons that hop from site to site in two dimensions while paying an energy cost U whenever any two of them happen to occupy the same site at the same time. Precisely these two elements appear in a model Hamiltonian that the British solid-state theorist John Hubbard had written down over twenty years earlier to discuss ferromagnetism.[19] It was Anderson's brilliant insight to propose that the Hubbard model was relevant to cuprate superconductivity.

Hubbard's model was notoriously difficult to analyze and Anderson did not have a new idea to do so. He also did not know how to write down mathematical formulas for his RVB wave

[19]   J. Hubbard, "Electron Correlations in Narrow Energy Bands," *Proceedings of the Royal Society of London A* **276**, 238–57 (1963); *This Week's Citation Classic* **22**, June 2 1980, p. 84.

functions. On the other hand, an opportunity to chat about the problem with high-quality theorists was just around the corner. In the second week of January, he was scheduled to present the keynote address at a big conference devoted to many-body phenomena in Bangalore, India.

The Bangalore meeting proved to be spectacularly useful to Anderson. Most importantly, he listened to a talk by his old Bell Labs colleague, Maurice Rice, who reported the results of computer calculations from his research group at the Swiss Federal Technical Institute. They found that the spin directions in a one-dimensional Heisenberg antiferromagnet are almost identical to the spin directions in a one-dimensional metal *if one removes the pieces of the metal's wave function where two electrons occupy one lattice site.*[20] The latter, as Rice and Anderson understood, would happen automatically in Nature if the Mott–Hubbard energy U was very large.

Rice's talk sparked the biggest eureka moment of Anderson's career. He realized abruptly that his RVB wave function must be none other than a *two-dimensional BCS superconductor wave function with the pieces removed from it where two electrons occupy one lattice site.* This was something he could write down explicitly. He flew to California (to honor a commitment to visit Caltech for a semester), excitedly phoned his Princeton research group, and then spent two weeks writing up a manuscript for submission to *Science* magazine. Joyce had never seen him so euphoric.

In early February, Phil combined a road trip to Stanford (to convince Ken Arrow to co-organize the first economics workshop at the nascent Santa Fe Institute) with a visit to give a talk at the Institute for Theoretical Physics in Santa Barbara. *Science* had rushed his article into print and it was conveniently available to

[20] C. Gros, R. Joynt, and T.M. Rice, "Antiferromagnetic Correlations in Almost-Localized Fermi Liquids," *Physical Review B* **36** 381–93 (1987). These authors state their aim was to confirm and extend earlier work by T.A. Kaplan, P. Horsch, and P. Fulde, "Close Relation between Localized-Electron Magnetism and the Paramagnetic Wave Function of Completely Itinerant Electrons," *Physical Review Letters* **49**, 889–92 (1982).

everyone at the ITP when he arrived.[21] That evening, a young theorist named Steve Kivelson told his wife, "I heard the most exciting talk today. Now I know what I'm going to be doing for the next year."[22] Little did he realize he would still be working on the problem thirty years later!

For many physicists, the months between the December 1986 Materials Research Society meeting and the 1987 March Meeting were a blur of non-stop activity and coffee consumption.[23] Experimenters raced to make chemical variants of the Bednorz–Müller material in hopes of finding a superconductor with an even higher value of T_C. These efforts bore fruit spectacularly at the end of February with the discovery that yttrium-barium copper oxide (YBCO) became a superconductor at $T_C = 93K$.[24] This was three times the transition temperature of LaBaCuO and well above the boiling point of liquid nitrogen.

Anderson flew back to Princeton from Caltech right before the March Meeting began. He was surprised to learn that two members of his research group (senior visitor Ganapathy Baskaran and PhD student Zou Zou) had already written an RVB paper and added his name to the author list. Starting with the Hubbard model, this was the first attempt to translate Phil's novel ideas into a quantitative theory.[25] Today, this paper is noteworthy mostly because it introduces the idea that the cuprates differ from BCS superconductors because, among other things, thermal

[21] P.W. Anderson, "The Resonating Valence Bond State in La_2CuO_4 and Superconductivity," Science 235, 1196–8 (1986).

[22] Interview of Steven Kivelson by the author, March 3, 2015.

[23] Robert M. Hazen, The Breakthrough. The Race for the Superconductor (Summit Books, New York, 1988); Bruce Schechter, The Path of No Resistance: The Story of the Revolution in Superconductivity (Simon and Schuster, New York, 1989).

[24] M.K. Wu, J.R. Ashburn, C.J. Torng, P.H. Hor, R.L. Meng, L. Gao, Z.J. Huang, Y.Q. Wang, and C.W. Chu, "Superconductivity at 93 K in a New Mixed-Phase Y-Ba-Cu-O System at Ambient Pressure," Physical Review Letters 58, 908–10 (1987).

[25] G. Baskaran, Z. Zou, and P.W. Anderson, "The Resonating Valence Bond State and High T_C Superconductivity," Solid State Communications 63, 973–6 (1987).

energy disorders the rigid locking of the phase variable illustrated in Figure 10.1 while Cooper pairs are still present.

At the March Meeting itself, Anderson was given the honor of presenting the first ten-minute talk at the high-T_C marathon session. Most of the 3000 physicists watching live or on the TV monitors were probably disappointed. The world-famous Nobel laureate spent over half his allotted time presenting a mash-up of known experimental results. He used one sentence to connect T_C for the cuprates to an antiferromagnetic exchange constant and a second sentence to vaguely define his resonating valence bond model. The names Mott and Hubbard never left his lips. In a great rush at the end, he tried (with little success) to summarize recent results from his Princeton group and to explain some ideas contributed by a trio of "Santa Barbarians" who had been inspired by his talk there.[26]

After the Thursday morning press conference, Anderson learned to his dismay that most of the physicists he spoke with (and all of the journalists) did not seem to appreciate that he had identified the origin of superconductivity in the cuprates. There was more interest in the four theory talks that had followed him at the marathon session. He was incredulous. Each of those talks discussed cuprate superconductivity using arguments he knew were hopelessly wrong.

His incredulity turned to cold fury a day later when he picked up a copy of the *New York Times*. Two photographs accompanied a front-page article about the marathon session. One showed the Bell Labs experimentalist Robert Cava holding samples of the new superconductors. The other featured a theorist from Northwestern University named Art Freeman. Freeman—a former PhD student of John Slater—approached the cuprates

---

[26] The three young theorists inspired by Anderson's talk at the Institute for Theoretical Physics were former Anderson senior thesis student Daniel Rokhsar, former Anderson PhD student James Sethna, and former Schrieffer postdoc Steve Kivelson.

using large-scale computer calculations. Anderson could not believe that the newspaper of record in the United States high-lighted theoretical work on the cuprates using a person who solved the Schrödinger equation in a manner that explicitly omitted all the physical effects he was convinced were vital to understanding the new superconductors.

The next nine months produced a tidal wave of scientific papers devoted to high-temperature superconductivity (HTS). The editors of the most prestigious physics journal at the time, *Physical Review Letters*, suspended their normal editorial policies and used a small review panel (not including Anderson) to decide which of the hundreds of HTS manuscripts they received (fully one-eighth of all submissions to the journal) should be published.[27]

Experiments suffered from the fact that people who had never worked with superconductors tried to publish measurements of dubious value.[28] Worse—and not appreciated at the time—it was quite easy to make low-quality samples of the superconducting cuprates but quite difficult to make high-quality samples.[29] The result was an initial experimental literature where it was hard to distinguish reliable results from unreliable ones.

Working in theory during this period was very much like drinking water from a fire hydrant.[30] Everyone knew that a Nobel Prize was at stake and preprints arrived daily from competitors. Anderson and his first group of HTS researchers at Princeton submitted one paper per month between March and December of 1987.[31]

---

[27] Private communication with Reinhardt Schuhmann, Managing Editor, *Physical Review Letters*.

[28] Interview of Frances Hellman by the author, March 16, 2016.

[29] By 'high-quality sample' we mean a single crystal that is substantially free of defects and impurities.

[30] Interview of Shivaji Sondhi by the author, May 4, 2016.

[31] Besides Baskaran and Zou, PhD students Theodore Hsu and Joseph Wheatley worked on the cuprates with Anderson.

Theorists displayed an irresistible compulsion to take whatever ideas and methodology were known to them and apply them to the cuprates. "Like Cinderella," said the British-American theorist Sebastian Doniach, "everyone brought out their glass slipper to see if it would fit."[32] Scientific luminaries with modest or no background in superconductivity like Edward Teller, Linus Pauling, and the particle physics Nobel Prize winner, Tsung-Dao Lee all fell victim to this syndrome.[33] Sir Nevill Mott quipped that there were as many HTS theories as HTS theorists.[34]

The Lost Decade

In 1997, Princeton University Press published Anderson's third book, *THE Theory of Superconductivity in the High-T_C Cuprates*. The capitalization of all the letters in the first word accurately reflected the author's sincere, albeit immodest, opinion of the work he and his collaborators had done over the preceding ten years. It is all the more remarkable, then, that he was forced to repudiate most of its contents almost immediately after its publication.

The book begins with a *mea culpa*:

> The glow of optimism in 1987 when we seemed to see an almost immediate conclusion to the problem of high T_C in the resonating valence bond idea faded rapidly and much of what we wrote and said then was false and misleading.

A problem from the outset was that Anderson's RVB was an interesting and exciting *idea* but it was very far from a coherent *theory*

---

[32] Quoted by David Campbell, interview by the author, March 4, 2015.

[33] Edward Teller, "The Chemistry of Superconductivity," in *World Congress on Superconductivity*, edited by C.G. Burnham and R.D. Kane (World Scientific, Singapore, 1988), pp. 303–10; Linus Pauling, "Influence of Valence, Electronegativity, Atomic Radii, and Crest-Trough Interactions with Phonons on the High-Temperature Copper Oxide Superconductors," *Physical Review Letters* **59**, 225–7 (1987); T.D. Lee, "Possible Relevance of Soliton Solutions to Superconductivity," *Nature* **330**, 460–1 (1987).

[34] Nevill Mott, "Is There an Explanation?," *Nature* **327**, 185–6 (1987).

with true predictive power. Even before his *Science* article appeared in print, he and his research group began tinkering with approximations to the Hubbard model that yielded RVB-type insulating and superconducting states as solutions. The tinkering continued in order to keep their predictions in accord with the ever-growing number of experimental results. This led them into a thicket of unfamiliar concepts (charge-spin separation, flux phases, gauge theories, holons, Luttinger liquids, plaquettes, quantum critical points, slave bosons, spinons, etc.) which made it difficult for physicists *not* working directly in HTS to follow what was going on.

Anderson's enthusiasm was very great at a 1989 summer workshop at Cargèse, a resort on the west coast of the Mediterranean island of Corsica. He brought along his entire Princeton team (Figure 14.4) and his report of the status of RVB theory had the perhaps unintended effect of revealing exactly how far away from

Figure 14.4 The Princeton HTS research group (left to right). Zou Zou, Joe Wheatley, Shou-Dan Liang, Ganapathy Baskaran, Benoit Doucot, Phil Anderson, Ted Hsu, and Xiao-Gang Wen. Cargèse, Corsica, summer of 1989. Source: Philip Anderson.

his original conception his thinking had drifted.[35] For example, he had concluded that a single doped CuO_2 layer is *never* a superconductor, nor even a conventional metal, although it is a conductor. He had also concluded that the cuprates are superconductors *only* because electron pairs tunnel between adjacent CuO_2 layers, not unlike what happens in a Josephson junction (Figure 10.2).

The second chapter of Anderson's HTS book uses the phrase "central dogma" (borrowed from the theoretical biologist Francis Crick) to mean an assertion "determined by logical deduction from the entirety of the experimental facts which tends to constrain the structure of any theoretical description." To strive for such assertions seemed entirely appropriate for a data-whisperer like Anderson—a scientist whose entire career demonstrated an enormous respect for experiment and a passion for listening carefully to what measurements have to say.

Anderson identified six central dogmas for HTS. We do not list them because the theoretical edifice they were meant to support almost immediately collapsed under the weight of new experimental evidence (see "Collapse and Revival" later in this chapter). Ironically, this fate was actually a testament to Anderson's theory. Unlike most other entrants in the HTS sweepstakes, his theory made explicit predictions that could be falsified in the laboratory.[36] When that falsification occurred, it forced him to abandon a decade of research and return to square one.

Discord

The radical originality and "first-out-of-the-box" nature of Anderson's RVB theory—not to mention his enormous stature

[35] P.W. Anderson, "Problems and Issues in the RVB Theory of High T_C Superconductivity," *Physics Reports* **184**, 195–206 (1989).

[36] Anderson's original *Science* article predicted that La_2CuO_4 would possess the disordered spin structure of his RVB state. Subsequent experiment showed that it was instead a simple Néel antiferromagnet. It turned out that the theory accommodated this flaw quite easily.

in the community—put him at the center of nearly every discussion of HTS from the very beginning.[37] For reasons many speculate about, but none can ever know, Anderson took an odd, proprietary attitude toward the theory of HTS. Participants recall a Gordon Conference where he announced that "all the other theorists should leave the room, I am the only one here who should talk to the experimentalists." Worse, he dismissed as "nonsense" and "folly" the work of other theorists who proposed mechanisms for cuprate superconductivity different from his own, even as his own ideas changed over time. Small wonder that some referred to RVB as "really vague bullshit."[38]

The Institute for Theoretical Physics sponsored a program devoted to HTS fairly early on. Anderson elected not to attend when the ITP Director at the time, Bob Schrieffer, and his advisory board declined to follow Phil's suggestions about the program content. Later, the genial Schrieffer proposed his own theory for the cuprates and he and Anderson published a reasonably civil dialog on the subject.[39]

Old Anderson friends David Pines and Doug Scalapino (a collaborator twenty-five years previously) did not fare so well. Anderson showered criticism on them because their research favored an electron pairing mechanism based on so-called "spin fluctuations."[40] The steady stream of workshops and conferences devoted to HTS provided Phil with plenty of opportunities to criticize them. The broader physics and scientific communities became aware of the dispute when the principals exchanged letters in *Physics Today* and *Discover* magazine devoted a feature article

[37] Citations to Anderson's original *Science* article exceeded 10,000 in 2020.

[38] M.R. Norman, "Unconventional Superconductivity" in *Novel Superfluids*, Volume 2, edited by K.H. Bennemann and J.B. Ketterson (Oxford University Press, Oxford, UK, 2014), pp. 23–79.

[39] Philip W. Anderson and Robert Schrieffer, "A Dialogue on the Theory of High T_C," *Physics Today* **44**(6), 54–61 (1991).

[40] P.W. Anderson, "A Re-Examination of Concepts in Magnetic Metals: the 'Nearly Antiferromagnetic Fermi Liquid'," *Advances in Physics* **46**(1), 3–1 (1997).

to the issue.[41] More than ten years later, a *Science* magazine article on the same subject by Anderson prompted an exchange between him and Scalapino in the electronic Letters section of the website of *Science* magazine.[42]

Anderson regarded Robert Laughlin as a worthy competitor on the strength of the latter's work in the early 1980s on the quantum Hall effect. Laughlin had some novel ideas about RVB and he was not a shrinking violet about promoting them. Attendees at one early HTS meeting recall an animated exchange where the echoing of Laughlin's booming voice repeatedly drowned out the remarks of the soft-spoken Anderson. In the end, the thirty-nine year old Laughlin and the sixty-six year old Anderson could only agree to disagree.

A few years later, Laughlin expressed disillusionment with Anderson and other physicists working on the cuprate problem in the form of an epic poem of over 1000 lines called *Hiawatha's Valence Bonding*.[43] This was surprising to some because, after sharing the 1998 Nobel Prize for his theory of the quantum Hall effect, Laughlin graciously credited Anderson for an off-hand remark which jump-started his thinking about the problem.[44]

Employing the same trochaic tetrameter Henry Wadsworth Longfellow used to write his *Song of Hiawatha*, Laughlin savaged many of Anderson's ideas, including his central tenet that Mott insulator physics was crucial to understanding the cuprates:

---

[41] Letter to the Editor from Philip W. Anderson and replies from David Pines and Douglas J. Scalapino, "In Explaining High T_C, Is d-Wave a Washout?," *Physics Today* **47**(2), 11–15,120 (1994); Tim Folger, "Call Them Irresistible," *Discover*, September 1995. Accessed December 27, 2019 from https://www.discovermagazine.com/the-sciences/call-them-irresistible.

[42] P.W. Anderson, "Is There Glue in Cuprate Superconductors?," *Science* **317**, 1705–7 (2007); eLetter exchange between D.J. Scalapino (December 5 2007) and P.W. Anderson (December 10, 2007), https://science.sciencemag.org/content/316/5832/1705/tab-e-letters.

[43] R. B. Laughlin, "Hiawatha's Valence Bonding," *Annals of Improbable Research*, May–June 2004, pp. 8–20.

[44] Robert B. Laughlin, Nobel Autobiography, 1998.

> There was also the assertion / Running rampant through the
> theory / That the essence of the cuprates / Was Coulombic insula-
> tion, / Which, on close inspection, turned out / No one could
> define precisely / With a few concrete equations, / But was none-
> theless a concept / People thought they comprehended / Like the
> fancy secret contents / Of competing brands of toothpaste / That,
> of course, are total fictions / Made up at lunch by ad guys.

Observers watched with dismay as theoretical HTS devolved into a
contact sport. The field acquired a reputation for unpleasant and
bitter argument for which Anderson and a few others must bear
responsibility. Some younger physicists reacted by avoiding the subject altogether. Others were marked by their association with their
research supervisors.[45] It is fact that none of Anderson's PhD students from this period made their careers in academic physics. In the
judgment of one observer, "Like World War I, there was a sacrifice of
young talent for no reason other than to defend or elaborate a
firmly held position. It was the end of old kings and kingdoms."[46]

Collapse & Revival

The centerpiece of Anderson's 1997 book is an *interlayer tunneling
theory* of cuprate superconductivity. The theory did not survive
long because high-quality experiments contradicted one of its
key assumptions and one of its key predictions. The assumption
(first made by BCS) was that the cuprate Cooper pair wave function possessed "s-wave" symmetry.[47] Anderson thought this was
reasonable because the cuprates were full of impurities and his
own "dirty superconductor" theorem from 1959 explained why
non-magnetic impurities had no effect on conventional superconductors that possessed this symmetry (Chapter 9).

---

[45] Interview of Douglas Scalapino by the author, March 17, 2015.
[46] Author correspondence with Gabriel Aeppli, August 22, 2017.
[47] A Cooper pair wave function with s-wave symmetry does not change its
algebraic sign when it is rotated in the plane of the layers.

For obvious reasons, Anderson was also predisposed to believe experiments done at Bell Labs and some early measurements from the Labs pointed to s-wave pairing.[48] Unfortunately for Phil, experimental evidence steadily accumulated that showed that the Cooper pairs of the cuprates possessed "d-wave" symmetry.[49] There is irony here because Anderson was very familiar with exotic Cooper pair wave functions. He and his student Pierre Morel had used them in their 1961 theory of superfluid ^{3}He.

Interlayer tunneling theory also made a specific prediction about the balance of energy in the cuprates. Careful experiments designed to test this prediction did not provide confirmation of Anderson's ideas. Interestingly, the editors of *Science* magazine permitted him to publish an alternative interpretation of the data back-to-back with the original experimental report.[50] It took a follow-up experimental paper to bury the theory once and for all.[51] There is irony here also because one of the co-authors of both experimental papers, Kathryn Ann Moler, was at the time a postdoctoral fellow at Princeton in close communication with Anderson.

[48] D. R. Harshman, L. F. Schneemeyer, J. V. Waszczak, G. Aeppli, R.J. Cava, B. Batlogg, L. W. Rupp, E.J. Ansaldo, and D.L. Williams, "Magnetic Penetration Depth in Single-Crystal $YBa_2Cu_3O_{7-\delta}$, *Physical Review B* **39**, 851–4 (1989).

[49] A Cooper pair wave function with d-wave symmetry changes its algebraic sign when it is rotated by $90°$ in the plane of the layers. Classically, the two electrons of the pair orbit one another. C.C. Tsuei and J.R. Kirtley, "Pairing Symmetry in Cuprate Superconductors," *Reviews of Modern Physics* **72**, 969–1016 (2000).

[50] Kathryn A. Moler, John R. Kirtley, D.G. Hinks, T.W. Li, and Ming Xu, "Images of Interlayer Josephson Vortices in $Tl_2Ba_2CuO_{6+\delta}$," *Science* **279**, 1193–6 (1998); Philip W. Anderson, "c-Axis Electrodynamics as Evidence for the Interlayer Theory of High-Temperature Superconductivity," *Science* **279**, 1196–8 (1998).

[51] A. A. Tsvetkov, D. van der Marel, K. A. Moler, J. R. Kirtley, J. L. de Boer, A. Meetsma, Z. F. Renk, N. Koleshnikov, D. Dulic, A. Damascelli, M. Gruninger, J. Schutzmann, J.W. van der Eb, H. S. Somal, and J.H. Wangk, "Global and local measures of the intrinsic Josephson Coupling in $Tl_2Ba_2CuO_6$ as a Test of the Interlayer Tunneling Model," *Nature* **395**, 360–2 (1998).

Anderson showed admirable aplomb in the face of high-quality data that contradicted his theory. He accepted defeat, sloughed off ten years of work, and returned to his original RVB ideas. An important nudge in this direction came from a research group at the Tata Institute for Fundamental Research in Mumbai, India. They began with Anderson's first guess for the RVB wave function, replaced his s-wave pairing by d-wave pairing, and showed that numerical calculations of the many-electron wave function led to many properties of the cuprates that agreed well with observations.[52]

Anderson contacted a dozen people he believed dwelled in what he called his "Big Tent" of supporters and asked them to join him as co-authors of a paper designed to outline a "back-to-basics" or "plain vanilla" version of RVB based on d-wave pairing. Only five chose to do so.[53] It mystified and disappointed him that the published paper did not alter the views of physicists who had not previously found merit in his RVB ideas. He nevertheless experienced a renewed sense of purpose and worked hard on plain vanilla RVB until his wife fell ill in August 2009. From that moment on, Phil devoted most of his time and energy to caring for her.

Anderson's Princeton colleague and good friend, the condensed matter experimentalist, Nai Phuan Ong, was instrumental in encouraging him during this final burst of research in superconductivity. Another trusted sounding board was V.N. Muthukumar. One day, Muthukumar asked Anderson why he had so quickly abandoned his original RVB ideas and spent ten years chasing gauge theories, flux phases, interlayer tunneling, and other esoterica. According to Muthukumar, "Phil was silent for a moment

---

[52] Arum Paramekanti, Mohit Randeria, and Nandini Trivedi, "Projected Wave Functions with High Temperature Superconductivity," *Physical Review Letters* **87**, 217002:1–4 (2001).

[53] P.W. Anderson, P.A. Lee, M. Randeria, T.M. Rice, N. Trivedi, and F.C. Zhang, "The Physics Behind High-Temperature Superconducting Cuprates: the 'Plain Vanilla' Version of RVB," *Journal of Physics: Condensed Matter* **16**, R755–R769 (2004).

and then looked at me and said, 'because I was surrounded by people who thought I was God'."[54]

This self-awareness mostly, but not completely, answers the question. We have seen the importance to Anderson of bouncing his ideas back and forth with other theorists. This strategy served him well at Bell Labs where he spent over thirty years sparring with extremely talented theoretical peers. One of his last Theory Group colleagues recalled that:

> I started [at Bell Labs] in January 1979 and was put in an office right next to Phil. I learned years later that this was intentional. Phil always thrived when other people argued with him, told him he was wrong about things, and generally pushed back on his ideas. By the time I got there, most people were pretty intimidated by him and they recognized that a new hire would be less so.[55]

Accomplished theorists near his own age played a similar role in his best work not done at Bell Labs.[56]

By contrast, there was virtually no theoretical condensed matter physics going on at Princeton (besides Anderson himself) when the HTS volcano erupted in 1987. By necessity, Phil engaged almost entirely with PhD students and postdocs. An exception was the Tamil theorist Baskaran (twenty-five years younger than Phil) who, if anything, exceeded Anderson in his enthusiasm for Anderson's ideas. As talented and argumentative as this group may have been, they could not provide the kind of intellectual resistance Phil needed to distinguish his good ideas from his questionable ones. Interesting hypotheses tended to morph into unquestioned axioms.

A peculiar bit of Anderson psychology also played a role. In early 1988, his former PhD student Gabriel Kotliar and his former Bell Labs colleague Maurice Rice, separately produced quantitative

---

[54] Author correspondence with V.N. Muthukumar, April 7, 2016.

[55] Interview of Daniel S. Fisher by the author, July 1, 2016.

[56] Sam Edwards and Elihu Abrahams argued back-and-forth with Anderson as they worked on the spin glass and weak localization problems, respectively.

theories of superconductivity that generalized Anderson's original conception of RVB.[57] Both found that the lowest energy super-conducting state had d-wave symmetry rather than the s-wave symmetry favored by Phil. Anderson's response was to pretend that their work did not exist; he cited neither paper for fifteen years. This is consistent with his career-long aversion to developing important insights discovered by others, particularly when they deviated from his way of thinking. Therefore, it is hardly a stretch to count Anderson's prejudice against d-waves as an additional driving force for him to turn away from RVB at a relatively early stage.

The Legacy of RVB

At the present time, there is no universally accepted theoretical understanding of the origin of superconductivity in the cuprates. This leaves in doubt the ultimate legacy of Anderson's resonating valence bond approach for HTS. Some believe the basic idea is very relevant; others believe it is not relevant at all. A view less connected to a belief system is that there is some range of doping where an RVB picture is most natural and some range of doping where other points of view are most natural.

Most theorists today accept Anderson's fundamental claim that the phonon mechanism of superconductivity plays little or no role in the cuprates. They also accept his brilliant insight that the relevant physics is some manifestation of the strong Coulomb repulsion between electrons. The Hubbard model (or some generalization of it) remains the starting point for many theorists and hope springs eternal that some unforeseen pen-and-paper approach to this model will appear and produce insight. In the meantime, most observers believe that any truly quantitative understanding of

---

[57] Gabriel Kotliar and Jialin Liu, "Superexchange and d-Wave Super-conductivity," *Physical Review* **38**, 5142–5 (1988); F.C. Zhang, C. Gros, T.M. Rice, and H. Shiba, "A Renormalised Hamiltonian Approach to a Resonant Valence Bond Wave Function," *Superconducting Science and Technology* **1**, 36–46 (1988).

superconductivity in the cuprates must await improvements in algorithms for computer-based numerical calculations.

RVB or no-RVB, Anderson deserves credit for another observation. Namely, the normal (non-superconducting) state of the cuprates differs in some fundamental way from the normal state of the classic low-temperature superconductors. Phil coined the term "strange metal" to distinguish the cuprates from conventional metals—the latter being well-described by Lev Landau's Fermi liquid theory (Chapter 9). For a while, Phil convinced himself that any conducting phase of the two-dimensional Hubbard was "strange" in this way and it excited him that this might be an important new contribution to many-body theory. It took some effort by others to show he was wrong. Indeed, it was a characteristic of Anderson that, even when he was wrong, he was usually wrong in some subtle and interesting way that opened up entirely new areas of research for others.

Quantum magnetism is an example. This field was the context for Anderson's original invention of RVB as a guess for the ground state of the two-dimensional, antiferromagnetic Heisenberg model. Subsequent theory (motivated by his application of RVB to the cuprates) showed that his guess was wrong.[58] However, these events inspired theorists and experimentalists to begin searching for other models and other physical systems where a more general form of RVB known as a "quantum spin liquid" is indeed the true ground state. At this writing, the jury is still out on this question.[59]

Finally, what of the cuprates themselves? Have they fulfilled the promise of the excitement generated at the 1987 Woodstock of Physics meeting? The answer is no. The cuprate superconductors are brittle ceramics like the materials from which toilet bowls and

[58] Sudip Chakravarty, Bertrand I. Halperin, and David Nelson, "Two-Dimensional Quantum Heisenberg Antiferromagnet at Low Temperatures," *Physical Review* **39**, 2344–71 (1989).
[59] See, e.g., J. Knolle and R. Moessner, "A Field Guide to Quantum Spin Liquids," *Annual Review of Condensed Matter Physics* **10**, 451–72 (2019).

ashtrays are made. This makes them difficult to form into flexible wires capable of passing high electric currents as the most important large-scale commercial applications require.[60] Eventually, this experimental/engineering problem will be solved, just as the corresponding theoretical problem will be solved.

[60] See the contributions to *100 Years of Superconductivity*, edited by H. Rogalla and P.H. Kes (CRC Press, Boca Raton, FL, 2012).

15

Four Facts About Science

The young Phil Anderson did not enjoy his sophomore philosophy class at Harvard. The professor seemed not to know that elementary calculus resolved a famous paradox posed by the Greek philosopher Zeno.[1] As he grew older, Phil took the time to think seriously about philosophical problems in the context of his own science. Eventually, he developed a personal philosophy of science based, not on abstract theorizing, but on a few core beliefs and concrete matters like criteria for choosing problems and the methodologies he used to attack those problems.

Previous chapters have explored some of these issues already, particularly those that bear on "more is different," the fundamentality (or not) of different types of physics, broken symmetry, complexity, and emergence. Indeed, the last of these grew in importance in Anderson's mind to the point where he identified the "fundamental philosophical insight of the twentieth century" as:

> Everything we observe emerges from a more primitive substrate, in the precise meaning of 'emergent', which is to say obedient to the laws of the more primitive level, but not conceptually consequent from that level.[2]

---

[1] Interview of PWA by Kaylee Ding, summer 2015.

[2] Philip Anderson, "Historical Overview of Twentieth Century Physics," in *Twentieth Century Physics*, volume III, edited by Laurie M. Brown, Abraham Pais, and Sir Brian Pippard (Institute of Physics and the American Institute of Physics, Bristol and New York, 1995), pp. 2017–32. Immediately following Anderson's essay is an essay by Steven Weinberg called "Nature Itself." Harking back to the position he took on the Superconducting Super Collider, Weinberg argues that "this has truly been the century of the triumph of reductionism."

This chapter examines some other aspects of Anderson's philosophy of science. His oft-mentioned inscrutability turns out not to be an impediment because his wife edited virtually all of his non-technical writing.[3] Joyce, the former English major, was a stickler who demanded clarity in his all-prose pieces.

Anderson's original retirement fantasy was to give up physics and write about early medieval history.[4] The 1986 discovery of high-temperature superconductivity obliterated that plan. Writing became part of his future anyway because the editor of *Physics Today* solicited him to become a regular contributor to a new opinion column in the magazine called *Reference Frame*. About half of the fifteen pieces he wrote for this column over the next five years discuss the sociology, structure, and methodology of physics.[5]

A broader audience became aware of Anderson's thinking between 1994 and 2006 when he published twenty-five reviews of science-related books in the *Times Higher Education Supplement*, a London-based weekly news magazine.[6] During the same period, he contributed twenty book reviews to *Science, Nature, Physics Today, Physics World*, and *American Scientist*. He reprinted some of these reviews in a collection of essays for the scientifically literate called *More and Different: Notes from a Thoughtful Curmudgeon*.

More and Different also includes thumbnail sketches of several well-known physicists, a few short popularizations, and discussions of subjects ranging from his personal scientific history to the politics of science, futurology, the science wars, and the

[3] Author correspondence with Susan Anderson.

[4] P.W. Anderson, *More and Different: Notes from a Thoughtful Curmudgeon* (World Scientific, Hackensack, NJ, 2011), pp. v–vi.

[5] The other half of Anderson's *Reference Frame* columns provide a personal history of the spin glass problem. These are reprinted in P.W. Anderson, *A Career in Theoretical Physics* (World Scientific, Singapore, 1994).

[6] These reviews were solicited by Andrew Robinson, the literary editor of the *Supplement*. Robinson's father was a University of Oxford physicist who befriended Anderson during visits he made to Bell Labs beginning in the 1950s.

meaning of physics. Reviewers found the essays insightful and readable, although occasionally marred by pettiness.[7]

A 1994 essay Anderson wrote for the *Daily Telegraph* newspaper of London provides a window into his philosophical thinking about science.[8] Its title, "Four Facts Everyone Ought to Know about Science," came from a talk he gave to liberal arts students at the University of Oxford as a contribution to the "two cultures" controversy that the chemist C. P. Snow and the literary critic F. R. Leavis began thirty-five years earlier.[9]

The original two-cultures issue was the supposed lack of respect humanists afforded scientists and engineers. Later, the debate expanded to include the need to preserve the moral resources provided by a liberal arts education in a world increasingly defined by scientific and technological advances.[10] In his *Daily Telegraph* piece, Anderson presented his thoughts about the nature of science, the characteristics of its practitioners, and how the subject should be communicated to the public.

Anderson's stated goal was to help his readers understand "how science really works." This was necessary because, in his view, the public suffered from a constant barrage of inaccurate or misleading information about the aims, methods, and results of

---

[7] N. David Mermin, "Essays on the Edge," *Physics Today* **65**(1), 44 (2012); Philip Phillips, "Review of 'More and Different'," *Physics in Perspective* **15**, 118–26 (2013).

[8] P.W. Anderson, "Four Facts Everyone Ought to Know about Science," *Daily Telegraph*, August 31, 1994, p. 14. See also, Andrew Zhang and Andrew Zangwill, "Four Facts Everyone Ought to Know about Science: The Two-Cultures Concerns of Philip W. Anderson," *Physics in Perspective* **20**, 342–69 (2018). The *Daily Telegraph* is a national newspaper similar to the *Wall Street Journal* in the United States.

[9] C. P. Snow, *The Two Cultures and the Scientific Revolution* (Cambridge University Press, Cambridge, 1959); F. R. Leavis, *Two Cultures?: The Significance of C. P. Snow* (Chatto & Windus, London, 1962).

[10] See, for example, Roger Kimball, "The Two Cultures Today," *The New Criterion* **12**(6), 10–16 (1994), 10; Guy Ortolano, *The Two Cultures Controversy: Science, Literature and Cultural Politics in Postwar Britain* (Cambridge University Press, Cambridge, 2009).

scientific research. He thought this could have grave consequences because our society depends so heavily on science-driven technology.

Anderson framed his essay in the form of four facts about science:

1. Science is not democratic.
2. Computers will not replace scientists.
3. Statistical methods are misused and often misunderstood.
4. Good science has aesthetic qualities.

It is possible to quibble with this list, offer alternatives, or debate whether some of them are even facts. The position taken here is to accept them as given and focus on their origin, their role in Anderson's thinking, and how subsequent developments may have affected their significance.

Science is not Democratic

Anderson begins with the simple declaration that science is not a search for consensus or compromise amongst rival theories of natural phenomena. Instead, scientists compare the predictions of theories with experimental observations in a ruthless search for explanations that are clear, unequivocal, and uncompromising. Democracy plays no role in deciding scientific issues because "there is only one explanation for every scientific phenomenon" presented to us by Nature.

This particular fact headed Anderson's list just as he was struggling to convince his colleagues of the correctness of his interlayer tunneling theory of high-temperature superconductivity. At a conference devoted to that subject, the chairperson jocularly asked the conferees to vote by a show of hands for the theory they thought would prove correct in the end. Anderson was not present, but he was outraged when he heard about it later. This was not because he lost the vote (which he did), but because it was beyond the pale to decide a scientific matter by vote, even in jest.

To Anderson, science is inherently elitist and undemocratic because "we pay scientists to discover the truth."[11] This position is consistent with his stance as a scientific realist. The core of this philosophical position is that the natural world exists objectively, independent of human perception and thought.[12] Scientific theories are true if they in some way provide a faithful account of that world. When a theory uses the word "crystal," the scientific realist presumes that the world includes objects that possess the properties assigned to a "crystal" in the theory. The same is true of atoms, quarks, and electrical resistance.

Anderson rejected the notion that science proceeds by the independent advance of each of its subdisciplines.[13] To him, the logical structure of modern scientific knowledge is not an evolutionary tree or a pyramid, but a highly interconnected web (Figure 15.1).[14] It is the seamless connectivity of the web, the mutual support the disciplines provide for each other, that is responsible for the great strength of modern science.[15]

Despite being undemocratic, science responds flexibly to empirical facts. A single experimental observation may spark a radical revision in a particular scientific field. However, the architecture of the web guarantees that any such revision typically

[11] P.W. Anderson, "When Scientists Go Astray," in *More and Different: Notes from a Thoughtful Curmudgeon* (World Scientific, Hackensack, NJ, 2011), pp. 204–17.

[12] Anjan Chakravartty, "Scientific Realism," in *The Stanford Encyclopedia of Philosophy* (Summer 2017), edited by Edward Zalta, https://plato.stanford.edu/archives/sum2017/entries/scientific-realism/. Accessed August 4, 2018.

[13] Philip W. Anderson, "More is Different—One More Time," in *More is Different: Fifty Years of Condensed Matter Physics*, ed. N. Phuan Ong and Ravin N. Bhatt (Princeton University Press, Princeton, NJ, 2001), p. 7.

[14] Philip W. Anderson, "Science: A 'Dappled World' or a 'Seamless Web'?," *Studies in the History and Philosophy of Modern Physics* 32(3), 487–94 (2001).

[15] Philip W. Anderson, "On the Nature of Physical Laws," *Physics Today* 43(12) 9 (1990). Anderson's web differs considerably from a web discussed by philosopher W.V. Quine. Quine depicts scientific knowledge (actually all knowledge) as a web or fabric of belief which "impinges on experience only along its edges." W.V. Quine, "Two Dogmas of Empiricism," in *From a Logical Point of View*, 2nd edition (Harper & Row, New York, 1951), pp. 20–46.

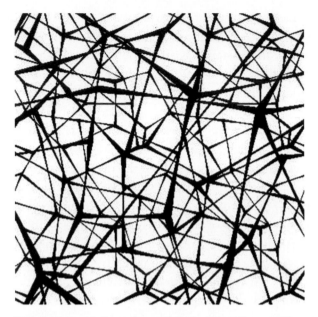

Figure 15.1 Anderson's "seamless web" of science. Source: Shutterstock.

induces only small changes, if any, in the workings of the many mature fields to which it is connected. It is this interconnectedness which protects the web from attacks from the outside. Real damage would require extensive slashing through its interior.

Anderson was keen for his readers to distinguish scientific truth from pseudoscientific untruth. He argued that the structure of the web itself facilitated seeing the falsity of subjects like homeopathy, cold fusion, and creation science because they inevitably contradicted evidence supplied from elsewhere in the web. He urged vigilance nonetheless. If a popular vote ever decided a scientific fact, the seamless web of interconnectedness would begin to fray, science would begin to lose its value as a repository of truth, and the door might open for pseudoscience to flourish.

Anderson opposed any attempt to distract from, or interfere with, the transmission of accurate and legitimate scientific information to the public. He identified the main distractors as (1) uncritical science journalists, (2) agenda-driven political pundits,

(3) religious fundamentalists, (4) social critics of science, and (5) publicity-seeking professional scientists.

Anderson despaired of the lack of critical acumen displayed by most science journalists. He never forgot the Krebiozen affair (Chapter 2) where the Chicago press gave equal weight to the negative opinion of the American Medical Association and the positive opinion of the promoters of this supposedly cancer-curing substance. The sociologist of science, Dorothy Nelkin, studied similar examples and concluded that because:

> they apply naïve standards of objectivity, reporters often deal with scientific disagreement simply by balancing opposing views. This approach does little to enhance public understanding of the role of science.[16]

Science reporting on the internet and social media are vastly more prominent today than they were in 1994 when Anderson assembled his list of facts. The current situation likely would appall him because, much more so than traditional science reporting, new media make it possible for "fake science" to flourish along with other forms of post-truth where facts do not matter.[17] The alleged connection between autism and vaccination is but one example.[18]

A 2005 book review gave Anderson the opportunity to excoriate agenda-driven political pundits for their efforts to corrupt the methods of scientific debate.[19] He took particular issue with

---

[16] Dorothy Nelkin, *Selling Science: How the Press Covers Science and Technology* (W.H. Freeman, New York, 1987), p. 68.

[17] Lee McIntyre, *Post-Truth* (MIT Press, Cambridge, MA, 2018), Chapters 2 and 6.

[18] Nan Li, Natalie Jomini Stroud, and Kathleen Hall Jamieson, "Overcoming False Causal Attribution: Debunking the MMR-Autism Association," in *The Oxford Handbook of the Science of Science Communication*, edited by Kathleen Hall Jamieson, Dan Kahan, and Dietram A. Scheufele (Oxford University Press, New York, 2017), pp. 433–43.

[19] Philip W. Anderson, "No Facts, Just the Right Answers," *The Times Higher Education Supplement*, December 23, 2005, p. 24.

neo-conservatives in the United States who provided intellectual cover for the prejudices of voting blocs whose favor they curried. He charged them with organizing town meetings where politically-motivated "experts" on issues like creationism and climate change could debate their scientific opponents on an equal footing. He worried actively that laypeople might mistakenly conclude that a simple vote among such experts is the best way to decide a scientific dispute.

Religious fundamentalism alarmed Anderson because it creates doubt about the veracity of science. He believed that the surrender to faith required to accept, say, the creation story in the Bible, is the reason that many people in the United States have a strong bias against the study of evolution and modern biology.[20] That being said, the atheist Anderson did not denounce religion unreservedly. It was, rather, a "deeply embedded evolved behavior pattern rather than as an intellectually justified, or necessarily useful concept."[21]

Postmodern and constructivist critics offer an anti-realist view of the scientific enterprise. The former claim that the aesthetic and cultural prejudices of scientists imply that science has no intrinsic truth value at all.[22] The latter focus on issues like careerism and professional prestige as the principal drivers of scientific decision making.[23] Anderson has no illusions that scientists are free from the social pressures that constrain the members of any group, but he finds no evidence to support the sweeping assertions of some social critics of science. A European scientist reported this remarkable obfuscation:

[20] Philip W. Anderson, "New Testaments," *The Sciences* **39**(4), 3 (1999). This is a response to Margaret Wertheim, "The Odd Couple," *The Sciences* **39**(2), 38–43 (1999).

[21] Philip W. Anderson, "Rebellious Thoughts," *Physics World*, April 2007, p. 42.

[22] Barry Barnes, David Bloor, and John Henry, *Scientific Knowledge: A Sociological Analysis* (University of Chicago Press, Chicago, 1996).

[23] Harry Collins and Trevor Pinch, *The Golem: What Everyone Should Know about Science* (Cambridge University Press, Cambridge, UK, 1993).

The profound and correct statement "At each level of complexity, new properties appear" was misread to imply that these properties were conjured rather than preexistent. Using this inaccuracy, Anderson's ideas were claimed to ascribe magical aspects to science.[24]

Finally, Anderson criticized self-promoting scientists who oversell the benefits that their research will bring to society. His observations of overhyping and overpromising in the early days of cuprate superconductivity provided all the evidence he needed that such behavior is rampant. The general practice continues to this day.[25]

By asserting that science is not democratic, Anderson aimed to warn his audience that the process of sorting fact from fiction extends to science. For him, any confusion on this point had serious and undesirable consequences, not only for the epistemic value of science, but also for the well-being of society.

Computers will not Replace Scientists

The *Oxford English Dictionary* identifies a computer as a machine that rapidly carries out arithmetic or logical operations so a user can efficiently perform complex calculations, store, manipulate, and communicate information, and control or regulate other devices. With this definition, Anderson confidently asserted that computers would never replace scientists because they lack the "creativity, serendipity, and lateral thinking" needed to produce significant science.

The trigger comment for this "fact" was a remark made by the late Stephen Hawking at his inaugural public lecture as Lucasian Professor of Mathematics at the University of Cambridge. The

---

[24] Author correspondence with Carlo Di Castro, August 5, 2016.

[25] See, e.g., Les Johnson and Joseph E. Meany, *Graphene: The Super-strong, Super-thin, Super-versatile Materials That Will Revolutionize the World* (Prometheus Books, Amherst, NY, 2018).

title of Hawking's lecture was "Is the End in Sight for Theoretical Physics?" He answered this question negatively by outlining several open questions in particle physics, field theory, and gravitational research. The final paragraph of Hawking's presentation struck Anderson forcefully:

> At present, computers are a useful aid in research, but they have never directed human minds. However, if one extrapolates their recent rapid rate of development, it would seem quite possible that they will take over altogether in theoretical physics. So maybe the end is in sight for theoretical physicists, if not for theoretical physics."[26]

It is possible that Hawking was joking.[27] If not, he was probably responding to years of exaggerated claims from the artificial intelligence (AI) subfield of computer science.[28] Many of these claims descend from Alan Turing's 1950 "imitation game" where a human interrogator is tasked to distinguish between two unseen interlocutors—one human and one digital—using only the typewritten responses each provides to the interrogator's questions.[29]

By 1990, the philosopher John Searle could explain to the readers of *Scientific American* that there were two types of AI.[30] "Strong AI" contended that an appropriately programmed

[26] S.W. Hawking, "Is the End in Sight for Theoretical Physics?," *Physics Bulletin* **32** (1), 15–17 (1981).

[27] By 1980, the degenerative disease amyotrophic lateral sclerosis had impaired Hawking's speech to the point where he asked his graduate student, Christopher Pope, to deliver the Lucasian inaugural lecture for him. Hawking did not share with Pope his reason for including the final paragraph, but Pope reports that Hawking "often liked to finish off a talk in a slightly jokey or whimsical way." Private communication with Christopher Pope, July 1, 2018.

[28] Alok Aggarway, "The Birth of Artificial Intelligence and the First AI Hype Cycle," 2018. https://www.kdnuggets.com/2018/02/birth-ai-first-hype-cycle.html. Accessed July 24, 2018.

[29] Daniel Crevier, *AI: The Tumultuous History of the Search for Artificial Intelligence* (Basic Books, New York, 1993), pp. 22–5; Margaret A. Boden, *Mind as Machine. A History of Cognitive Science* (Clarendon Press, Oxford, 2006), Section 10.iii.a.

[30] John R. Searle, "Is the Brain's Mind a Computer Program?," *Scientific American*, January 1990, pp. 26–31.

computer really is a mind, in the sense that computers given the right programs could be literally said to understand and have other cognitive states. "Weak AI" focused only on using the computational abilities of computers to study the mind the way one studies weather, economics, or molecular biology.

Anderson had no truck with strong AI. It was perfectly clear to him that the mind does things that are beyond the capabilities of mere computing machines. But he also saw danger with weak AI, despite the fact that the digital computer was well entrenched in nearly every research subfield of physics by 1970.[31] He shared the concern of the British plasma physicist Keith Roberts, who sounded the alarm at a 1971 summer school devoted to computing: "It is sometimes thought that computers will eventually kill theoretical physics; all that one will need to do is to program the equations and press the button in order to get a numerical answer. This is very far from being the case."[32]

Anderson went further. He told his *Daily Telegraph* readers that "consciously or unconsciously, it can be all too easy to adjust the input to achieve the desired output. Biased, meaningless, or even faked computations are just as common as the corresponding malfeasances in the biomedical sciences." He did not offer a specific example of either form of misbehavior.

More so than others trained in the pre-computer era, Anderson (b. 1923) maintained a career-long suspicion of the use of computers in physics. He granted that some practitioners used this tool to do creative and interesting physics (like his former colleagues Bill McMillan and Volker Heine) but he railed tirelessly against people whom he felt turned off their brains when they turned on the computer. His bogey man in this regard was John Slater, the brilliant theoretical physicist we first met in Chapter 6.

---

[31] Sidney Fernbach and A. H. Taub, *Computers and Their Role in the Physical Sciences* (Gordon and Breach, New York, 1970).

[32] K. V. Roberts, "Computers and Physics," in *Computing as a Language of Physics* (International Atomic Energy Agency, Vienna, 1972), pp. 3–26.

Slater did important work on the theory of many-electron systems in the early days of quantum mechanics.[33] After 1960, he devoted his considerable resources to obtaining numerical solutions of an approximate version of the many-electron Schrödinger equation. A visitor to his group was struck by the contrast between Slater's zeal for obtaining results for wave functions and energy levels for specific material systems and his relative disinterest in using them to extract broad qualitative trends.[34]

Anderson made clear his attitude about people like the post-1960 Slater in a 1980 essay published in the French language magazine *La Recherche*.[35] Halfway in, he alludes to the classic reductionist statement of Paul Dirac (quoted in Chapter 5) that quantum theory described "much of physics and the whole of chemistry" but that an exact application of that theory to any real many-body problem was impossibly difficult.[36] Unfortunately, rather than paraphrase Dirac accurately (including his desire for the development of practical methods of approximation), Anderson quotes him to say, "everything else is chemistry" and characterizes his tone as derisive, which it is not. This device allows him to fume that creative non-computational condensed matter physicists (like himself) could be driven to extinction by scientists who suffer from what he calls the "Dream Machine" syndrome—a malady which presumes that computing is sufficient to understand every interesting physical phenomenon.

---

[33] John C. Slater, *Solid State and Molecular Theory: A Scientific Biography* (Wiley-Interscience, New York, 1975); S. S. Schweber, "The Young John Clarke Slater and the Development of Quantum Chemistry," *Historical Studies in the Physical and Biological Sciences* **20**(2) 339–406 (1990).

[34] Interview of Volker Heine by the author, July 11, 2015.

[35] P.W. Anderson, "La Grande Illusion des Physiciens," *La Recherche* **11**, 98–102 (1980). This is a translated and heavily edited version of the manuscript Anderson submitted titled "The Great Solid State Physics Dream Machine." The original English version circulated widely at the time.

[36] P.A.M. Dirac, "Quantum Mechanics of Many-Electron Systems," *Proceedings of the Royal Society of London A* **123**, 714–33 (1929).

The *La Recherche* article reflects an epistemology where computers play little or no role. To gain knowledge of the world, says Anderson, the theoretical physicist should identify inconsistencies between experiment and theory, construct a simple model Hamiltonian for the situation of interest (90 percent of the task, he says), and use that Hamiltonian to extract a consistent physical picture. Earlier chapters testified to Anderson's success applying this philosophy to learn about antiferromagnetism, disorder-induced localization, magnetic impurities in metals, BCS super-conductivity, the Kondo effect, and spin glasses.

Anderson's method foundered only with cuprate supercon-ductivity. This was not because he did not identify a suitable model Hamiltonian—the Hubbard model—but because none of the pen-and-paper methods of theoretical physics he used to attack it satisfied the theoretical community as a whole. Even greater skepticism accompanied every attempt he made to replace the Hubbard model with a more tractable model Hamiltonian. The direct numerical attacks on the Hubbard model favored by his old friend Doug Scalapino and others were galling to him precisely because, in his view, they operated with an entirely different idea about what it meant to understand the physics.

How well does Anderson's 1994 assertion about computers hold up in light of twenty-first century developments in artificial intelligence? One recent achievement is that a computer program now handily defeats the best human players of Go—Anderson's favorite board game.[37] This is non-trivial because, unlike the strategies computers use to defeat human chess champions, success at Go requires a sophisticated pattern recognition strategy.

[37] David Silver, Julian Schrittwieser, Karen Simonyan, Ioannis Antonoglou, Aja Huang, Arthur Guez, Thomas Hubert, Lucas Baker, Matthew Lai, Adrian Bolton, Yutian Chen, Timothy Lillicrap, Fan Hui, Laurent Sifre, George van den Driessche, Thore Graepel, and Demis Hassabis, "Mastering the Game of Go without Human Knowledge," *Nature* **550**, 354–9 (2017).

In 2009, *Science* magazine published a paper that used a computer analysis of a large trove of raw observational data to "discover" two mathematical laws of motion relevant to a pendulum.[38] These laws were well known to classical physics, but they were far from apparent from just a cursory inspection of the data. Anderson and fellow physicist Elihu Abrahams responded with a letter to *Science*'s editor.[39] In their view, the paper's authors were "seriously mistaken about the nature of the scientific enterprise, particularly what theorists do and the meaning of physical law."

Anderson and Abrahams conceded that computers might contribute to normal science where one "simply fleshes out the consequences of existing scientific paradigms."[40] However, they could see no mechanism for a computer to "create Kuhnian scientific revolutions and thereby establish new physical laws." The latter, as Anderson had told his *Daily Telegraph* readers fifteen years earlier, was the exclusive domain of human theorists.

The Go and pendulum examples came from the domain of *machine learning*, a relatively new field where computers infer patterns from enormous sets of data and use those patterns to perform tasks without being explicitly programmed to do so.[41] Anderson's dim view of this activity reinforces his *Daily Telegraph*

---

[38] Michael Schmidt and Hod Lipson, "Distilling Free-Form Natural Laws from Experimental Data," *Science* **324**, 81–5 (2009). The *double pendulum* studied in this paper consists of two pendula connected in series.

[39] Philip W. Anderson and Elihu Abrahams, "Machines Fall Short of Revolutionary Science," *Science* **324**, 1515–16 (2009). Abrahams and Anderson published together seven times, the Gang of Four weak localization paper being the most notable.

[40] Thomas S. Kuhn, *The Road Since Structure*, edited by James Conant and John Haugeland (University of Chicago Press, Chicago, 2000).

[41] Peter Donnelly et al. "Machine Learning: the Power and Promise of Computers that Learn by Example," Royal Society White Paper, April 2017. Available at https://royalsociety.org/machine-learning. Accessed January 10, 2020.

message in a manner we can illustrate using the story of Tycho Brahe.[42]

Brahe was a sixteenth-century Danish nobleman whose collection of astronomical observations constituted the first Big Data set in history. We remember Brahe because he engaged the young Johannes Kepler to analyze his trove. Kepler inferred two mathematical formulas from this data, which together implied that the planets move around the Sun in elliptical orbits.

From Anderson's perspective, Kepler practiced machine learning and his work was an example of normal science. However, the human genius of Isaac Newton was needed to create a scientific revolution by inventing the subject of mechanics. Only then could Kepler's formulas be deduced from a logical foundation and fitted into the seamless web of scientific knowledge.

Statistical Methods are Misused and Often Misunderstood

Anderson's third fact that everyone should know about science involved statistics. He was not concerned with the well-known biases that lead all people to struggle with arguments based on statistical reasoning.[43] Instead, he wanted the literate public to know that most of his fellow scientists misunderstood (or at least lacked an appreciation of) an approach to statistical analysis called *Bayesian inference*.[44]

---

[42]   Sui Huang, "The Tension Between Big Data and Theory in the 'Omics' Era of Biomedical Research," *Perspectives in Biology and Medicine*, **61**(4), 472–88 (2018).

[43]   See, e.g., Nate Silver, *The Signal and the Noise: Why Some Predictions Fail—But Some Don't* (Penguin Press, New York, 2012); Daniel Kahneman, *Thinking Fast, Thinking Slow* (Farrar, Straus and Giroux, New York, 2013).

[44]   Sharon McGrayne recounts the colorful history of Bayesianism in *The Theory that Would Not Die: How Bayes' Rule Cracked the Enigma Code, Hunted Down Russian Submarines, & Emerged Triumphant from Two Centuries of Controversy* (Yale University Press, New Haven, CT, 2011).

There are two distinctly different approaches to statistical inference. The first applies to scientific questions that have a definite numerical answer. For example, "what is the temperature of the Universe?" or "what is the mass of the proton?" The job of the scientist is to deduce the desired numerical value from a statistical analysis of measured data. For these cases, Anderson recommended the "standard" (so-called frequentist) method of statistical analysis that is built into pocket calculators and spreadsheet programs.

More problematic to Anderson were situations where the question is, "how probable is it that a particular phenomenon actually occurs?" For this class of questions, the frequentist approach assumes that the phenomenon occurs with a non-zero probability and the task of the analysis is to estimate the numerical value of that probability. By contrast, the Bayesian approach allows explicitly for the possibility that the phenomenon of interest does not occur at all.

The two approaches do not always give the same answer and one cannot always tell which one is more correct based on the available information. After all, the absence of complete information is the reason one resorts to a statistical method in the first place. Nevertheless, all things being equal, Anderson preferred the Bayesian approach because "the epistemology of modern science seems to be basically Bayesian induction with a very great emphasis on its Ockham's razor consequences (a penalty for the use of unnecessary hypotheses). One is searching for the simplest schematic structure that explains all the observations."[45]

Bayesian inference copes with uncertain knowledge by learning from experience. It does this by making explicit use of any prior beliefs one might have about the issue of interest and then providing a mechanism to update those beliefs in the light of new

[45] Philip W. Anderson, "Science: A 'Dappled World' or a 'Seamless Web'?," *Studies in the History and Philosophy of Modern Physics* **32**(3), 487–94 (2001).

information.[46] The late Richard Feynman illustrated Bayesianism this way:

> Suppose there are two theories of some effect: Theory A and Theory B. Suppose also there is a test for the effect which involves dipping a strip of paper into a prepared solution. Theory A says nothing should happen. Theory B says the strip should turn blue. For some reason, you have a prior belief that Theory A is much more likely to be correct than Theory B. You now perform the test and the strip turns greenish. This is very unlikely if Theory A is true, but it is not impossible if Theory B is true. Using the rules of Bayesian inference, the observation of a greenish strip weakens your belief in Theory A and strengthens your belief in Theory B. This becomes your new "prior belief" when you perform a different test where Theory A and Theory B make different predictions. Bayesian analysis constantly makes use of new data to readjust our relative belief in Theory A and Theory B.[47]

Similar readjustments occur in machine learning.

Anderson was thinking about Bayesian inference because he had just published a *Physics Today* opinion piece on the subject in connection with two research papers which disturbed him.[48] One used new nuclear physics data to infer the existence of a previously

---

[46] Anderson has argued that a Bayesian analysis inevitably leads to a very small probability for the existence of God, no matter how large your prior belief in God might be. This so because the Ockham's razor character of the method reduces the predicted probability for God every time one introduces a new parameter into the analysis. The problem is that the number of these new parameters can be very large. Does God have a long grey beard? Is God benevolent? Is God malicious? Does God impose dietary restrictions? See Anderson's response to the 2006 Edge. org question, "What is Your Dangerous Idea?, https://www.edge.org/responses/what-is-your-dangerous-idea."Accessed August 2, 2018.

[47] This is a paraphrase of a paragraph in Richard P. Feynman, *The Meaning of It All: Thoughts of a Citizen-Scientist* (Addison-Wesley, Reading, MA, 1998), 67–8. The contents of this book were transcribed posthumously from audio tapes of three public lectures given in 1963.

[48] Philip W. Anderson, "The Reverend Thomas Bayes, Needles in Haystacks, and the Fifth Force," *Physics Today* **45**(1), 9–11 (1992).

unknown elementary particle.[49] The other re-analyzed a famous experiment in the history of physics and inferred the existence of a previously unknown force in the Universe.[50] Both garnered a great deal of publicity and launched an avalanche of work by others. Eventually, the weight of accumulated experimental evidence led the physics community to reject both claims.[51]

In his *Physics Today* piece, Anderson stated that a Bayesian analysis at the outset would have revealed that both phenomena were implausible. This would have saved the physics community the time, money, and effort it cost to mount the many experiments needed to establish implausibility using standard, non-Bayesian statistical methods.[52]

The reason Anderson chose to discuss statistical analysis in a mass circulation newspaper becomes clearer after learning about an event that occurred at Princeton a decade earlier. In 1980, Robert Jahn, the Dean of Princeton's School of Engineering and Applied Science, established a laboratory for the experimental study of paranormal and psychokinetic behavior. In one test, researchers dropped steel balls through a series of channels and asked subjects to use their minds to try to influence the trajectories taken by the balls.[53] Over a number of years, Jahn and his

[49] J. J. Simpson, "Evidence of Heavy-Neutrino Emission in Beta Decay," *Physical Review Letters* **54** (17), 1891–3 (1985).

[50] Ephraim Fischbach, Daniel Sudarksy, Aaron Szafer, Carrick Talmadge, and S. H. Aronson, "Reanalysis of the Eötvös Experiment," *Physical Review Letters* **56**(1), 3–6 (1986).

[51] Douglas R. O. Morison, "The Rise and Fall of the 17-keV Neutrino," *Nature* **366** 29–32 (1993). Allan Franklin, *The Rise and Fall of the Fifth Force: Discovery, Pursuit, and Justification in Modern Physics* (American Institute of Physics, New York, 1993).

[52] For a critique of Anderson's assertion, see Allan Franklin and Ephraim Fischbach, *The Rise and Fall of the Fifth Force: Discovery, Pursuit, and Justification in Modern Physics*, 2nd edition (Springer, Heidelberg, 2016).

[53] Steven Schultz, "Robert Jahn, Pioneer of Deep Space Propulsion and Mind-Machine Interactions Dies at 87," Princeton University Office of Engineering Communications, https://www.princeton.edu/news/2017/11/30/robert-jahn-pioneer-deep-space-propulsion-and-mind-machine-interactions-dies-87. Accessed July 31, 2018.

coworkers reported that his subjects' intentions influenced their results in a way that deviated from pure chance.[54]

Anderson felt strongly that Jahn's results were not statistically significant. That fact, Brian Josephson's embrace of the paranormal as a research subject (Chapter 10), and the general fascination of the public with astrology, telepathy, and psychic spoon-bending practically compelled him to exhort his *Daily Telegraph* readers to be skeptical when told that an allegedly "scientific study" came to a conclusion that violated their common sense.

Anderson's message stripped to its bones was simply that Bayesian inference is common sense expressed in mathematical form. He hoped the lay public would hold accountable those who commission or disseminate the results of statistical studies and insist that they impose quality control on the scientists who produce them.

Good Science has Aesthetic Qualities

Anderson's final fact about science aimed to dispel the image of the scientist as a white-coated automaton devoid of passion. Making new science is a creative act and scientists respond to aesthetic principles just like writers, artists, and musicians. Practitioners know the difference between a beautiful piece of science and an ugly piece of science. This is not a novel point of view. But with his characteristic contrariness, Anderson suggested a set of aesthetic principles for science that differed considerably from those traditionally offered for this subject.

If Plato was the first person to connect knowledge to beauty, Johannes Kepler was typical of his time when he cited the beauty and perfection of the Divine as the inspiration for his heliocentric

[54] Douglas J. Matzke and Loren L. Howard, "A Review of Psychical Research at SRI and Princeton University," Technical Report EE 85004, January 1985, http://www.matzkefamily.net/doug/papers/princeton_research.pdf. Accessed July 31, 2018.

model of the solar system.[55] To him, "God laid out the world so that it might be best and most beautiful and finally most like the Creator."[56] In the secular twentieth century, most physicists transferred their conception of beauty from the Divine to ideas like unification and symmetry.[57] This led them to favor theories endowed with mathematical elegance and permitted them to quote John Keats, "beauty is truth, truth beauty—that is all ye know on earth, and all ye need to know."[58]

Some scientists concluded that, like a company too large to fail, there can be theories too beautiful to be wrong. Albert Einstein, when asked how he would feel if a recent experiment had failed to confirm his theory of general relativity, replied "I would have to pity the dear Lord, the theory is correct anyway."[59] The quantum pioneer Paul Dirac expressed a similar sentiment when he declared that "it is more important to have beauty in one's equations than to have them fit experiment."[60] Today, some regard an approach to particle physics called supersymmetry as the leading candidate for a theory that is too beautiful to be wrong.[61] Others

[55] Nickolas Pappas, "Plato's Aesthetics," in *The Stanford Encyclopedia of Philosophy* (Fall 2017), ed. Edward N. Zalta, https://plato.stanford.edu/archives/fall2017/entries/plato-aesthetics/. Accessed August 2, 2018.

[56] Johannes Kepler, *Harmonices Mundi*, Book III, Chapter 1, 1618. Quoted in Judith V. Field, "Astrology in Kepler's Cosmology," in *Astrology, Science, and Society. Historical Essays*, edited by Patrick Curry (Boydell Press, Woodbridge, UK, 1987), p. 123.

[57] Peter Atkins, *Galileo's Finger: The Ten Great Ideas of Science* (Oxford University Press, Oxford, 2003), Chapter 6; Frank Wilczek, *A Beautiful Question: Finding Nature's Deep Design* (Penguin Press, New York, 2015).

[58] Ian Stewart, *Why Beauty is Truth: A History of Symmetry* (Basic Books, New York, 2007), Chapter 16; Frank Wilczek, *A Beautiful Question: Finding Nature's Deep Design* (Penguin Press, New York, 2015). The quotation from Keats is from "Ode on a Grecian Urn," *Annals of Fine Arts*, January 1820.

[59] Quoted in Ilse Rosenthal-Schneider, *Reality and Scientific Truth: Discussions with Einstein, Von Laue, and Planck* (Wayne State University Press, Detroit, MI, 1980), p. 74.

[60] P. A. M. Dirac, "The Evolution of the Physicist's Picture of Nature," *Scientific American*, May 1963, p. 47.

[61] Dan Hooper, *Nature's Blueprint: Supersymmetry and the Search for a Unified Theory of Matter and Force* (Harper Collins, New York, 2008).

reserve that distinction for a particular expression of supersymmetry called string theory.[62] It does not seem to bother enthusiasts that one makes predictions that disagree with experiment (supersymmetry) while the other makes no testable predictions at all (string theory).[63]

Phil Anderson's set of aesthetic principles for science have nothing to do with beauty in the sense used above.[64] Criteria like naturalness, elegance, and symmetry do not obviously bear on the scientific mandate of objectivity.[65] Indeed, for Anderson, a beautiful theory does not require beautiful mathematics at all. His model is the fine arts.

Art exists in context and one can only judge the presence or absence of beauty against the substrate of this context. The paintings by Da Vinci and Caravaggio speak to us through a substrate of religious symbolism. The paintings by Picasso and Rothko do this through their abstract qualities of color and form. Similarly, a work of art achieves beauty if it communicates multiple layers of meaning. Anderson offers the example of *The Wasteland* by T. S. Eliot. In this poem, there is a "gorgeous use of language" at the surface level, "a sense of despair at the moral emptiness of the modern world" at a deeper level, and "a series of references to myth" at an even deeper level.

Abstracting from these examples, Anderson offered four aesthetic principles for theoretical physics. First, a beautiful theory must

[62] Lee Smolin, *The Trouble with Physics: The Rise of String Theory, The Fall of Science, and What Comes Next* (Houghton Mifflin, Boston, 2006).

[63] Not all philosophers of science regard the absence of experimental predictions as a disqualifying vice for a scientific theory. See, e.g., Richard Dawid, *String Theory and the Scientific Method* (Cambridge University Press, Cambridge, 2013).

[64] Using the aesthetic criterion of "a proper conformity of the parts to one another and to the whole" while still showing "some strangeness in their proportion," an eminent astrophysicist nominated Einstein's classical theory of gravitation as "probably the most beautiful of all existing physical theories." Subrahmanyan Chandrasekhar, "Beauty and the Quest for Beauty in Science," *Physics Today* **32**(7), 25–30 (1979).

[65] Sabine Hossenfelder, *Lost in Math: How Beauty Leads Physics Astray* (Basic Books, New York, 2018), p. 2.

deal with reality as its substrate. This means the theory must address some aspect of the physical world that is amenable to study by experiment. Pure mathematical constructions untethered to observations need not apply.

Second, a beautiful theory must exhibit craftsmanship of a sort that reflects its creator's intentions and his/her efforts to maximize the potential of the theory. From this point of view, the trial-and-error method of invention used by Thomas Edison lacks beauty. By contrast, Johann Sebastian Bach took the existing technique of counterpoint and used his great genius to craft musical compositions of striking beauty.

Third, Anderson expects a beautiful theory to exhibit maximal cross-reference. Not only must the theory possess multiple levels of meaning, it must exhibit a large breadth of reference and application in the physical world. The most beautiful theories contribute maximally to the seamless web of interconnectedness that is fundamental to his architecture of science. Or as he put it to his *Daily Telegraph* readers:

> How wide and far-reaching are its implications? How subtle and unexpected are the connections? How deeply does it delve into Nature?

Ockham's razor returns when Anderson demands that a beautiful theory obey a principle of simplicity. This does not mean that its mathematics should be particularly elegant or economical. Rather, there should be a core idea that leads to all the essential consequences of the theory. Anderson emphasizes that a beautiful theory must provide the maximum amount of information about the real world using a minimum number of ideas.

An outstanding example of scientific beauty in the mode of Anderson is the deduction of the atomic structure of the DNA molecule by Francis Crick and James Watson. On the one hand, this work displayed exquisite craftsmanship in the brilliant detective work which led them to the double helix. On the other hand, the structure itself suggested the mechanism needed to replicate genetic material and eventually the molecular mechanism of heredity.

By Anderson's criteria, supersymmetry and string theory are not very beautiful. The predictions of the former have not been observed and the latter lacks a physical substrate altogether. By the same standard, the Nobel Prize-winning model Hamiltonians proposed by Anderson for disorder-induced localization, magnetic impurities in metals, and the spin glass *are* beautiful because they include only the essentials abstracted from experiment while exhibiting the full richness of the phenomena he hoped to understand.

Chapter 9 used the aesthetic principles elaborated here to evaluate the beauty of the BCS theory of superconductivity. Anderson is quick to point out that BCS was *not* built using beauty as an a priori criterion to help guide its formulation. Its beauty became apparent only afterward with the realization that it spoke directly to experiment, it exhibited a high degree of craftsmanship, it informed topics as diverse as liquid ^{3}He, atomic nuclei, and neutron stars, and it possessed one simple idea: analyze the phase of the macroscopic quantum wave function.

But where does the beauty of BCS reside, exactly? Anderson locates it in the creative tension between theory and experiment and in the seamless web of its connections to other subjects. This answer—which applies equally well to the concept of broken symmetry—is interesting because virtually no commentator besides Anderson speaks about the connection with experiment when extolling the beauty of a piece of theoretical physics.[66] As for the seamless web, Anderson likens its construction to the construction of one of the great medieval cathedrals of Europe he

---

[66] A grudging exception is "it also must be admitted…that a concept that provides widespread empirical unification will thereby acquire aesthetic value." This comment appears in Stanley Deser, "Truth, Beauty, and Supergravity," *American Journal of Physics* **85**(11), 810–11 (2017).

so loved.[67] In both cases, the edifice that results has enormous grace, power, and endurance.

And finally, what of high-temperature superconductivity? Aesthetics surely played a role in some of the more high-minded theoretical models that were proposed to explain this puzzle. For his part, Anderson began with a powerful idea, wandered away in thrall with a new theory, and finally returned to his original insight. All the while, he was nettled by the appearance and re-appearance of theories he thought were both ugly and untrue. More than ten years after the cuprates were discovered, a sum-mary of the theoretical situation slipped past the editors of *Physical Review Letters* when they permitted a technical article about the cuprates to conclude:

> The tragedy of beautiful theories, Aldous Huxley once observed, is that they are often destroyed by ugly facts. One perhaps can add that the comedy of not so beautiful theories is that they can-not even be destroyed; like figures in a cartoon they continue to enjoy the most charming existence until the celluloid runs out.[68]

[67] P.W. Anderson, "Some Ideas on the Aesthetic of Science," Lecture given at the 50th Anniversary Seminar of the Faculty and Science and Technology, Keio University, Japan, May 1989. Reprinted in P.W. Anderson, *A Career in Theoretical Physics* (World Scientific, Singapore, 1994), pp. 569–83.

[68] B.K. Chakraverty, J. Ranninger, and D. Feinberg, "Experimental and Theoretical Constraints on Bipolaronic Superconductivity in High T_c Materials: An Impossibility," *Physical Review Letters* **81**, 433–5 (1998).

16

Conclusion

Phil Anderson was eighty years old when he and his co-authors published the "plain vanilla" version of his RVB model of super-conductivity. The same year (2004), he oversaw the publication of the second edition of a selection of his research papers, *A Career in Theoretical Physics*. He also read with great interest an article claiming to see the first experimental evidence for superfluidity in the solid phase of ^{4}He.[1] This odd-sounding combination had been predicted years earlier and the possibility that such a *supersolid* existed delighted Anderson. It was another quantum many-body problem to which he could apply his still sharp theoretical skills. Over the next decade, he wrote six articles exploring the phe-nomenology of a supersolid. Unusually for him, the fact that other physicists were unable to reproduce the results of the ori-ginal experimetal report did not deter him from the belief that the phenomenon would eventually be seen in the laboratory.

A life-changing event overtook the Anderson family in August 2009 when Joyce suffered a near-fatal stroke.[2] Phil was away at a physics meeting in western Canada. Emergency services rushed Joyce to the hospital, but it took Phil two days to get home. Daughter Susan shuttled back-and-forth between Boston and New Jersey every week for a month. Seven weeks passed before Joyce returned to the Frank Lloyd Wright prairie-style house she had lived in for over two decades. Phil and Susan moved her bed

---

[1] E. Kim and M.H.W. Chan, "Probable Observation of a Supersolid Helium Phase," *Nature* **427**, 225–7 (2004).

[2] Author correspondence with Susan Anderson, August 19, 2016.

Figure 16.1 A view from the window of the living room of the last house owned by Phil and Joyce Anderson. Source: Susan Anderson.

into the living room and encouraged a Norwegian forest cat to keep her company. A large window allowed her to look out on the beautiful ten acre property she had spent so many hours maintaining (Figure 16.1).

Phil curtailed his travel dramatically and organized his life around caring for Joyce. Her ability to read was not seriously impeded and keeping her supplied with the daily *New York Times* and the weekly *New Yorker* magazine became important rituals. Phil's own health was quite good for an eighty-five year old man. A pacemaker controlled his heart rhythm, but besides a bout with hepatitis in 1976 and colon surgery in 1981, he had always been a remarkably healthy and vigorous person. More than a few younger physicists who hiked with him or played tennis with him were embarrassed by their inability to keep up.

Susan's relationship with her parents had always been prob-lematical. A comfortable truce was easy to maintain as long as she lived and worked in Boston. Joyce's stroke changed the landscape and Susan eventually settled into a pattern of driving down to New Jersey every two weeks to organize her parents' finances, manage Joyce's aides, etc. She discovered she loved her father and

Figure 16.2 Philip Warren Anderson in 2016 at age 92. Source: Peter Badge/Lindau Nobel Laureate Foundation.

real warmth flowed between them, perhaps for the first time in their lives.

Anderson slowly returned to spending time at his Princeton office. He graduated his final PhD student in 2010 and he published his collection of essays, *More and Different: Notes from a Thoughtful Curmudgeon*, in 2011. He attended seminars that interested him, and talked physics and gossiped about physicists with visitors and colleagues. Even as a nonagenarian he continued to write: a review of a biography of the iconoclastic mathematical physicist Freeman Dyson in 2013 and an update about superconductivity and the Higgs mechanism in 2015 (Figure 16.2). His "Four Last Conjectures" of 2018 concerned supersolids, the cuprates, and the "dark energy" mystery of contemporary astrophysics.[3] The tone of all these contributions—literate, didactic, and personal—differed little from the tone of his PhD thesis written seventy years earlier.[4]

---

[3] Available at https://arxiv.org/abs/1804.11186. Accessed March 18, 2020.

[4] Philip Anderson, "An Iconoclast's Career," *Physics World*, March 2013, pp. 62–3; Philip W. Anderson, "Higgs, Anderson and all that," *Nature Physics* **11**, 93 (2015); Philip W. Anderson, "Four Last `Conjectures'," arXiv:1804.11186.

At a certain point, Anderson's physical frailty prevented him from traveling to Princeton. Nevertheless, his spirit remained positive and he enjoyed dining with visitors at the retirement community to which he and Joyce had moved in 2013. He also maintained regular e-mail contact with a variety of friends, former colleagues, and former students. His cognitive skills were still good when he celebrated his ninety-sixth birthday on December 13, 2019. His health took an irreversible turn for the worse after the New Year. At the beginning of March, the Japanese quince Anderson had given to Ryogo Kubo in 1954 failed to flower for the first time, withered, and died.[5] Phil succumbed to pneumonia three weeks later.

Who We Know

The Introduction of this book described Anderson as one of the most accomplished and influential physicists of the second half of the twentieth century.[6] Later chapters demonstrated why his name resonates not only with physicists, but also with scholars in other fields. Nevertheless, the abstruse nature of Anderson's research achievements makes it difficult to communicate the magnitude of his influence on his own community.

An analogous figure might be John Maynard Keynes (1883–1946), a person whose influence on his field of economics was equally technical and profound, but whose name happens to be known to the general public. Keynes founded modern macroeconomics and his academic work provides the theoretical underpinning for government intervention in national economies. Not all economists agree with Keynesian analysis, but none can ignore his *General*

[5] Private communication with Hiroto Kono.

[6] Using a metric based on citations and references in published research journals, one data scientist named Philip Anderson and Edward Witten as the most creative physicists of the second half of the twentieth century. Steven Weinberg was a close third. José M. Soler, "A Rational Indicator of Scientific Creativity," *Journal of Informetrics* 1, 123–30 (2007).

Theory of Employment, Interest, and Money. A recent characterization of this book as a "masterwork" of "passion driven by intuition" that is "at times obscure, tedious, and tendentious" could, in the minds of many, be applied equally well to Anderson's *Basic Notions of Condensed Matter Physics.*[7]

The public knows Keynes' name because pundits and politicians occasionally praise or damn an economic policy as "Keynesian." College students know somewhat more about him because general survey courses (Economics 101) routinely discuss his importance. By contrast, survey courses barely exist to inform curious liberal arts students about quantum science and technology. This may change as those technologies penetrate more and more deeply into our personal and working lives. When such courses proliferate, non-physicists will begin to recognize the name Philip Anderson.

A Man in Full

Phil Anderson was a child of the Depression who had the good fortune to grow up in a financially stable and nurturing academic home. He absorbed a liberal political outlook from his parents, excelled in academics and athletics, and attended Harvard on a scholarship. His class graduated early to contribute to the war effort and he served his country as a radar engineer before returning to Harvard to earn his PhD in physics.

Naturally shy, Anderson hung out with non-physics graduate students when he wasn't using theoretical methods he learned from Julian Schwinger to solve a thesis problem set by John Van Vleck. By more good luck, he met and married his life-partner Joyce eighteen months before Van Vleck secured him a job at Bell Labs. He launched his career just when his new employer and his country were at their most confident and expansive.

---

[7] Benn Steil, "The Economic Engineer," *Wall Street Journal*, May 9–10, 2020, p. C7.

Solid-state physics and Anderson grew up and flourished together. He never studied the simplest solids, but made a conscious decision to focus on the puzzles presented by disordered solids, macroscopic quantum systems, and many-body phenomena. Mentoring by senior theorists like William Shockley, Gregory Wannier, Conyers Herring, Charles Kittel, and John Bardeen proved invaluable.

Anderson treated experimental data as his private secret weapon. He used that data to identify issues and construct novel theories in a way that few other physicists could. In the decade of the fifties alone, he discovered disorder-induced wave localization, recognized the importance of symmetry breaking in magnetism, and explained why many antiferromagnets do not conduct electricity. Most of his colleagues did not recognize the importance of these topics at the time. Eventually, the enthusiasm of Nevill Mott and some striking successes with the BCS theory of superconductivity put his colleagues on notice that they should heed any problem that Anderson deemed worthy of attention.

Within the solid-state physics community, Anderson was renowned for his signature methodology: engage deeply with experiment, be alert to "anomalies" that defy current understanding, strip a problem to its essentials, invent a model Hamiltonian incorporating those essentials, and analyze the model in just enough detail to extract the physics. It is a method that is often imitated but rarely replicated with the skill of the master. Over time, his answers to some of the most profound questions in his field earned him awards and accolades from his professional colleagues, recognition by learned societies in his own and other countries, and a share of a Nobel Prize for Physics.

Many physicists win awards and even Nobel Prizes. Not often, however, does the work of a physicist open a truly new and unforeseen area of research. This happened to Anderson three times: after his discovery of disorder-induced localization; after his invention (with Sam Edwards) of the spin glass model; and after his invention of the resonating valence bond (or quantum spin liquid) description of a many-body system.

Anderson influenced the broader scientific community in many ways. First, he resurrected the forgotten concept of emergence and explained how phenomena at any scale must be consistent with, but generally cannot be derived from, phenomena at smaller scales. Second, he recognized the importance of broken symmetry as a fundamental, discipline-crossing principle and argued how it can drive emergent properties like rigidity. Third, he identified and exploited a set of conceptually unifying principles which now permeate the research and pedagogical literature of condensed matter physics. Fourth, he brilliantly defended his field (the physics of the very many) against claims that it was less fundamental than elementary particle research (the physics of the very small) or cosmological research (the physics of the very distant). Finally, he championed the idea that complexity is not a hindrance to problem-solving, but an invitation to attack problems from entirely novel points of view.

Anderson unquestionably made profound contributions to science. Yet, he often felt compelled to remind others of his achievements. He quietly helped friends and colleagues in need, but he could be gruff, arrogant, and dismissive of those he did not respect. His intuition was off-scale but his communication skills were not. These traits made him inspirational to some and inscrutable to others.

Anderson believed fiercely in civil liberty, refused to affiliate with organizations that required a security clearance, and was nothing if not outspoken. Many physicists felt he was speaking for them when he testified before Congress that the United States could not justify the cost to single-handedly build and maintain the Superconducting Super Collider requested by the particle physics community. This led some others to blame him (unfairly) when Congress voted to cancel the project.

Anderson was widely read, loved Nature deeply, and cherished the love of his wife and her commitment to his professional success. Every year, the couple spent a month in England and thereby renewed their love of that country. They raised one child, Susan, although Phil left most of her upbringing to Joyce. Susan and her parents were distant for some years, but she and her father grew close in later years after they became Joyce's primary caregivers.

Like all people, Anderson suffered setbacks. His research career was almost stillborn when his refusal to pursue a PhD in nuclear physics led to a single academic job offer from a college with a large teaching load and no graduate program. He was deeply disappointed in 1973 when the Nobel Prize for Physics went to Brian Josephson and two experimenters. He regarded himself as just as deserving as they. This angst dissolved a few years later when he won his own share of a Nobel Prize. Nevertheless, the external trappings of success were never as important to him as the recognition of his professional colleagues for his scientific achievements. This trait explains why he tirelessly demanded his priority in all scientific matters.

Anderson worked to produce a theory of high-temperature superconductivity for the last twenty-five years of his professional career. At almost every moment during those years, he thought he possessed the answer to this great mystery. It disappointed him deeply that his colleagues did not accept his explanations and their behavior was a source of great frustration and unhappiness for him.

Many of his colleagues, in turn, were disappointed by him. During this final phase of his research career, they felt he had come to rely too much on his vaunted intuition. They were dismayed also by the combative attitude he and a few others brought to the subject. The siren call of a possible second Nobel Prize proved too strong and he adopted a proprietary attitude about the theory. You were either with him or against him in what became (for him) a struggle against the forces of ignorance.

A certain apartness was a constant in his life. The earnest scholarship boy from the Midwest did not mix with the prep school crowd at college. The graduate student from a small town did not join the mostly big-city cohort who studied theoretical nuclear physics with Julian Schwinger. The theorist who solved problems for the experimental magnetism group at Bell Labs was not the one who pushed management to create a separate theory department. He published the vast majority of his scientific papers as a single author and he never followed theoretical trails blazed by other theorists.

Anderson positively influenced a number of brilliant people (including at least three future Nobel Prize winners), but he pushed several of them away when they deviated from his message. The last decades of his professional life found him intellectually isolated from many of his theoretical colleagues and unable to convince them of the correctness of his theory for high-temperature superconductivity.

For fifty years, Phil Anderson was one the brightest stars in the firmament of theoretical physics. He did no experiments and he was not involved with applied problems directly. Nevertheless, his conceptual formulations profoundly influenced a broad swath of the physics world and beyond. His impact will be felt by generations of future scientists.

Appendix
Highlights of the Scientific Career
of Philip W. Anderson

1949 Pressure Broadening of Molecular Absorption Lines

1950 Superexchange in Antiferromagnets

1951 A Model for Ferroelectricity

1952 Spontaneous Symmetry Breaking in Antiferromagnets

1958 Disorder-Induced Localization
Gauge Invariance of Bardeen–Cooper–Schrieffer theory
Collective Excitations in Superconductors

1959 Theory of Dirty Superconductors
Mott Repulsion as the Origin of Superexchange

1960 Origin of Superfluidity in Helium Three
Soft Modes in Ferroelectrics

1961 Local Magnetic Moments in Metals

1962 Flux Creep in Superconductors

1963 Observation of the DC Josephson Effect
Concepts in Solids (book)
Tunneling Spectra for Superconductors
A Mechanism of Mass Generation for Elementary Particles

1966 Superfluid Flow Properties of Helium Four

1967 Infrared Catastrophe in Fermi Gases
National Academy of Sciences

1975 Pulsar Glitches

1970 Renormalization Group Approach to the Kondo Effect

1972 Thermal Properties of Glasses
More is Different critiques reductionism and champions emergence

1973   Resonating Valence Bonds in Insulators

        Spin Fluctuations in Helium Three

1977   Topological Theory of Defects

1975   Negative U Centers in Amorphous Semiconductors
        Theory of Spin Glasses

1977   Nobel Prize for Physics (1/3 share)

1979   Scaling Theory of Localization

1981   Emergence and Life

1984   *Basic Notions of Condensed Matter Physics* (book)
        Triplet Pairing in Heavy Fermion Superconductors
        Chemical Pseudopotentials

1985   Helps launch the Santa Fe Institute

1987   Resonating Valence Bond Theory of High-T_C
        Superconductivity

1988   *The Economy as an Evolving Complex System* (book)
        Strange Metals

1993   Congressional Testimony Opposing the Superconducting
        Super Collider

1997   THE *Theory of Superconductivity in the High-T_C Cuprates* (book)

2004   Plain Vanilla Theory of High-T Superconductivity

2006   Theory of the Pseudogap in the Cuprates

2012   Supersolidity

Horizontal lines divide Anderson's career into three periods (see Chapter 12)

Glossary

ADIABATIC CONTINUITY The replacement of a many-body system by a simpler system which captures much of its behavior in an average manner

ADVANCED LIGHT SOURCE An electron accelerator at Lawrence Berkeley National Laboratory used to produce x-rays for materials science research

ANDERSON LOCALIZATION A phenomenon that arrests wave propagation in a disordered medium

ANTIFERROMAGNET A system where nearest neighbor spins tend to point in opposite directions

AI Artificial intelligence in connection with computers

BAYESIAN INFERENCE A form of statistical analysis where prior information and updating with evolving knowledge are important features

BCS Acronym for John Bardeen, Leon Cooper, and Robert Schrieffer, and also for the first successful microscopic theory for the elemental superconductors which these authors originated

BOSE-EINSTEIN CONDENSATION A situation when all the particles in a many-boson system occupy the lowest energy state of the system at very low temperature

BOSON A quantum particle with the property that any number of them can occupy a given quantum state. The many-particle wave function of a collection of bosons does not change sign when any two of them exchange locations

BROKEN SYMMETRY A situation where a system no longer appears the same after some operation has been performed on it

CERN The European center for accelerator-based particle physics research

COLLECTIVE EXCITATION An excited state of many-particle systems where all the particles move together in an orchestrated fashion

COMPLEXITY Situations where the interactions amongst many constituents leads to emergent behavior

CONDENSED MATTER PHYSICS The study of matter in its solid and liquid phases

CONDUCTION BAND The completely unoccupied energy band of a semi-conductor that lies lowest in energy amongst all the unoccupied bands

CONDUCTOR A material capable of passing an electric current

COOPER PAIR Two electrons bound together (like a diatomic molecule) which move without electrical resistance through a superconductor

COULOMB INTERACTION The electric force between two charged objects

CRITICAL TEMPERATURE The temperate at which a phase transition (magnetic to non-magnetic, superconducting to non-superconducting) occurs

CRYSTAL A solid composed of a repeated stacking of a fixed motif of atoms

CUPRATE SUPERCONDUCTORS A class of ceramic materials (many oxides) that exhibit superconductivity at much higher temperature than all elemental superconductors

DELOCALIZED An electron or wave function that can spread out over an arbitrarily large region of space

DIFFRACTION Describes the outgoing waves when a single incoming wave strikes an object with sharp edges

DIFFUSION A spreading out process whereby one or many particles take a succession of steps, changing direction randomly after every step

DIRTY SUPERCONDUCTOR A superconductor with non-magnetic impurities

DISORDER Any disruption of regularity, particularly crystalline regularity

DONOR An impurity atom in a semiconductor capable of donating an electron to the host crystal

DONOR LEVEL The energy of the electron in a donor atom that can be donated to the host crystal

DOPING The addition of impurities to a system to achieve a desired behavior

ELECTRON GAS A fictitious system composed of electrons moving unobstructed through a uniform "jelly" of positive charge so the entire system has zero net electric charge

ELECTRON–PHONON INTERACTION The change in energy of an electron due to its Coulomb attraction to positive ions making small-amplitude vibrational excursions away from their usual lattice positions

EMERGENCE The notion that the behavior of a complex system is consistent with the behavior of its constituents but cannot be predicted merely by knowing their properties and the forces between them

ENERGY BAND An interval of energy within which the spacings between the allowed energies of an electron in a solid are extremely small

ENERGY GAP An interval of energy wherein no allowed energies for an electron in a solid are found

EXCHANGE The quantum description of the interaction between two particles with spin

FERMI ENERGY The energy of the highest occupied state of a collection of fermions, e.g., electrons

FERMI LIQUID THEORY A parameterized theory of the electron gas due to Lev Landau that takes full account of the Coulomb repulsion between all pairs of electrons

FERMION A quantum particle with the property that only one can occupy a given quantum state. The many-particle wave function of a collection of fermions changes sign when any two of them exchange locations

FERROELECTRIC A system capable of generating an internal electric field at low temperature

FERROMAGNET A system where nearest-neighbor spins tend to point in the same direction

GAUGE INVARIANCE A property of a theory which guarantees conservation of charge for a system of particles and electromagnetic fields

GAUGE SYMMETRY The unchanging predications made by a theory when one changes the value of an angle-type variable in the theory

GINZBURG—LANDAU THEORY A phenomenological theory of superconductivity due to Vitaly Ginzburg and Lev Landau that preceded the microscopic theory of BCS

Green function (see many-particle Green function)

HAMILTONIAN A mathematical representation of the total energy of a system

HARTREE—FOCK APPROXIMATION An approximate treatment of a many-electron system that takes account of electron exchange only

HEISENBERG MODEL A mathematical model of spins on a lattice used to account for the thermal properties of crystalline ferromagnets and antiferromagnets

HIGGS PARTICLE A particle associated with a theory which describes how certain elementary particles acquire a mass

HUBBARD MODEL A total energy expression for electrons in a solid that hop from site to site and pay an electrostatic energy cost when two electrons occupy one site

IFF Identify friend or foe: a radar system designed to distinguish Allied aircraft form Axis aircraft during World War II

INSULATOR A material that cannot pass an electric current

INTERLAYER TUNNELING THEORY A failed theory of high-temperature superconductivity based on the tunneling of Cooper pairs between layers of the cuprate materials

INVERSION An event when the apical atom of a pyramid-shaped molecule moves from one side of the molecule to its mirror-image position

JOSEPHSON EFFECT A quantum phenomenon whereby Cooper pairs tunnel across a small gap

KONDO EFFECT A low-temperature minimum in the electrical resistance of a dilute magnetic alloy

KONDO MODEL A mathematical model of the interaction of one fixed impurity spin with many neighboring conduction electrons spins

KREBOIZEN A bogus substance alleged to promote the cancer-fighting ability of the human body

LATTICE A periodic arrangement of points that locate the center of the unit cells of a crystal

LIQUID CRYSTAL A fluid system where the constituent molecules tend to arrange themselves in a crystal-like fashion

LOCALIZED An electron or wave function that is confined to a small region of space

MACROSCOPIC Large enough to be visible with the naked eye

MAGNETIC MOMENT A vector which characterizes the strength and orientation of the magnetic field produced by an object

MAGNETIC RESONANCE An experimental technique where a microwave frequency is tuned to a characteristic frequency of the system studied

MAGNETISM Physical effects induced by or influenced by a magnetic field

MANY-BODY PHYSICS The study of many (quantum) particles interacting with each other

MANY-PARTICLE GREEN FUNCTION A mathematical object used to calculate the properties of a many-body system without direct appeal to the many-particle wave function

MICROSCOPIC Too small to be visible with the naked eye

MICROWAVES Electromagnetic waves with wavelength between one meter and one millimeter

MODEL A physical or mathematical description of a system, usually in much simplified form

MOTT INSULATOR A solid where the electrostatic repulsion between two electrons on any atomic site prevents them from hopping from atom to atom and thus prevents electrical conduction

NONLINEAR A mathematical problem or system where the output is not proportional to the input

OCKHAM'S RAZOR The maxim that one should make no more assumptions to solve a problem than are absolutely necessary

ORBITAL ANGULAR MOMENTUM A vector quantity that characterizes the orbital motion of one particle about a given center

PARTICLE PHYSICS The study of subatomic particles and the forces between them

PERCOLATION The motion of a liquid or gas through a porous medium

PERTURBATION THEORY A systematic method to study how small changes in the Hamiltonian of a system affects its properties

PHASE TRANSITION The transformation of a system between two readily distinguishable states as a function of a controllable parameter

PHENOMENOLOGY Study of the behavior of a physical system

PHONON A quantum particle of atomic vibrational motion

PHOTON A quantum particle of electromagnetic radiation

PSEUDOSPINS Non-spin variables that behave in many ways like spins

QUANTUM FIELD THEORY The generalization of the quantum mechanics of particles to deal with fields, e.g., the electromagnetic field

QUANTUM HALL EFFECT The characteristic behavior of two-dimensional electron system subject to an external voltage and an external magnetic field

QUANTUM MECHANICS The mathematical theory which describes single particles and many-particle systems at the microscopic scale

QUASIPARTICLE The object produced when an isolated particle acquires new properties by turning on its interactions with all other particles

RADIATION Electromagnetic waves

RANDOM PHASE APPROXIMATION An approximate treatment of a many-particle system that includes electron exchange (like Hartree–Fock) plus certain long-distance and short-distance aspects of the Coulomb force

REDUCTIONISM The belief that a complex, many-particle phenomenon can be understood entirely from the properties and interactions amongst its constituents

RENORMALIZATION GROUP A mathematical scheme which relates the behavior of a physical system at one scale to its behavior at a different scale

RESONANCE The situation when two oscillating systems in contact have the same frequency

RESONATING VALENCE BOND A many-body wave function composed of a sum of pair wave functions where the spins of the pairs are antiparallel

RVB (see resonating valence bond)

SCALING Increasing or decreasing a quantity in a systematic way and observing the changes produced in other quantities

SCHRÖDINGER EQUATION An equation whose solution gives the quantum wave function for a system

SEMICONDUCTOR A solid where the highest occupied allowed energy level of the electrons is separated by an energy gap from the lowest unoccupied allowed energy level of the electrons

SINGLET A wave function of a pair of particles whose spins point in opposite directions

SOLID-STATE PHYSICS The study of solids, the form of matter where atoms occupy fixed positions in space with respect to each other

SPATIAL CORRELATION A relationship between two objects at different points in space

SPECTRAL LINE A feature resulting from the emission or absorption of electromagnetic radiation in a very narrow frequency range

SPECTROSCOPY Study of the interaction of objects with electromagnetic waves as one varies the wavelength of the wave

SPECTRUM Range or breadth, particularly of the frequency or wavelength of electromagnetic waves

SPIN A vector property of an electron, proton, neutron, etc. which characterizes the magnitude and orientation of the intrinsic magnetic field it produces

SPIN FLUCTUATIONS Transient wave-like motion of a collection of spins invoked by some theories of high-temperature superconductivity

SPIN GLASS A magnetic alloy where atomic spins freeze in random directions at low temperature

SPONTANEOUS SYMMETRY BREAKING A special case of symmetry breaking where a system possesses the ability to restore the symmetry by itself

STANDARD MODEL A theory of particle physics describing the weak and electromagnetic forces

SUPERCONDUCTING SUPER COLLIDER A large proton accelerator that was begun and then abandoned by the United States government

SUPERCONDUCTIVITY A low-temperature phenomenon where many metals and alloys exhibit zero electrical resistance

SUPEREXCHANGE An indirect interaction between two spins

SUPERFLUIDITY A low temperature phenomenon where a fluid (like liquid helium) loses its viscosity

SUPERSOLID A hypothetical state of matter where a solid exhibits some of the characteristics of a superfluid

SUSCEPTIBILITY The response of an object to an electric or magnetic field

SYMMETRY The situation where a system appears the same after some operation has been performed on it, e.g., rotation or translation

TRANSISTOR A semiconductor device used to switch or amplify electronic signals

TUNNEL EFFECT A quantum phenomenon where microscopic particles pass through a barrier they could not surmount classically

UNIT CELL A geometrical object containing a fixed motif of atoms which, when periodically repeated in space, produces a perfect crystal

VACUUM TUBE A device which formerly performed the electrical amplifying and switching functions of a transistor

VALENCE BAND The completely filled energy band of a semiconductor that lies highest in energy of all the filled bands

VECTOR A mathematical object entirely defined by a magnitude and a direction, often represented geometrically by a directed arrow

WAVE FUNCTION The quantity used in quantum mechanics to describe an object in Nature

WOODSTOCK OF PHYSICS Nickname of the nighttime session of the 1987 March Meeting of the American Physical Society devoted to high-temperature cuprate superconductivity

Index